Proteins: A Guide to Study by Physical and Chemical Methods

PROTEINS

A Guide to Study by Physical and Chemical Methods

RUDY H. HASCHEMEYER
Cornell University Medical School

AUDREY E. V. HASCHEMEYER
Hunter College of the City University of New York

A WILEY-INTERSCIENCE PUBLICATION

JOHN WILEY & SONS, New York · London · Sydney · Toronto

Library of Congress Cataloging in Publication Data

Haschemeyer, Rudy Harm, 1930–
Proteins: a guide to study by physical and chemical methods.

"A Wiley-Interscience publication."
Includes bibliographical references.
1. Proteins. 2. Biological chemistry—Technique.
I. Haschemeyer, Audrey E. V., 1936– joint author.
II. Title.

QP551.H37 574.1′9245 72–13134
ISBN 0-471-35850-9

Printed in the United States of America

10 9 8 7 6 5 4 3 2 1

Preface

Proteins comprise the structural building blocks as well as most of the functional machinery of living organisms. There are few areas of biological or biomedical research today that do not demand a working knowledge of the chemical and physical properties of these complex macromolecules. Thus the advanced study of proteins beyond the level of the general biochemistry course has become increasingly essential. This book evolved from many years of teaching courses in proteins to graduate students in biochemistry and biology and to advanced medical students. To meet the needs of these students, and of researchers in all branches of the biomedical sciences, we have attempted to present a comprehensive and critical guide to the physical and chemical methods available for the study of proteins. An evaluation of the current state of knowledge of protein structure, including the important areas of protein folding and conformational equilibria, is also presented.

The book is organized for a two semester course or, if physical methods alone are to be emphasized, a one semester course based on Part II. A general theoretical background is presented for each of the physical methods discussed to indicate both the potential and the limitations of the technique in its application to macromolecules. More rigorous derivations of equations are presented where we have found them to be valuable in teaching the material. Primary emphasis, however, is given to the usefulness of the various techniques for real problems involving proteins and their contributions to our overall understanding of protein structure and function. To facilitate access to further information on each topic a listing of important general references and reviews, and selected references of interest in the current literature is provided.

We wish to express our appreciation to the colleagues and students who offered valuable criticism and advice on portions of the manuscript. Special thanks are due to Dr. Robert W. Woody for his review of the complete manuscript. All responsibility for the opinions expressed herein and any errors that may occur, however, rests with the authors. We will welcome communications from interested scientists and students.

RUDOLPH HASCHEMEYER
AUDREY H. HASCHEMEYER

New York, New York
February 1973

Contents

Proteins: A Guide to Study by Physical and Chemical Methods

PART ONE

I Introduction

In all areas of biological and medical research today there is increasing need for knowledge about proteins. These complex macromolecules, with particle weights ranging from the thousands to the millions, comprise in essence the working machinery of life. They cannot claim the central position held by the nucleic acids as the carriers of heredity, yet certain proteins are responsible for the control of expression of hereditary information from its first transcription from the gene to its final translation into new polypeptide chains. Hundreds of proteins have been identified in the category of enzymes, catalyzing a myriad of complex biochemical reactions. Others are responsible for the basic structural framework of living organisms; in higher animals these structural proteins include collagen of bones, cartilage, and tendons; keratin of hair and nails; elastin of blood vessels and ligaments; and myosin of muscle.

The name protein (Greek, *proteios*, of the first rank) was first used in 1838 by Mulder following a suggestion by Berzelius. Mulder was also among the first to do a systematic study of the elemental composition of proteins. Most proteins were found to contain 50 to 55% carbon, 6 to 7% hydrogen, 20 to 23% oxygen, and 12 to 19% nitrogen. Protein determinations based on nitrogen (assuming an average content of 16%) came to be used for analysis of tissues and food samples. Sulfur (0.2–3.0%) was found to occur in proteins, as was phosphorus in some cases (as high as 3%). Trace elements identified in certain proteins (e.g., 0.34% iron in hemoglobin) permitted calculation of minimum molecular weights. These results gave the first indication that proteins have large molecular weights compared with other organic substances known at that time. Decomposition to smaller molecular weight units could be achieved through hydrolysis catalyzed by acids, alkalis, or certain biological preparations (containing proteolytic enzymes). During the late 1800s, amino acids were identified as the basic building units of proteins. Eventually 20 different amino acids were shown to occur as components of most proteins; a number of others were found in special cases.

Analysis of the more common amino acids proved that all but proline (which is actually an imino acid) have structures consisting of a carbon atom (the α-carbon) and four substituent groups: a carboxyl group, an amino group, a hydrogen atom, and an R group which differs among the various amino acids. Pasteur had shown in 1851 that amino acids are optically active, and this property was soon correlated with the asymmetry of the α-carbon atom resulting when all four substituents of the α-carbon are different (glycine is not optically active because R = hydrogen). Although amino acids were found to differ in the direction in which they rotate polarized light, it was eventually established that all amino acids which occur in proteins have the same configuration (denoted L) with regard to the arrangement of groups about the α-carbon atom.

The nature of the bond which links amino acids together in proteins was elucidated independently by Emil Fischer and Franz Hofmeister in 1902. They proposed that water is eliminated between the α-carboxyl group of one amino acid and the amino group of another to produce an amide linkage. The condensation of two amino acids to form a peptide is illustrated by the general equation.

$$\underset{\text{Amino acid}}{NH_2\overset{\overset{\displaystyle R}{|}}{C}HCOOH} + \underset{\text{Amino acid}}{NH_2\overset{\overset{\displaystyle R'}{|}}{C}HCOOH} \rightarrow \underset{\text{Peptide}}{NH_2\overset{\overset{\displaystyle R}{|}}{C}HCO\text{—}NH\overset{\overset{\displaystyle R'}{|}}{C}HCOOH} + H_2O \tag{1-1}$$

Proteins isolated from different sources and by different techniques were found to vary considerably in their properties (e.g., solubility), and in 1908 an attempt was made to develop a scheme of classification. First, two general groups were differentiated: the simple proteins, which yield only amino acids upon hydrolysis, and the conjugated proteins, which contain prosthetic groups or other substances that are released upon hydrolysis. The proteins categorized as "simple" (though they are certainly quite complex structurally) were further subdivided into several groups. These include the albumins which are readily soluble in water; the globulins which are insoluble or sparingly soluble in water but are soluble in dilute neutral salts; the glutelins which are soluble in dilute acid or alkali; the prolamines which are soluble in 70 to 80% alcohol but insoluble in either water or absolute alcohol alone; the albuminoids or scleroproteins which are fibrous insoluble animal proteins; the histones which are basic proteins containing a high percentage of the basic amino acids (lysine, histidine, and arginine) and are soluble in water; and the protamines which are low molecular weight, basic proteins and are soluble in water. The conjugated proteins are classified according to their nonprotein moieties or prosthetic groups, which may be nucleic acid (nucleoproteins), carbohydrate (mucoproteins and glycoproteins), lipid (lipoproteins), highly colored prosthetic groups (chromoproteins such as the hemoglobins, cytochromes, and flavoproteins), or metals (metalloproteins, a group that includes many enzymes). Although the division between groups is not sharp (as between albumins and globulins) and despite the fact that proteins may belong to more than one group (e.g., hemoglobin is both a chromoprotein and a metalloprotein), some of the terminology of this classification scheme is still in use.

Early attempts at protein fractionation were largely limited to methods that took advantage of the different solubility properties of proteins. Egg albumin from egg white was successfully crystallized by Hofmeister in 1889, but most proteins proved difficult to purify or to crystallize. In the 1940s the development of chromatography led to improved methods of purification of proteins and analysis of their components. Separation of acylated amino acids on the basis of their different partition coefficients between water and an immiscible organic solvent was achieved by Neuberger in 1938. In 1941 Martin and Synge used silica gel column chromatography for the separation of acylated amino acids, and in 1944 Consden, Gordon, and Martin obtained separation of free amino acids by paper chromatography. These methods made it possible to determine the amino acid composition of proteins quickly and conveniently. The techniques of chromatography were later extended to include separation of proteins themselves by either absorption, ion exchange, or molecular sieving.

With the automation of chromatographic methods, rapid and convenient determination of amino acid composition of extremely small amounts of proteins became possible. Table 1-1 shows the amino acid composition of a variety of proteins. Most proteins have fairly similar amino acid compositions even though they differ greatly in physical and biological properties. Thus an important next step was the determination of the order of the amino acids in the polypeptide chains of the protein. Sanger showed in 1947 that the compound 1-fluoro-2,4-dinitrobenzene could be reacted with free α-amino groups at the end of polypeptide chains to form a linkage stable to acid hydrolysis. This reaction permitted the identification of the *N*-terminal amino acid residues of proteins and contributed to the first sequence determination, that of the protein hormone insulin. Insulin was shown to consist of two chains joined by disulfide bonds, one containing 21 amino acid residues, the other containing 30 residues (Fig. 1-1). Since that time the complete or partial sequences of hundreds of proteins have been determined.

A major contribution to the study of the three-dimensional structure and physical properties of proteins was the development of the ultracentrifuge

Table 1-1 Amino acid composition of proteins [from a compilation by D. M. Kirschenbaum, *Anal. Biochem.*, **44,** 159 (1971)].

Amino acid	Glycine	Alanine	Valine	Leucine	Isoleucine	Proline	Serine	Threonine	Aspartic acid	Glutamic acid	Half-cystine	Methionine	Lysine	Arginine	Histidine	Phenylalanine	Tyrosine	Tryptophan	Amide ammonia	MW × 10^{-3}
Albumin (bovine)	15	44	35	58	14	28	26	32	54	80	36	4	58	23	17	26	19	2	27	65
Carbonic anhydrase B (porcine)	18	20	16	23	15	17	29	11	34	21	0	0	23	6	14	9	11	6	17	30
Casein (human)	3	7	19	26	13	39	9	9	11	39	0	3	11	3	5	5	7	1	32	24
Chymotrypsinogen B (porcine)	22	22	25	17	11	14	28	17	20	15	10	2	6	8	3	8	4	13	—	26
Creatine kinase (human)	66	48	46	64	25	45	41	31	73	75	7	19	66	32	28	26	20	6	53	81
Deoxyribonuclease A (bovine)	9	23	25	23	12	9	30	15	34	20	4	4	9	12	6	12	16	3	—	31
Enolase (salmon)	96	104	61	73	61	37	56	37	121	95	14	14	91	34	22	33	28	3	—	106
Ferredoxin (*Chromatium*)	5	3	6	3	6	5	4	6	8	16	9	1	2	2	2	0	3	0	—	10
β-Galactosidase (*E. coli*)	84	94	72	112	44	71	69	70	122	145	17	21	27	73	36	45	34	32	—	135
Glyceraldehyde-3-*P* dehydrogenase (human)	134	127	98	78	73	56	79	80	162	81	8	28	102	38	39	55	45	—	—	140
Glycerol-3-*P* dehydrogenase (rabbit)	30	25	25	24	22	12	9	11	22	33	9	5	22	7	7	12	3	2	—	30
Invertase (*Neurospora*)	158	124	122	114	57	93	141	137	189	130	9	20	55	54	26	67	63	44	—	187
Methionyl-tRNA synthetase (*E. coli*)	114	143	94	136	80	78	83	64	158	171	24	38	94	78	35	77	49	33	—	173
Penicillinase (*B. cereus*)	20	33	18	20	17	9	13	24	34	24	—	4	21	11	4	8	11	5	28	32
Prolactin (ovine)	11	9	10	22	11	11	15	9	22	22	6	7	9	11	8	6	7	2	—	23
Rhodopsin (bovine)	21	25	15	20	10	16	15	25	19	25	6	8	8	6	4	22	10	—	—	28
D-Serine dehydratase (*E. coli*)	45	47	27	46	17	14	27	20	31	49	5	12	19	17	11	18	13	4	—	46
Thrombin (bovine)	30	18	19	33	13	18	18	16	32	35	8	4	26	24	8	14	11	10	—	34
Thyroglobulin (human)	394	383	310	476	155	347	484	283	400	697	244	62	195	275	80	259	110	117	549	670
Urease (jackbean)	64	61	45	57	51	41	39	45	72	62	11	19	39	31	20	19	16	5	45	75

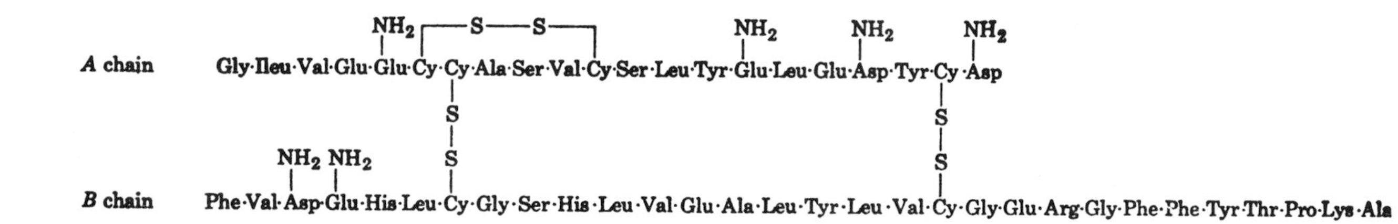

Fig. 1-1 The primary structure of beef insulin. (From F. Sanger, in *Currents in Biochem. Res.*, D. E. Green, ed., Wiley, New York, 1956.)

by Svedberg in the 1930s. From observations on rates of migration of proteins under high centrifugal fields, Svedberg and his associates demonstrated that different proteins have characteristic molecular sizes and shapes. Since this early work the technique of ultracentrifugation has been developed to become a powerful tool for many biological purposes. The use of electrophoresis for separation and characterization of proteins on the basis of migration in an electric field was introduced by Tiselius in 1933. The application to proteins of other physical techniques, such as the measurement of light scattering, viscosity, and diffusion, followed shortly thereafter.

During the past 10 or 20 years tremendous strides have been made in elucidating the three-dimensional structure of proteins. Structural analysis of crystals of amino acids and small polypeptides led to the determination of the spatial configuration of the peptide bond, and Pauling and Corey's

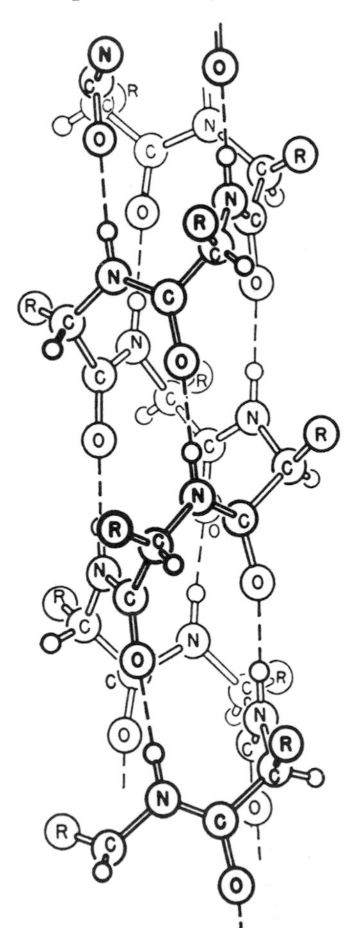

Fig. 1-2 The right-handed α-helix. (From R. B. Corey and L. Pauling, *Proc. Intern. Wool Textile Research Conf.*, Australia, Part B, 1955.)

proposal in 1951 of the α-helix as a basic model for polypeptides (Fig. 1-2). In the α-helix the polypeptide backbone is folded in such a way that the carbonyl group of one amino acid is hydrogen-bonded with the amide hydrogen of an amino acid further up the helix. Further investigation has largely substantiated these ideas, and it is now known that many proteins contain α-helical segments as well as other types of repeating hydrogen-bonded structures. Attempts to determine the complete three-dimensional structure of a protein by means of X-ray crystallography began in the middle 1930s and have reached an impressive productivity today. The first structure solved at high resolution was that of sperm whale myoglobin by J. C. Kendrew and his associates (1961), and these results were effectively applied to the study of a much larger but related molecule, horse hemoglobin, by M. F. Perutz's group. In the case of myoglobin, the positions of all non-hydrogen atoms in the molecule have been determined to yield the structure shown in Fig. 1-3. Other structures solved at high resolution in the late 1960s

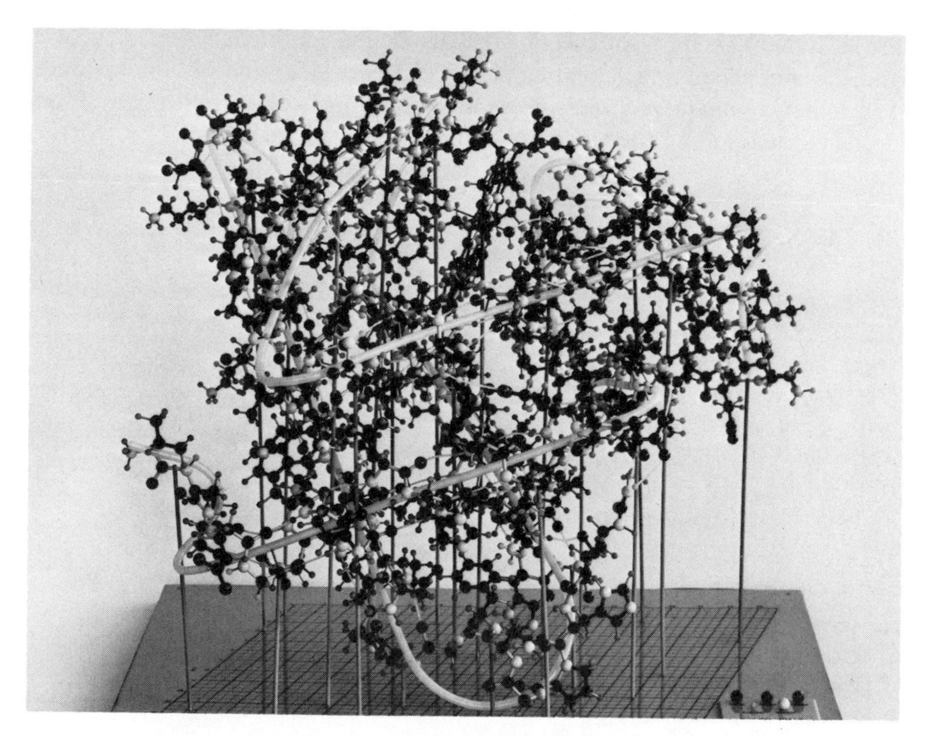

Fig. 1-3 Sperm whale myoglobin—a wire atomic model with a white cord marking the path of the α-helices. (Courtesy of J. C. Kendıew.)

include lysozyme, ribonuclease, carboxypeptidase, and chymotrypsin. It can be expected that the list will continue to grow slowly. However, experimental limitations, as well as the enormous investment of time and equipment required for each structure solution, indicate that we still have to rely on other methods for studying protein structure for a long time.

It is now generally accepted that the major portion of the three-dimensional architecture of a protein must remain intact for biological activity. In ribonuclease, for example, amino acid residues from both ends of the chain must be in close proximity for enzymatic activity to exist. Although it has been demonstrated that the genetic code and protein synthesizing system of a cell contain the information necessary to determine the proper amino acid sequence of a protein, no cellular mechanism (e.g., enzymatic "foldases") has yet been found to fold a protein into its "native" configuration. This folding is apparently accomplished by intramolecular and intermolecular noncovalent interactions acting cooperatively to produce the functional conformation of the protein in its cellular environment. This hypothesis is supported by the observation that a number of proteins can be denatured to a completely unfolded state and then renatured *in vitro* to the biologically active state. The types of forces involved in the maintenance of structure and the manner in which folding is achieved are not yet fully understood. The following chapters discuss these and other areas of current research on protein structure and the techniques employed for these studies.

REFERENCES

For a general background on proteins and for additional historical information, the following are recommended:

J. S. Fruton and S. Simmonds, *General Biochemistry*, 2nd ed., Wiley, New York, 1958.

E. S. West, W. R. Todd, H. S. Mason, and J. T. Van Bruggen, *Textbook of Biochemistry*, 4th ed., Macmillan, New York, 1966.

A. L. Lehninger, *Biochemistry*, Worth Publishers, New York, 1970.

H. E. Schultze and J. F. Heremans, *Molecular Biology of Human Proteins*, Vol. 1, American Elsevier, New York, 1966.

The major continuing monographic series dealing with proteins are:

H. Neurath, *The Proteins*, 2nd ed., Vols. 1–5, Academic Press, New York, 1963–1970.

Advances in Protein Chemistry, Vols. 1–26, Academic Press, New York, 1946–1972.

S. P. Colowick and N. O. Kaplan, Eds., *Methods in Enzymology*, Vols. 1–26, Academic Press, New York, 1955–1972.

P. D. Boyer, Ed., *The Enzymes*, 3rd ed., Vols. 1–7, Academic Press, New York, 1970–1972, and earlier editions.

Advances in Enzymology, Vols. 1–36, Wiley-Interscience, New York, 1941–1972.

II Protein Composition

On the basis of our present knowledge of protein chemistry and of protein biosynthesis, we may divide the components of proteins into two major classes. The first class comprises the amino acids that occur in proteins in peptide linkage as shown in Eq. 1-1. This group includes the 20 types now recognized as genetically coded as well as those produced by special reactions that occur after a precursor is incorporated into the polypeptide chain(s) of the molecule. In the former category, however, we do not include amino acids that, like *N*-formylmethionine, may be found to have only a transient existence in polypeptides during biosynthesis. The second class of components occurring in proteins includes a variety of substances other than amino acids that may be covalently linked or bound by strong noncovalent forces. Proteins containing substances of this type are termed conjugated proteins.

Since the amino acids constitute the entire composition of all simple proteins and a significant part in conjugated proteins, we present a brief review of pertinent physical and chemical properties of these substances. For a more detailed consideration of the subject and for references to the original literature, several treatises are available.[1–4]

1 AMINO ACIDS

The 20 amino acids that form the basic building blocks of proteins and their genetic codon assignments are listed in Table 2-1. All except proline have the general formula and spatial configuration shown in Fig. 2-1, where R represents the side chain which varies among the different amino acids. The structure is shown in the absolute L-configuration and in the ionized form which predominates in aqueous solution at neutral pH. All of these substances are

Table 2-1 Structures of the 20 amino acids that occur in proteins as a direct result of genetic coding. The standard abbreviation for each amino acid is given, as well as the currently established codon assignments [referring to the nucleotide sequence of messenger ribonucleic acid, where A = adenylic acid, C = cytidylic acid, G = guanylic acid, U = uridylic acid, and (N) = any of these nucleotides].

Aliphatic side chains (apolar)

Glycine (Gly)
[GG(N)]

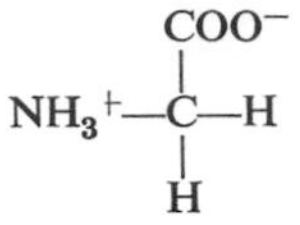

COO^- / $NH_3^+—C—H$ / H

L-Alanine (Ala)
[GC(N)]

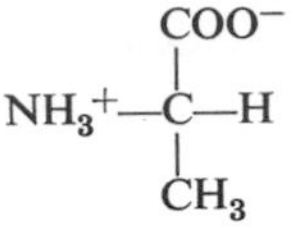

COO^- / $NH_3^+—C—H$ / CH_3

L-Valine (Val)
[GU(N)]

COO^- / $NH_3^+—C—H$ / CH / CH_3 CH_3

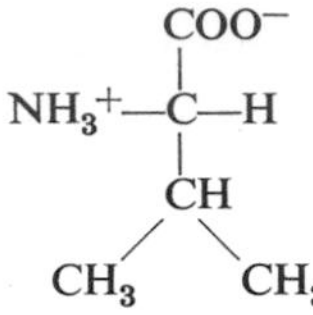

L-Leucine (Leu)
[CU(N), UUA, UUG]

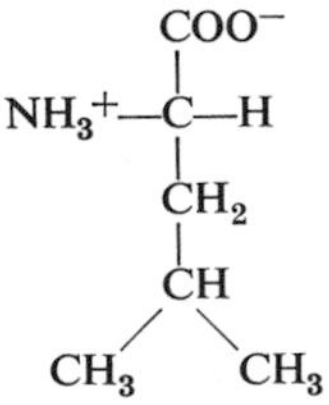

COO^- / $NH_3^+—C—H$ / CH_2 / CH / CH_3 CH_3

L-Isoleucine (Ile)
[AUU, AUC, AUA]

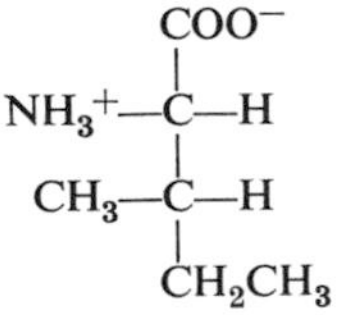

COO^- / $NH_3^+—C—H$ / $CH_3—C—H$ / CH_2CH_3

L-Methionine (Met)
[AUG]

COO^- / $NH_3^+—C—H$ / CH_2 / CH_2 / $S—CH_3$

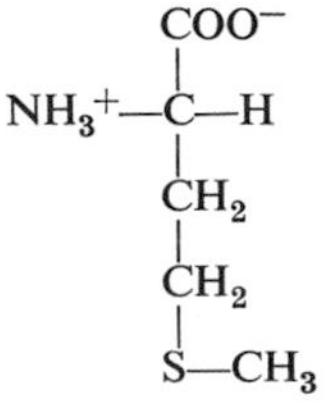

Heterocyclic and aromatic side chains (apolar)

L-Proline (Pro)
[CC(N)]

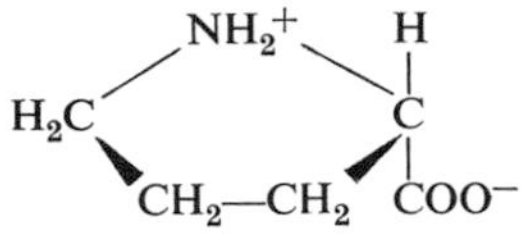

NH_2^+ H / H_2C C / $CH_2—CH_2$ COO^-

L-Phenylalanine (Phe)
[UUU, UUC]

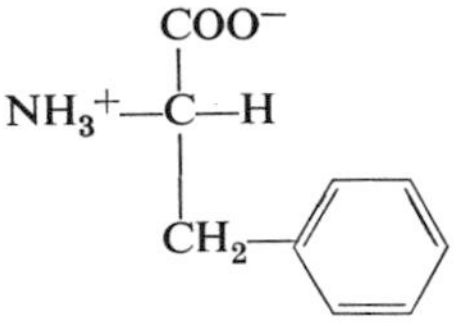

COO^- / $NH_3^+—C—H$ / $CH_2—$

L-Tryptophan (Trp)
[UGG]

COO^- / $NH_3^+—C—H$ / $CH_2—$ / N / H

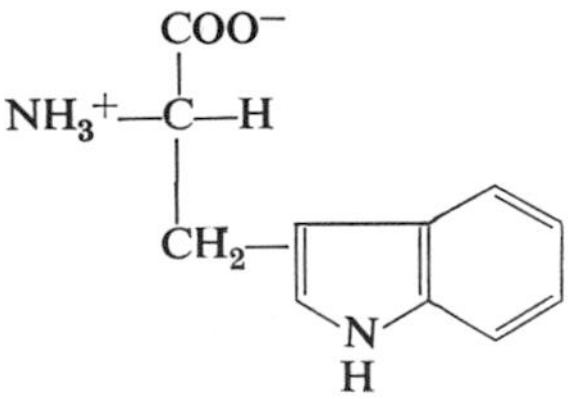

L-Tyrosine (Tyr)
[UAU, UAC]

COO^- / $NH_3^+—C—H$ / $CH_2—$ —OH

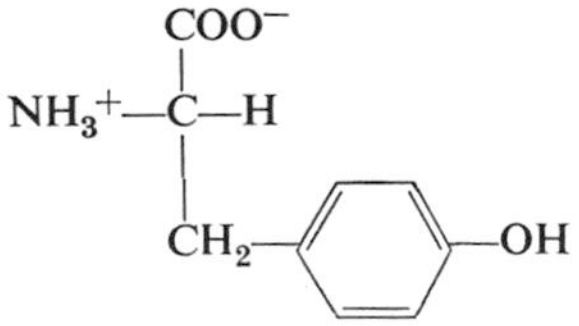

Table 2-1 (Cont.)

Polar aliphatic side chains

L-Serine (Ser)
[UC(N), AGU, AGC]

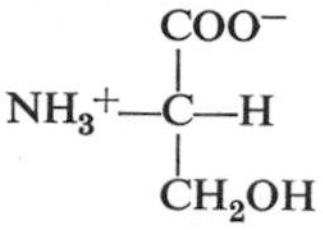

COO^-
$NH_3^+—C—H$
CH_2OH

L-Threonine (Thr)
[AC(N)]

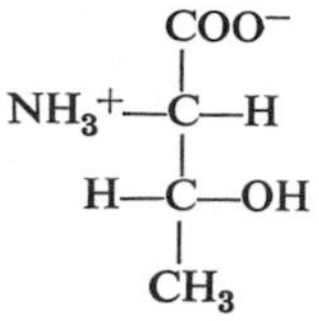

COO^-
$NH_3^+—C—H$
$H—C—OH$
CH_3

L-Cysteine (Cys)
[UGU, UGC]

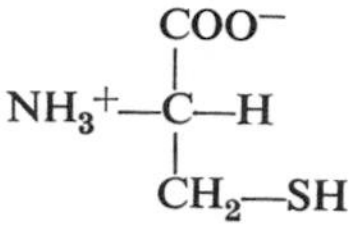

COO^-
$NH_3^+—C—H$
$CH_2—SH$

L-Asparagine (Asn)
[AAU, AAC]

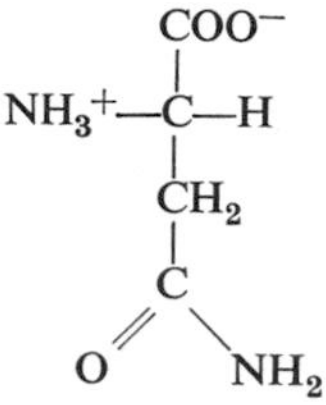

COO^-
$NH_3^+—C—H$
CH_2
C
$O \quad NH_2$

L-Glutamine (Gln)
[CAA, CAG]

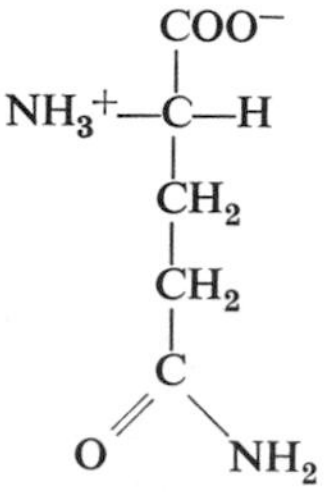

COO^-
$NH_3^+—C—H$
CH_2
CH_2
C
$O \quad NH_2$

Ionizable side chains

L–Aspartic Acid (Asp)
[GAU, GAC]

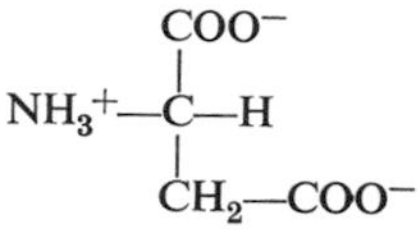

COO^-
$NH_3^+—C—H$
$CH_2—COO^-$

L-Glutamic Acid (Glu)
[GAA, GAG]

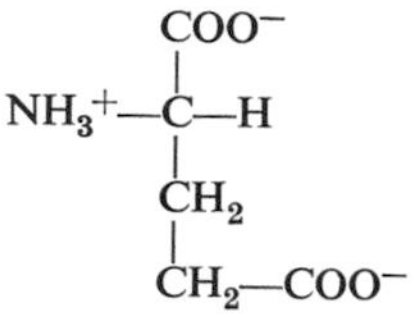

COO^-
$NH_3^+—C—H$
CH_2
$CH_2—COO^-$

L-Histidine (His)
[CAU, CAC]

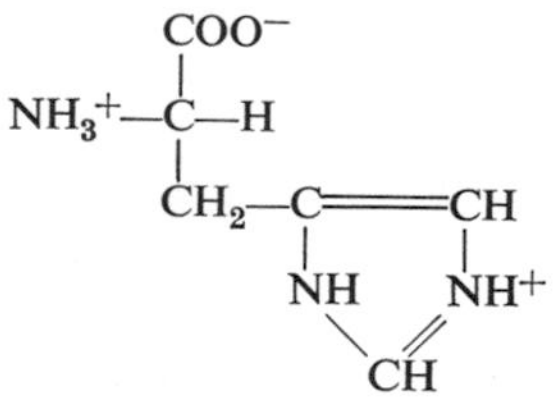

COO^-
$NH_3^+—C—H$
$CH_2—C═══CH$
$NH \quad NH^+$
CH

L-Lysine (Lys)
[AAA, AAG]

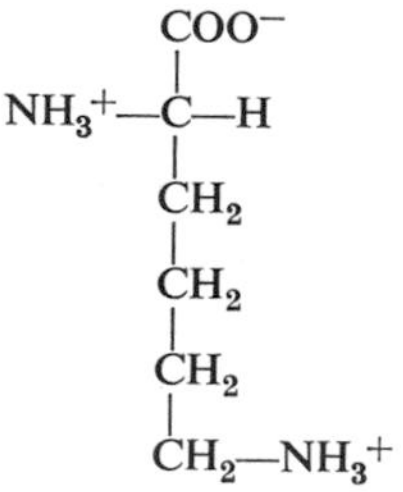

COO^-
$NH_3^+—C—H$
CH_2
CH_2
CH_2
$CH_2—NH_3^+$

Table 2-1 (Cont.)

L-Arginine (Arg)
[CG(N), AGA, AGG]

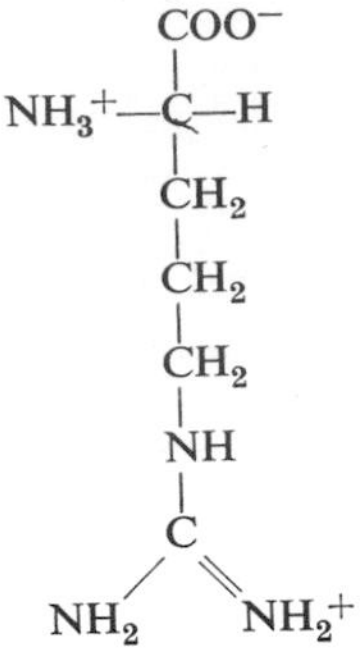

COO^-
NH_3^+—C—H
CH_2
CH_2
CH_2
NH
C
NH_2 NH_2^+

α-amino acids, that is, the amino group is in the α-position to the carboxyl function; the central carbon atom is called the α-carbon. In proline the α-nitrogen atom is part of a ring structure forming a secondary amine.

The 20 amino acids of Table 2-1 are grouped according to the physical properties of their side chains, particularly on the basis of their interaction with water. This classification is useful in relation to our later consideration of the three-dimensional structure of proteins. The primary distinction to be

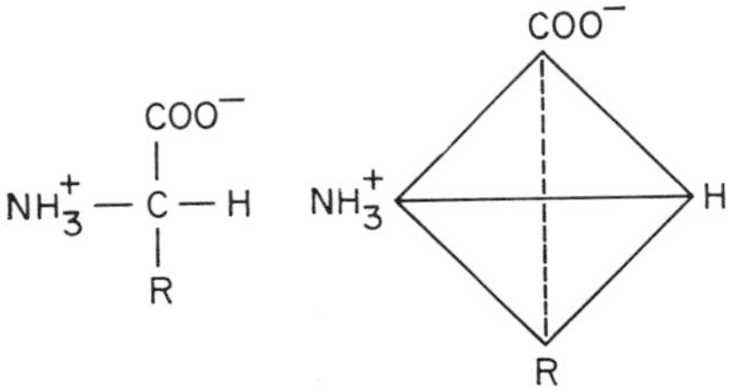

COO^-
NH_3^+ — C — H
R

COO^-
NH_3^+ H
R

Fig. 2-1 Schematic illustration of the structure of an α-amino acid in the L configuration: tetrahedral and planar representations. The tetrahedron is viewed with NH_3^+—H lying above the plane of the paper and COO^-—R lying below the plane of the paper.

made is between those amino acids whose side chains are relatively insoluble in water (alanine, valine, leucine, isoleucine, methionine, proline, phenylalanine, tryptophan, and tyrosine) and those whose side chains in almost all cases are ionized under the usual conditions of life and thus are strongly hydrophilic (aspartic acid, glutamic acid, lysine, and arginine). Histidine falls close to this latter group due to its conversion to a protonated form at about pH 7. All other amino acids fall between the two extremes.

The second category of amino acids in proteins includes those for which genetic coding has not been established. Most are apparently formed from

Table 2-2 Structures of some modified amino acids that occur in proteins. All are derivatives of amino acids in Table 2-1

L-Cystine (half-cystine in a polypeptide chain is denoted Cys)

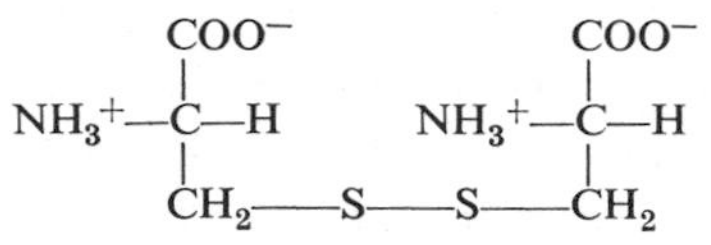

$$\begin{array}{ccc} COO^- & & COO^- \\ | & & | \\ NH_3^+\text{—}C\text{—}H & & NH_3^+\text{—}C\text{—}H \\ | & & | \\ CH_2\text{——}S & \text{——} & S\text{——}CH_2 \end{array}$$

L-Hydroxyproline (Hyp)

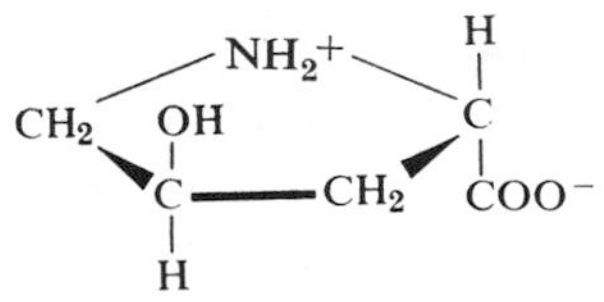

$$\begin{array}{ccccc} & & NH_2^+ & & H \\ & & & & | \\ CH_2 & OH & & & C \\ & | & & & | \\ & C & \text{——} CH_2 & & COO^- \\ & | & & & \\ & H & & & \end{array}$$

L-Hydroxylysine (Hyl)

$$\begin{array}{ccccc} & H & & & NH_3^+ \\ & | & & & | \\ H_2N\text{—}CH_2\text{—} & C & \text{—}CH_2\text{—}CH_2\text{—} & C & \text{—}H \\ & | & & | & \\ & OH & & COO^- & \end{array}$$

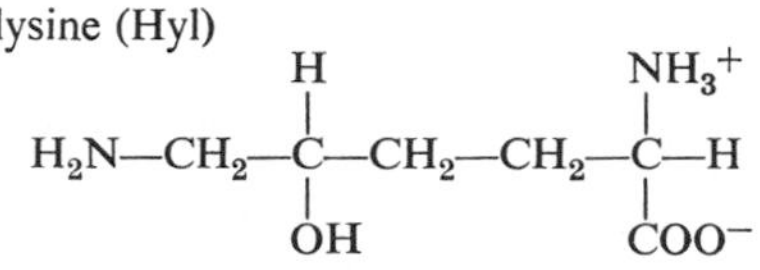

L-Thyroxine (often denoted T4)

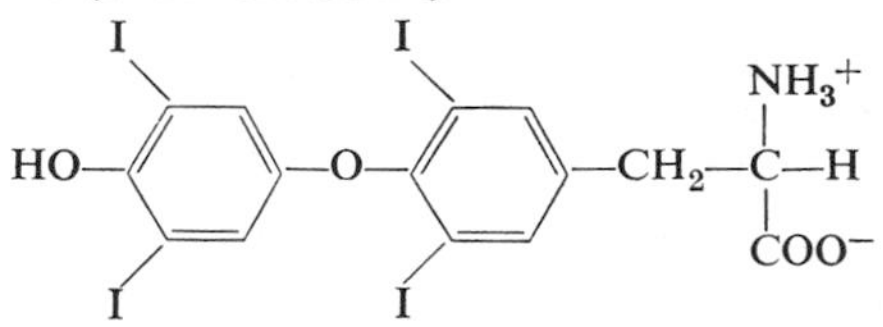

$$\begin{array}{c} I \quad\quad\quad\quad I \quad\quad\quad\quad\quad\quad NH_3^+ \\ HO\text{—}\langle\bigcirc\rangle\text{—}O\text{—}\langle\bigcirc\rangle\text{—}CH_2\text{—}C\text{—}H \\ I \quad\quad\quad\quad I \quad\quad\quad\quad\quad\quad COO^- \end{array}$$

precursors by special reactions that occur during or after polypeptide chain biosynthesis. The structures of some of these are shown in Table 2-2. One of the most important of these derived amino acid residues is cystine, formed by the oxidation of sulfhydryl groups of two cysteine residues in the same or different polypeptide chains. The term half-cystine refers to one cysteine residue participating in a disulfide bond, and is often used to express the total cysteine residue content of proteins, including both cysteine and cystine. Hydroxyproline and hydroxylysine, shown in Table 2-2, have been identified in only a single class of proteins (collagen, elastin). Thyroxine (3,5,3′,5′-tetraiodothyronine) occurs in the protein thyroglobulin as does a triiodo form (T3) and monoiodo- and diiodotyrosine. Other unusual amino acids that form through crosslinking reactions are desmosine and isodesmosine in elastin and di- and trityrosine in resilin. Unusual crosslinkages appear to be a characteristic of many structural proteins. The crosslinks of collagen are discussed in Chapter 16.

Another type of modified amino acid is that representing a simple derivative of one of the amino acids of Table 2-1. Hydroxyproline and hydroxylysine, illustrated in Table 2-2, are among these. Others such as phosphoserine and phosphothreonine are produced by phosphorylation of the hydroxyl function of serine and threonine. Phosphorylation may occur in conjunction with enzyme function (e.g., in phosphoglucomutase and other phosphate-transfering enzymes) or may be mediated by another enzyme for purposes of control (e.g., in the conversion of muscle phosphorylase *b* into the more active phosphoserine-containing form phosphorylase *a* by the enzyme phosphorylase kinase). The occurrence of *O*-sulfotyrosine has also been established (e.g., in fibrinogen) and several enzymes containing adenylated amino acids are known (e.g., *Escherichia coli* glutamine synthetase).

1-1 Configuration of the Amino Acids

The Asymmetric Carbon Atom. Much of the uniqueness and specificity of proteins derives from the absence of symmetry in their primary structures. Later we shall see that the opposite is true at higher levels of organization and that symmetry is an important aspect of quaternary structures. Although the configuration of the α-carbon atom and the optical activity of amino acids are normally treated in basic biochemistry or organic chemistry, a brief review here may be helpful.

The sp^3 hybridization of the valency electrons of the carbon atom produces four equivalent bonding orbitals projected toward the corners of a regular tetrahedron. If the four substituents of the carbon atom are all different, as is the case for the α-carbon of most amino acids, the atom is said to be asymmetric. The four substituents can then be arranged in two different ways to obtain structures which are mirror images of one another. As shown below, for groups W, X, Y, and Z, the two possible structures are:

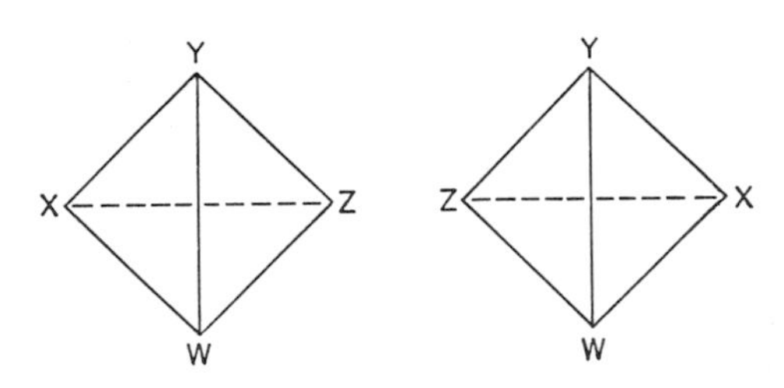

All other ways of arranging the substituents are equivalent, since they may be interchanged by simple rotation. The representation of these structures in two dimensions according to the Fischer convention (in which the tetrahedra are rotated so that W lies below the plane of the paper before projection into that plane) is as follows:

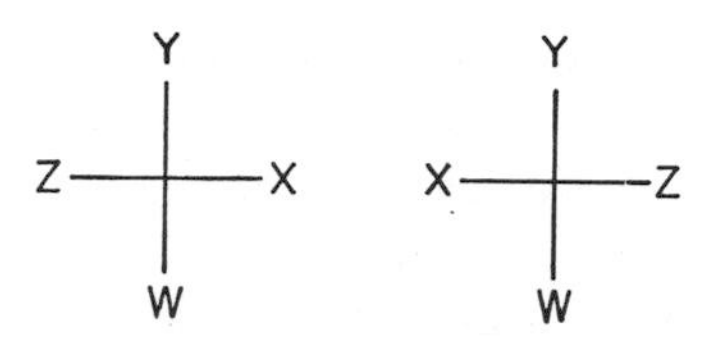

Two such compounds, which are mirror images of one another, are termed *enantiomorphs* (from the Greek *enantio*, meaning opposite). Enantiomorphs are identical in melting point, solubility, and other physical properties but can be distinguished by the fact that they rotate plane polarized light in opposite directions.

Optical Isomerism. All of the amino acids of Tables 2-1 and 2-2 (except glycine) possess at least one asymmetric center, the α-carbon atom, and exhibit optical activity, the ability to rotate the plane of polarization of linearly polarized light. Although optical rotation is not completely understood on a quantitative basis, theories have been developed which facilitate at least a qualitative understanding (cf. Chapter 9). A particularly helpful approach is the demonstration that a chromophoric group (any group of atoms in which an electronic transition occurs) lying in a potential field of suitable dissymmetry (such as that produced by the charge distribution of groups about an asymmetric carbon atom) has rotatory capacity. According to the general quantum mechanical equation for optical rotation, the molecular rotation of a substance consists of contributions from all of the absorption bands of the molecule. Contributions also result from coupling effects between chromophoric groups, but the relative importance of these various factors is still debated.

In the one-electron model of optical rotation developed by Condon, Altar, and Eyring (cf. Chapter 9), it is shown that rotation of the plane of polarization of plane polarized light will occur for an electron whose energy levels are perturbed by a potential field lacking planes of symmetry or a center of symmetry. When the tetrahedral structures of glycine and L-alanine are examined, a plane of symmetry (determined by the α-carbon, the carboxyl carbon, and the amino nitrogen) can be located in the glycine structure due to the presence of two identical substituents on the α-carbon atom. Hence, glycine molecules in solution (which can rotate to assume all possible orientations in space) do not exhibit optical rotation. For alanine, on the other hand, all substituents of the α-carbon atom are different and thus the potential field produced by the electron distribution in these groups has neither a plane of symmetry nor a center of symmetry. Thus alanine has the ability to rotate the plane of polarized light. This concept may be extended to understand qualitatively why certain conformations in proteins and polypeptides, such as the helical structure of poly-L-alanine, exhibit optical rotation differing from that expected for the sum of rotations of the individual amino acid residues (cf. Chapter 9).

When polarized light is passed through a solution containing an optically active material, the emerging resultant beam is still plane polarized light, but a small rotation of the plane has occurred. The degree of rotation may be determined with an apparatus as depicted in Fig. 2-2. A light source L is placed at the focal point of a lens such that light passing through this lens will be parallel. Preceding the lens is a polarizing prism or filter which passes the light waves that oscillate in one plane only. The polarized beam travels through the sample solution and then through a second polarizing prism (analyzer) from which the intensity of the emergent light can be determined with a photoelectric device or observed visually. In the absence of an optically active sample and with the polarizer and analyzer in a crossed (perpendicular) position, no light comes through. When an optically active sample is then placed between the polarizer and analyzer, rotation of the plane of polarization occurs, and the analyzer in the crossed position no longer blocks all light transmission. To obtain total darkness, the analyzer must be rotated, and the number of degrees of rotation is then noted. If the analyzer must be rotated clockwise as viewed by an observer looking toward the light source, the substance is said to have a positive optical rotation (dextrorotatory); if

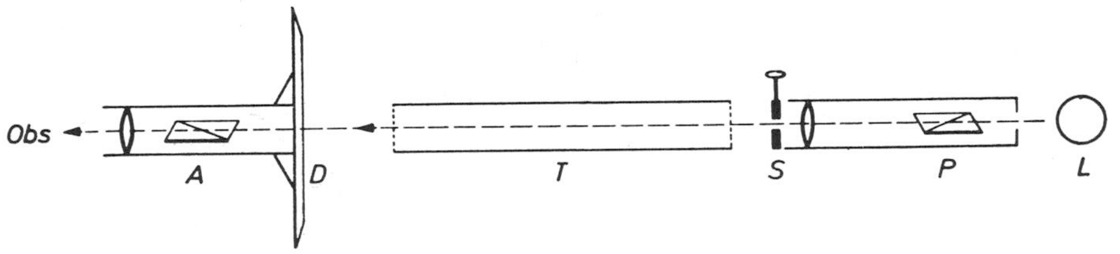

Fig. 2-2 Schematic illustration of a simple polarimeter for the measurement of optical rotation. L, light source; P, polarizer; S, slit; T, sample tube, and D, a graduated disc rotating with the analyzer A. (From B. Jirgensons, *Optical Rotatory Dispersion of Proteins and Other Macromolecules*, Springer-Verlag, New York, 1969.)

counterclockwise, the substance has a negative rotation (levorotatory). An equal mixture of the dextrorotatory form (given the symbol *d*) and the levorotatory form (given the symbol *l*) of a compound is termed a racemic mixture or *dl* form; it produces no net rotation.

Optical rotatory power is normally expressed in terms of the specific rotation $[\alpha]_\lambda^T$ given by the formula

$$[\alpha]_\lambda^T = \frac{\alpha}{lc}$$

where α is the observed rotation (degrees), l is the length of the sample tube (dm), and c is the concentration (g/ml). The rotation depends upon the wavelength of light used (λ), the temperature (T), and solvent, and these conditions must be specified. For example, $[\alpha]_D^{24}$ refers to rotation at 24°C, measured at the sodium D line ($\lambda = 589$ nm). The solution pH is an important factor in the case of the amino acids since their optical rotation is influenced by the ionic state of the ionizable groups.

D- and L-Amino Acids. It is clear from the above that although the direction (or sign) of optical rotation depends upon the arrangement of substituents about an asymmetric carbon atom, it does not yield information on absolute configuration. According to convention, the capital letters D and L are used to denote particular configurations. When absolute configuration is not known, optically active compounds are designated *d* or *l* depending upon the direction of rotation in aqueous solution at 589 nm. There is no correlation between absolute configuration, D or L, and observed direction of rotation, *d* or *l*; the structure shown earlier for L-alanine is, in fact, the dextrorotatory form under most conditions of measurement.

Although D-isomers of the amino acids are found in nature, all of the amino acids that occur in proteins have the L configuration. Resolution of amino acids (separation of optical isomers in a D,L mixture) may be accomplished in the laboratory by the use of specific enzymes that degrade one but not the other enantiomorph. This is particularly effective for acylated amino acids whose hydrolysis by certain hydrolytic enzymes is highly stereospecific. Column chromatographic procedures using optically active packing materials such as cellulose have also proved of moderate success in resolving racemic mixtures of amino acids.

Compounds having more than one asymmetric center may exist in as many as 2^n different configurations, where n is the number of asymmetric centers in the molecule. Some of these (diastereoisomers) will not be mirror images of one another. Threonine, isoleucine, hydroxyproline, and hydroxylysine each have two asymmetric carbon atoms and four possible configurations. Those of threonine are shown schematically below:

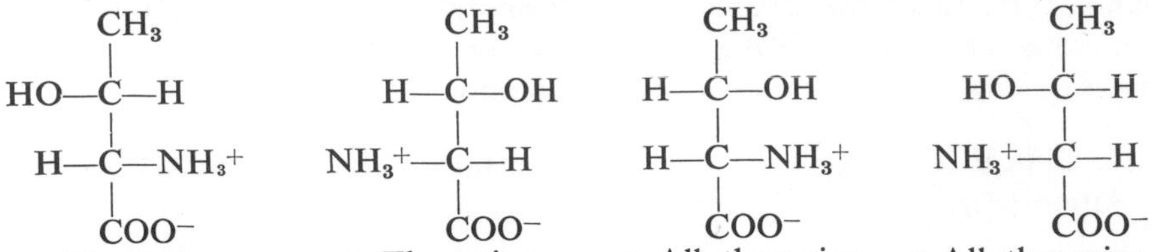

$$\begin{array}{cccc}
\begin{array}{c} CH_3 \\ | \\ HO-C-H \\ | \\ H-C-NH_3^+ \\ | \\ COO^- \end{array} &
\begin{array}{c} CH_3 \\ | \\ H-C-OH \\ | \\ NH_3^+-C-H \\ | \\ COO^- \end{array} &
\begin{array}{c} CH_3 \\ | \\ H-C-OH \\ | \\ H-C-NH_3^+ \\ | \\ COO^- \end{array} &
\begin{array}{c} CH_3 \\ | \\ HO-C-H \\ | \\ NH_3^+-C-H \\ | \\ COO^- \end{array} \\
\text{L-Threonine} & \text{D-Threonine} & \text{L-Allothreonine} & \text{D-Allothreonine}
\end{array}$$

Only L-threonine has been found to occur in proteins. Another interesting type of isomer is that illustrated by mesocystine, formed by disulfide linkage between mirror image isomers:

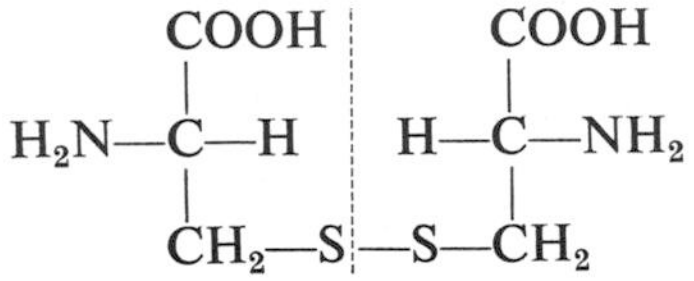

$$\begin{array}{cc|cc}
\text{L-Cysteine portion} & \begin{array}{c} COOH \\ | \\ H_2N-C-H \\ | \\ CH_2-S \end{array} & \begin{array}{c} COOH \\ | \\ H-C-NH_2 \\ | \\ -S-CH_2 \end{array} & \text{D-Cysteine portion}
\end{array}$$

This molecule has compensating asymmetric centers and therefore has no net optical rotation, since one half of the molecule exactly compensates for the other.

1-2 Acid-Base Properties of Amino Acids[1,2]

All amino acids contain an acidic —COOH group that can ionize to —COO⁻ and H^+ and a basic amino (or imino) group that can accept a proton to form the quaternary ammonium salt (e.g., $R-NH_3^+$). Substances which can behave as both acids and bases are said to be amphoteric. The dissociation of glycine may be represented as

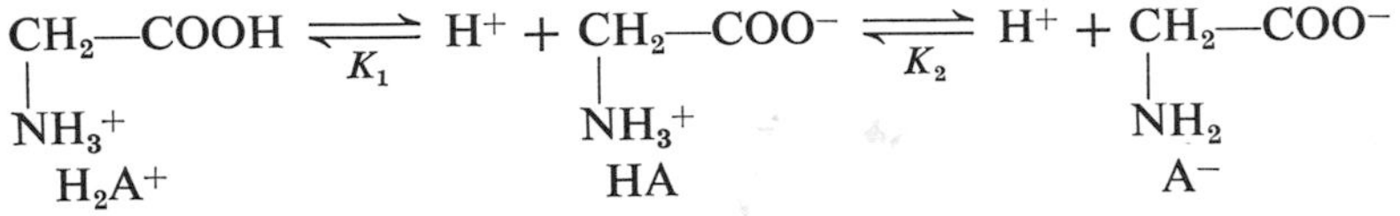

$$\begin{array}{c} CH_2-COOH \\ | \\ NH_3^+ \\ H_2A^+ \end{array} \underset{K_1}{\rightleftharpoons} H^+ + \begin{array}{c} CH_2-COO^- \\ | \\ NH_3^+ \\ HA \end{array} \underset{K_2}{\rightleftharpoons} H^+ + \begin{array}{c} CH_2-COO^- \\ | \\ NH_2 \\ A^- \end{array}$$

The species HA, which has no net charge, is an "inner salt" or zwitterion. Glycine has been shown to have this form in the crystalline state; glycine also crystallizes as a hydrochloride ($HOOC-CH_2-NH_3^+Cl^-$). The titration curve of the zwitterion form of glycine is shown in Fig. 2-3. When glycine is dissolved in water at 0.1 *M*, it exists primarily as the zwitterion, having a pH of about 6. Addition of acid causes the carboxylate group to titrate; the inflection point of the curve occurs at pH 2.4 (pK_1'). The $-NH_3^+$ group titrates upon addition of base with an inflection point at about pH 9.6 (pK_2'). By convention p*K*'s are numbered starting with the most acidic group dissociation as pK_1'.

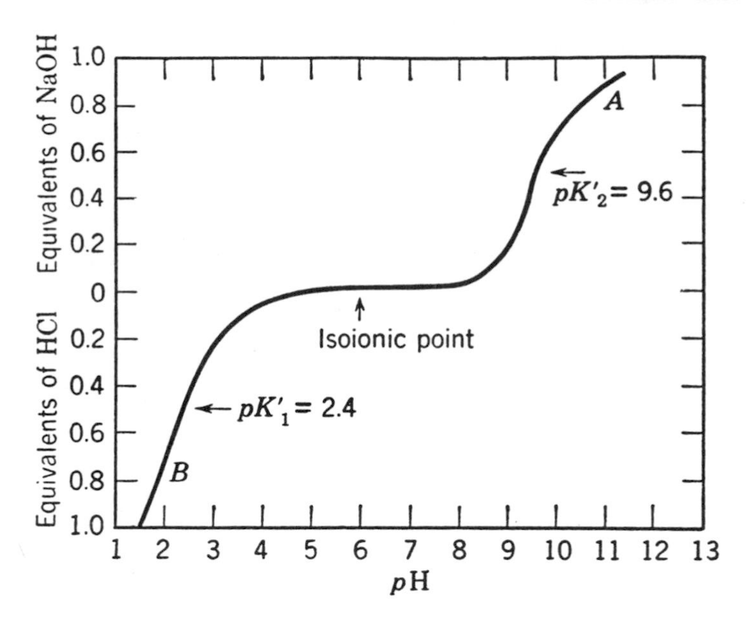

Fig. 2-3 Titration curve of glycine. (From J. S. Fruton and S. Simmonds, *General Biochemistry*, Wiley, New York, 1958.)

The titration behavior of the side chains of the ionizable amino acids are of considerable interest to protein studies. Figures 2-4*a* and 2-4*b* illustrate their contribution to the titration of histidine and aspartic acid in the free amino acid and dipeptide forms. The degree of ionization of the side chains of tyrosine and cysteine can be conveniently followed by measurement of ultraviolet absorption. Figures 2-5*a* and 2-5*b* show the changes in absorbance with pH associated with the ionization of these side chains. From the data for cysteine it is possible to obtain directly the titration curve for the sulfhydryl group, as shown in Fig. 2-6. These data are of particular interest in relation to the disulfide interchange reaction in proteins which depends on the presence of ionized sulfhydryl groups. Results of p*K* determinations in model compounds (substances which serve as simple chemical models for amino acid residues in proteins) and in proteins are presented in Chapter 10.

1-3 Reactions of Amino Acids[3–5]

The reactions discussed in this section represent some aspects of amino acid chemistry of particular interest in protein work. An elaboration of these and other reactions and references to the original literature may be found in the general references cited above.

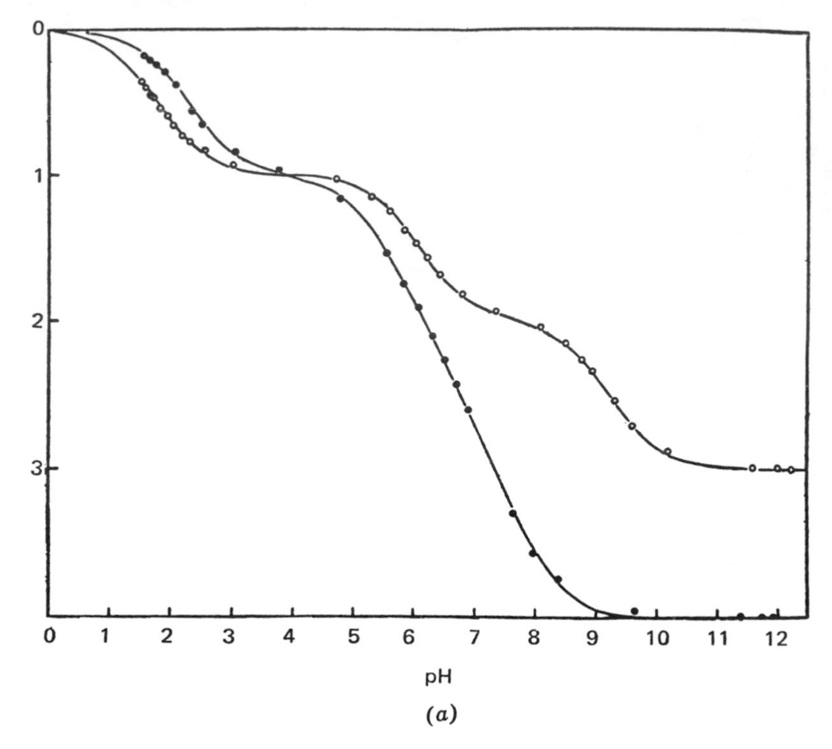

Fig. 2-4*a* Titration curves for histidine (open circles) and histidylhistidine (closed circles) based on the following ionization constants: histidine, $pK_1' = 1.77$, $pK_2' = 6.10$, $pK_3' = 9.18$; histidylhistidine, $pK_1' = 2.25$, $pK_2' = 5.60$, $pK_3' = 6.80$, $pK_4' = 7.80$. [From J. P. Greenstein, *J. Biol. Chem.*, **93,** 479 (1931).]

Amino acids undergo typical reactions for primary amines. One such reaction involves the action of nitrous acid on the α-amino group:

$$NH_2—CH_2—COOH + HNO_2 \rightarrow HO—CH_2—COOH + N_2 + H_2O$$

This reaction was used in 1912 by Van Slyke as the basis of his "nitrous acid" method for the estimation of amino acids by volumetric or manometric techniques. The reaction may be used to modify amino groups in proteins for studies of the role of those groups in relation to structure and function.

The α-amino group of amino acids reacts readily with a variety of acylating agents. Among these are the acid chlorides such as acetyl chloride (CH_3COCl), benzoyl chloride (C_6H_5COCl), and carbobenzoxy chloride ($C_6H_5CH_2OCOCl$). The reaction of a primary amine with acetyl chloride is

$$CH_3CO—Cl + NH_2—R \rightarrow CH_3CO—NH—R + HCl$$

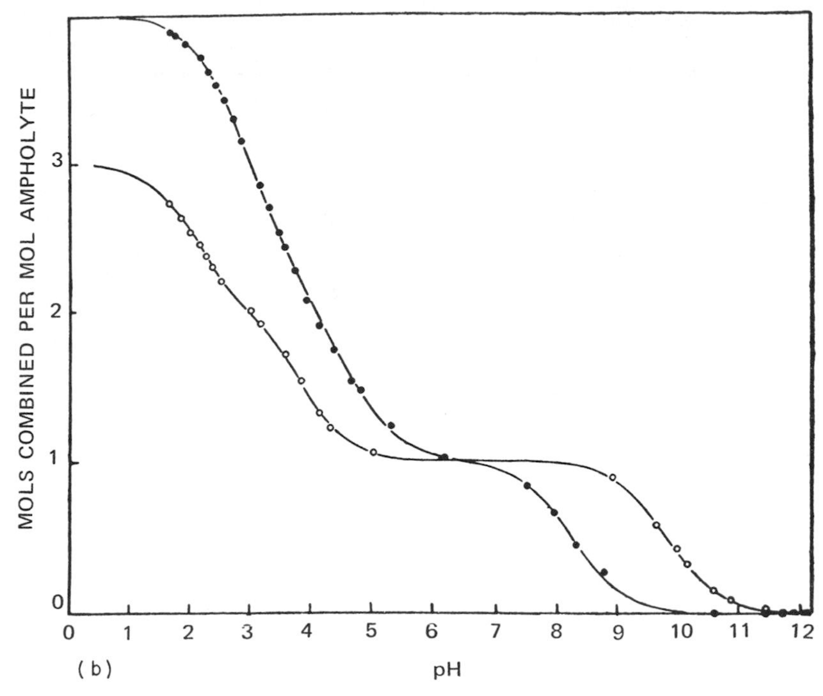

Fig. 2-4b Titration curves for aspartic acid (open circles) and aspartyl–aspartic acid (closed circles) based on the following ionization constants: aspartic acid, $pK_1' = 2.10$, $pK_2' = 3.86$, $pK_3' = 9.82$; aspartyl–aspartic acid, $pK_1' = 2.70$, $pK_2' = 3.40$, $pK_3' = 4.70$, $pK_4' = 8.26$. [From J. P. Greenstein, *J. Biol. Chem.*, **93**, 479 (1931).]

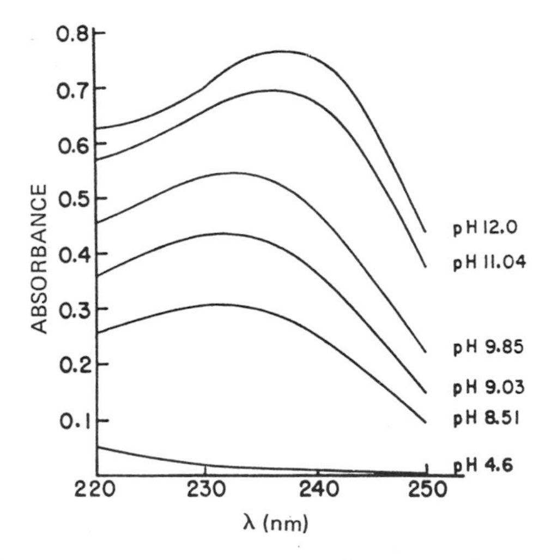

Fig. 2-5a Absorption spectra of cysteine solutions in the ultraviolet as a function of pH. At pH 12 to 13, where dissociation to—S^- is complete, the molar absorptivity is approximately 5000 M^{-1} cm^{-1} at λ_{max}. [From R. E. Benesch and R. Benesch, *J. Amer. Chem. Soc.*, **77**, 5877 (1955).]

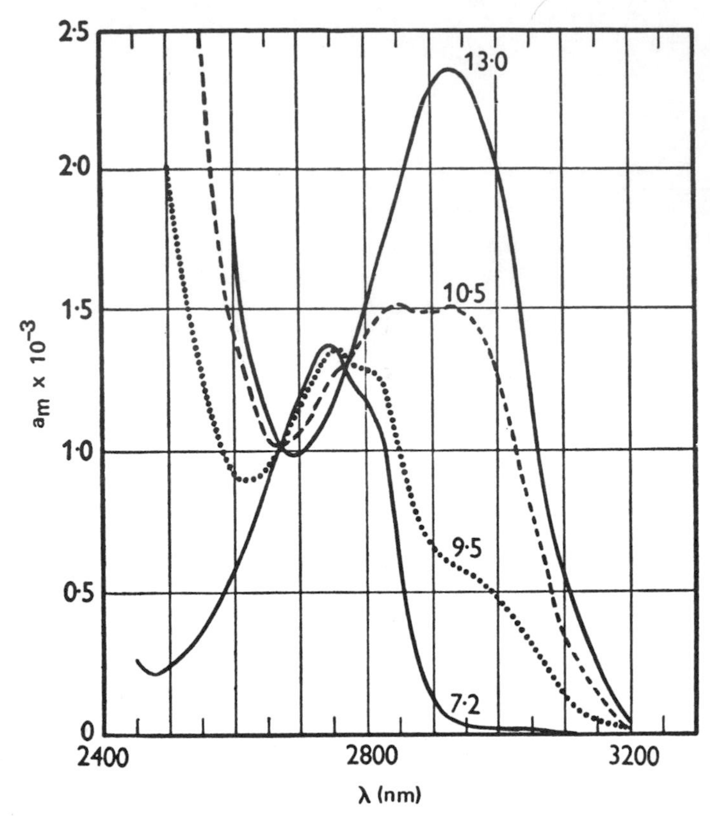

Fig. 2-5b Absorption spectra of aqueous tyrosine solutions at pH values in the region of the dissociation of the phenolic group. The apparent p*K* determined from these data is 10.0. [From D. Shugar, *Biochem. J.*, **52,** 142 (1952).]

Reactions of this type are of value in peptide synthesis (cf. Chapter 4). Several amino group reagents of importance in the determination of the sequence of amino acids and proteins, such as phenylisothiocyanate (C_6H_5NCS) and 1-fluoro-2,4-dinitrobenzene, are discussed in Chapter 4.

The amino group may react with various aromatic aldehydes such as benzaldehyde to form condensation products of the Schiff base type:

$$C_6H_5CHO + NH_2—R \rightarrow C_6H_5CH{=}N—R + H_2O$$

The ε-amino group of lysine is also subject to this type of reaction, and a linkage of this type with an aliphatic aldehyde is thought to be responsible for crosslinking of polypeptide chains in collagen (cf. Chapter 17).

The reaction of amino acids with formaldehyde produces the methylol derivative of the amino acid:

$$R—NH_2 + HCHO \rightarrow R—NHCH_2OH$$

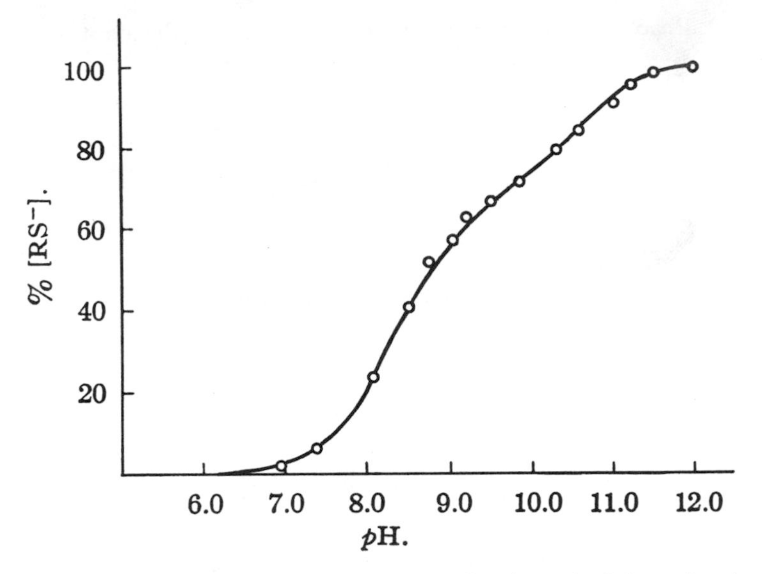

Fig. 2-6 Titration of the sulfhydryl group of cysteine determined from the ultraviolet absorption of the ionized form $[RS^-]$ at 232 to 238 nm (see Fig. 2-5*a*). [From R. E. Benesch and R. Benesch, *J. Amer. Chem. Soc.*, **77,** 5877 (1955).]

This compound reacts with an additional mole of formaldehyde to yield the dimethylol derivative:

$$R—NHCH_2OH + HCHO \rightarrow R—N(CH_2OH)_2$$

Reaction with formaldehyde and glutaraldehyde (a "double-headed reagent") is often carried out in tissues and in isolated protein preparations in order to fix or preserve structural features for subsequent microscopic examination. The reaction of amino acids with ninhydrin (triketohydrindene hydrate) is of special importance in biochemistry since it can be used for the quantitative determination of amino acids and is convenient for the detection of amino acids on chromatograms. Most amino acids react with ninhydrin to yield hydrindantin, carbon dioxide, ammonia, and the corresponding aldehyde:

$$C_6H_4{<}^{CO}_{CO}{>}C{<}^{OH}_{OH} + NH_2\overset{R}{\overset{|}{C}}HCOOH \rightarrow$$

Ninhydrin

$$C_6H_4{<}^{CO}_{CO}{>}C{<}^{H}_{OH} + RCHO + NH_3 + CO_2$$

Carbon dioxide formed in the reaction may be measured manometrically. A specific test for glycine is obtained by determination of formaldehyde produced. Reaction of ammonia with ninhydrin and hydrindantin yields an intense blue-violet color called Ruhemann's Purple:

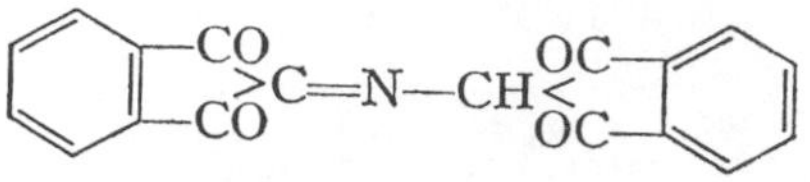

$$C_6H_4{<}\begin{matrix}CO\\CO\end{matrix}{>}C{=}N{-}CH{<}\begin{matrix}OC\\OC\end{matrix}{>}C_6H_4$$

A yellow product is obtained with proline and hydroxyproline under certain conditions. The ninhydrin reaction permits colorimetric analysis of peptides and other primary amines, as well as free amino acids, and is widely employed in chromatographic procedures.

Free α-carboxyl groups are esterified by reaction with alcohols such as methanol under acidic conditions:

$$RCOOH + CH_3OH \rightarrow RCOOCH_3 + H_2O$$

Reduction of free carboxyl groups or of esters produced in the above reaction can be achieved by a variety of reducing agents (e.g., lithium borohydride) to yield the corresponding amino alcohol.

Under suitable pH conditions, amino acids and peptides form complexes with cupric ions of the type

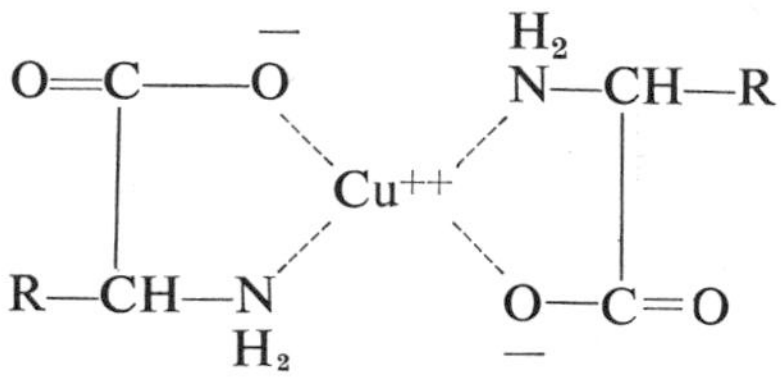

```
                 _             H2
   O═C─────O                N─CH─R
     |      ╲              ╱   |
     |        Cu++             |
     |      ╱              ╲   |
 R─CH─N                     O─C═O
      H2                     _
```

This reaction forms the basis of a sensitive spectrophotometric method for quantitation of free amino acids (cf. Chapter 3).

Reactions Used for Identification of Amino Acids. Automated chromatography is generally the method of choice for identification and quantitation of free amino acids, provided the necessary apparatus is available (cf. Chapter 4). However, many other methods exist for a variety of special applications, particularly where only one amino acid is to be assayed. Details of the following procedures may be found in Reference 3.

Two methods of analysis that have general applicability are the nitrous acid method and the ninhydrin reaction described above. Enzymatic methods based on the specificity of an enzyme for a particular amino acid are of value in the determination of certain of the amino acids; for example, in the decarboxylation of histidine, lysine, arginine, glutamic acid, aspartic acid, phenylalanine, and tyrosine by specific bacterial preparations. Bacterial strains that have a specific growth requirement for an amino acid are useful in detection of minute quantities of that amino acid.

Certain α-amino acids undergo specific reactions that may be used for analysis. Serine, threonine, and hydroxylysine, which have hydroxyl and amino groups on adjacent carbon atoms, are oxidatively cleaved by periodate, as shown below for serine:

$$HOCH_2CH(NH_2)COOH + HIO_4 \rightarrow OHCCOOH + HCHO + NH_3 + HIO_3$$

The formaldehyde produced from serine or hydroxylysine (acetaldehyde from threonine) or one of the other products (e.g., the ammonia) can be measured quantitatively. The aromatic amino acids tryptophan and tyrosine are conveniently detected and quantitatively measured by means of their specific absorption in the ultraviolet (cf. Chapter 9). The amide nitrogen of glutamine and asparagine may be assayed by analysis of ammonia released upon hydrolysis of the amide groups. One common procedure is the Conway microdiffusion technique. Additional information and literature citations may be found in References 3 to 6.

Specific color reactions are valuable in chromatographic procedures for detection of particular amino acids either in a free state or as components of peptides. The Ehrlich stain for indoles permits the location of tryptophan or tryptophan-containing peptides on chromatograms. A distinctive purple color is observed after treatment with *p*-dimethylaminobenzaldehyde in acidic acetone medium for a few minutes at room temperature. Imidazole compounds (histidine) can be detected by treatment with sulfanilic acid in alkaline sodium nitrate solution (Pauly reaction). A weak reaction is also given by phenolic compounds (tyrosine). An effective means for location of tyrosine or tyrosine-containing peptides is the red color formed with 1-nitroso-2-naphthol. Isatin reagent (indole-2,3-dione) is used for specific detection of proline and hydroxyproline on chromatograms.

The Sakaguchi reaction is used in analysis for arginine either as the free amino acid or in proteins. Arginine and other monosubstituted guanidines produce a deep red color upon treatment with 1-naphthol and sodium hypobromite (or sodium hypochlorite). A highly sensitive test for cysteine (and other SH-containing compounds) is obtained by reaction with nitroprusside solution (sodium nitroferricyanide). Detection of cystine is possible by pretreating with sodium cyanide in methanol. Polychromatic techniques are available for detection of as many as 20 amino acids on a single paper chromatogram. In one procedure a ninhydrin–cupric nitrate reagent is used; another involves treatment with diethylamine followed by ninhydrin. Experimental details for these colorimetric methods are given by Bailey.[7]

2 OTHER COMPONENTS OF PROTEINS

Proteins that occur in biological systems are often functionally associated with nonprotein substances either through noncovalent interactions or, in some cases, through covalent crosslinkages. In this section we consider briefly some of the categories of conjugated proteins that may be isolated as defined entities from biological systems.

Several important classes of macromolecules contain protein in intimate association with nucleic acids, both DNA and RNA. The association is noncovalent in nature. One important class of nucleoproteins is the viruses (complex viruses contain lipid and carbohydrate as well) in which the protein is thought to play several functional roles.[8] The protein component of viruses is generally external to the nucleic acid, thereby affording protection against nuclease digestion and functioning in a "packaging" role. Additional functions of proteins in viruses, including host range specificity and cellular penetration, are also evident in some viral systems. Another nucleoprotein, the ribosome, plays a leading role in protein synthesis. This RNA-containing macromolecule contains several types of polypeptide chains and both their structure and function are the subject of intensive current investigation.[9] Small basic proteins (and perhaps other proteins as well) are associated with the DNA genome in cells,[10] and both structural and control functions have been postulated for these proteins.

Proteins may be found conjugated to lipids in several important systems. An important case of this association in structural components (e.g., in cell membranes) has been difficult to study by techniques described here, because only extreme denaturing conditions have permitted the isolation of the protein (freed from the lipid). A more thoroughly studied and better defined class of lipid-conjugated proteins are the lipoproteins found in the blood stream of higher animals. In man, these may be divided into three groups, the high-density lipoproteins, the low-density lipoproteins, and the chylomicra. These macromolecules apparently serve important functions in transporting lipids to and from various body tissues. There is mounting evidence that the lipid is primarily attached by noncovalent interactions and that the lipoproteins may become larger or smaller in size depending on the amount and type of lipid available in the physiological medium[11]. As a result, considerable heterogeneity is found in these lipoproteins with respect to size, molecular weight, and density.

Glycoproteins are formed by conjugation of proteins with oligosaccharides through covalent linkages.[12] The carbohydrate is usually attached through *O*-glycosidic linkages to serine or threonine, or through a glycosylamine structure involving the reducing end of the carbohydrate and the amide group of asparagine, or less commonly glutamine. The carbohydrate portion is attached to the protein as one or more relatively short segments (e.g., molecular weight 520 to about 3500). The carbohydrate usually contains glucosamine and/or galactosamine and one or more of the monosaccharide units galactose, mannose, fucose, and sialic acid. The amount of carbohydrate in glycoproteins varies from just a few percent (e.g., fibrinogen) to over 50% for glycoprotein components of mucous secretions. Glycoproteins containing a high percent of carbohydrate (mucoproteins) have unusual solubility and hydrodynamic properties. The lubricating and clearing function of mucous glycoproteins is readily inferred, but the functional significance of the carbohydrate in other glycoproteins is still not certain. Proteins are also found to be associated with several highly charged polysaccharides such as chondroitin sulfate, heparin, and hyaluronic acid. Both covalent and noncovalent linkages have been reported.

A large number of proteins require small molecular weight cofactors to carry out their biological function. These cofactors are often strongly protein associated and are isolated along with the protein. Both covalent and noncovalent association is common. Among the many known examples are coenzyme and metal-containing enzymes; prosthetic groups like the heme in myoglobin, hemoglobin, and cytochrome *c*; and the metals and cofactors associated with the protein systems of electron transport. Other proteins are involved in more unique associations, such as with the ferric hydroxide core of ferritin. Further details and examples of these protein components may be found in most general textbooks of biochemistry.

REFERENCES

1. E. J. Cohn and J. T. Edsall, *Proteins, Amino Acids and Peptides as Ions and Dipolar Ions*, Hafner Publishing Co., New York, 1943; reprinted 1965.
2. J. T. Edsall and J. Wyman, *Biophysical Chemistry*, Vol. 1, Academic Press, New York, 1958.
3. A. Meister, *Biochemistry of the Amino Acids*, 2nd ed., Vol. 1, Academic Press, New York, 1965.
4. J. P. Greenstein and M. Winitz, *Chemistry of the Amino Acids*, Wiley, New York, 1961, 3 vols.
5. S. Blackburn, *Amino Acid Determination*, Marcel Dekker, Inc., New York, 1968.
6. C. H. W. Hirs, Ed., *Methods in Enzymology*, Vol. 11, Section I, "Amino acid analysis and related procedures," Academic Press, New York, 1967; also C. H. W. Hirs and S. N. Timasheff, Eds., *Methods in Enzymology*, Vol. 25, Academic Press, New York, 1972.
7. J. L. Bailey, *Techniques in Protein Chemistry*, 2nd ed., Elsevier, New York, 1967.
8. H. Fraenkel-Conrat, Ed., *Molecular Basis of Virology*, Reinhold, New York, 1968.

9. C. G. Kurland, "Ribosome structure and function emergent," *Science*, **169**, 1171 (1970).
10. R. H. Stellwagen and R. D. Cole, "Chromosomal proteins," *Ann. Rev. Biochem.*, **38**, 951 (1969).
11. A. V. Nichols, "Human serum lipoproteins and their interrelationships," *Adv. Biol. Med. Phys.*, **11**, 109 (1967).
12. A. Gottschalk and E. R. B. Graham, "The basic structure of glycoproteins," in *The Proteins*, 2nd ed., Vol. 4, Academic Press, New York, 1966. A. Gottschalk, *Glycoproteins*, Elsevier, New York, 1966.

III Fractionation and Analysis of Proteins[1–10]

Living systems contain thousands of genetically distinct proteins. Many occur in combination with other macromolecules and/or various small molecules. In most cases study of a particular protein requires identification of the protein (through a biological assay or by characteristic physical or chemical properties) and some degree of purification. The extent of purification required depends on the type of subsequent study to be undertaken. Very high purity is required for meaningful chemical analysis (amino acid composition, sequence determination) and for most kinds of physical studies, as in the determination of sedimentation and diffusion coefficients (although these physical parameters can be estimated in crude mixtures by the use of biological assays). The determination of molecular weight distribution by sedimentation equilibrium also requires highly purified and chemically homogeneous protein preparations (in solution); single crystals of suitable size are necessary for structure analysis by X-ray crystallography. Other types of physical methods (e.g., sedimentation velocity) are effective for mixtures of proteins or other macromolecules, provided the number of components is not too great.

1 A GENERAL APPROACH TO PROTEIN PURIFICATION

A large proportion of the literature of biochemistry is devoted to the means by which particular proteins are separated from the mass of proteins and other material present in living cells (or in the extracellular structures or fluids of organisms). In most cases the starting point for a purification procedure is the development of a biological assay (e.g., of enzymatic activity) that can be used to distinguish a particular protein from others and to estimate the total quantity of that protein in terms of activity units in the starting material (e.g., in a crude tissue homogenate). The choice of the assay will, of course, dictate the nature of the purified product. The general aim of purification is to increase the specific activity of the preparation (activity/unit weight) at each step, while minimizing losses of total activity.

Although there is no single protocol that will assure success in protein isolation, some key elements in most procedures can be identified. A typical case might be the isolation of an intracellular enzyme from a single-celled organism (e.g., bacteria) or from a fairly homogeneous tissue (e.g., liver) where the great bulk of tissue mass is associated with a single cell type. The first step normally is the liberation of the cell contents by disruption of the tissue in a suitable medium. It is inevitable that the organization of the cell must be destroyed and its contents diluted in order to obtain the desired enzyme activity in a convenient (usually soluble) form for assay and purification. The procedure used for cell breakage is likely to have an effect on the physical properties of the desired protein (e.g., the state of aggregation) in subsequent steps and the distribution of other substances released from the tissue. A medium that is isoosmotic with the intracellular fluid (cell sap) is often chosen in order to aid in preservation of subcellular organelles for subsequent separation. The gentlest possible method for breaking open the cells is generally desirable: for soft tissues, a hand-operated glass homogenizer might be tested first; where cell walls are present, a detergent or gentle grinding might be used.

The second phase of purification usually involves differential centrifugation which may be used to separate nuclei, mitochondria, microsomes or ribosomes, and other subcellular particles from the remaining soluble phase. The latter may be further fractionated by high-speed centrifugation to obtain enzymes present in high-molecular-weight aggregates or complexes. When the desired enzyme activity has been obtained in solution (e.g., as the centrifugal supernatant resulting from particle sedimentation or as an extract or suspension of a sedimented fraction), the investigator may attempt gross fractionation by one or more solubility methods (see Section 2 of this chapter). Methods such as ammonium sulfate fractionation or isoelectric precipitation have the advantage of simplicity and ease in scaling up for large preparations. For proteins that can tolerate them, these procedures are extremely valuable throughout a purification scheme both for elimination of unwanted mass and for concentration of the desired protein.

The next stage of protein purification is based on chromatographic procedures. A typical sequence will ordinarily involve steps of ion-exchange chromatography (e.g., DEAE–cellulose, CM–cellulose), molecular exclusion or gel chromatography (Sephadex, Bio-Gel), and possibly adsorption chromatography (e.g., hydroxyapatite). Affinity chromatography is valuable in special cases. Ion-exchange methods are particularly powerful and may be coupled with molecular exclusion. Prior determination of the pH stability of the enzyme activity indicates the permissible range for experimentation with ion-exchange procedures. The molecular basis of the various chromatographic methods is presented in Section 4 of this chapter. For small-scale preparations chromatography may be followed by preparative zone electrophoresis and isoelectric focusing (Section 3 of this chapter and Chapter 10). Where possible, crystallization of the protein may be carried out to obtain a product of high purity (although contaminants may still be trapped in the crystals).

Quantitation of protein yields may be carried out by a variety of methods (see Section 5 of this chapter). Analytical methods for assessment of purity include discontinuous (disc) electrophoresis in polyacrylamide gels and isoelectric focusing (Chapter 10). Biological assays may be made to test for possible contaminating enzymes. Sedimentation velocity centrifugation (Chapter 7) is often used as a test for physical heterogeneity; although not highly sensitive to impurities (of similar molecular weight), particularly in the case of lower-molecular-weight proteins, it is effective for aggregated species and complexes. Zone centrifugation (Chapter 7) may also be used in these cases. For proteins of high chemical purity, analysis of the distribution of aggregated forms (monomer, dimer, trimer, etc.) is possible by sedimentation equilibrium (Chapter 8).

One of the greatest problems in protein purification is due to the fact that throughout the procedure the investigator must subject the protein to the more-or-less deleterious effects of solution environments quite different from the living state. This is particularly critical in the case of proteins whose entire functional existence occurs within the highly controlled environment of the living cell (many secreted proteins, in contrast, can tolerate more environmental variation without loss of structure and function). Maintenance of high total protein concentration in solution is one way to try to mimic intracellular conditions and reduce the spontaneous denaturation of many proteins that occurs in dilute aqueous solutions. The role of water in relation to the noncovalent interactions responsible for protein three dimensional structure is considered in Chapter 6. Other ways to reduce the aqueous character of solutions during protein purification are the addition of other purified proteins (e.g., bovine serum albumin, gelatin) or of low-molecular-weight substances (e.g., glycerol, sucrose). Temperature is usually kept low during preparative procedures in order to minimize the effect of degradative enzymes that are likely to be present and active after homogenization of the tissue (although cold may cause loss of quaternary structure in some enzymes, cf. Chapter 15). Careful control of pH is advisable to avoid possibly irreversible conformational alterations accompanying titration of ionizable groups. Consideration must also be given to possible side reactions that may develop during purification on account of the necessary changes in solution media. Sulfhydryl proteins (those containing cysteine residues not linked by disulfide bonds) are particularly susceptible to oxidation reactions, and maintenance of a reducing environment is often advisable. An oxygen-free buffer containing a sulfhydryl reagent such as 2-mercaptoethanol may be used. In other cases, however, sulfhydryl compounds may inhibit activity of a protein. For enzymes the presence of the substrate is often helpful in protecting the native form of the protein against denaturation during preparative procedures. The presence of certain divalent ions (e.g., Mg^{++}, Ca^{++}) may contribute to stability of proteins or protein complexes; elimination of undesirable ions (particularly contaminating metal ions introduced from the water supply) by the use of a complexing agent (e.g., ethylenediamine tetraacetate) is sometimes required.

Another factor to keep in mind is the possibility that, even with meticulous care in handling, a protein may undergo structural changes during a purification sequence which cause the final molecular conformation to differ from the fully functional or native form of the protein, even though substantial activity is retained. Under these circumstances, other properties of the protein measured *in vitro* may not accurately reflect the behavior of these macromolecules in the living system, independently of the environmental alteration. Further attention will be given to such changes, or partial denaturation, in Chapter 15.

A special category of denaturation is that of dissociation in proteins containing multiple subunits. In cases where only the activity expressed by one type of subunit is measured during purification, other subunits may be inadvertently lost if dissociation occurs. Similarly, there may be cases where a protein possesses multiple functional forms *in vivo* or where more than one genetically distinct protein exhibits a particular biological activity. Special caution is then needed in extrapolating from the properties of a particular purified protein to the total function *in vivo*. The assay of total activity units during purification is helpful in this regard. The higher the final yield at the end of the process, the more likely it is that the protein obtained is the principal agent for that activity in the starting material. A source of uncertainty, however, is the true value of the initial activity. Assays made with aliquots of unfractionated tissue homogenates are affected by the medium and method used for disruption of the cells and the interaction of other components of the homogenate with ingredients of the standardized assay system. In early stages of purification, total activity may appear to increase as the result of the elimination of substances that interfere with the assay. If the highest total activity is taken as the base, it would be desirable to recover at least 50% of that amount at the end of the purification. A reasonable range achievable in practice is 5 to 20%; whether a selection of molecules has taken place depends on the nature of the purification steps. In some cases, particularly where an enormous amount of starting material is available, an investigator may concentrate on obtaining a high specific activity product even though the procedure is costly in terms of total recovery (e.g., only 1% of starting activity may be recovered). Identity between the final product and the protein responsible for activity in the crude homogenate may be tested by the use of a specific antibody prepared against the purified protein (see Reference 10). Physical properties may be compared by the determination of sedimentation and diffusion coefficients or size (by gel exclusion) based on activity in the unpurified tissue homogenate and based on total protein concentration in the purified material (cf. Chapter 7 and Section 4-4 of this chapter).

The above discussion is intended as a basic guideline for protein purification. A brief review of the most important methods used in preparative procedures is presented in the following sections. Specific details of experimental procedures are available in the cited references. The series *Methods in Enzymology* is a particularly valuable source of information in this area.

2 SOLUBILITY AND SOLUTION METHODS

A number of methods are available for fractionation of protein mixtures based upon solubility differences. The use of graded concentrations of ammonium sulfate (of high purity) is a common procedure, although other neutral inorganic salts are also effective. The precipitation of different proteins as the salt concentration is increased ("salting-out") is discussed further in Chapter 15. Differences in content and surface distribution of charged and polar groups probably play a role in determining the order of precipitation of different proteins as the concentration of ammonium sulfate is increased by steps.

Other methods of fractionating proteins by selective precipitation depend on the variation of temperature or pH or the addition of organic solvents. Some proteins may be selectively precipitated from solution by long periods in the cold; conversely, certain highly stable proteins may be retained in solution at a high temperature where contaminating substances are heat-denatured and coagulated. This method has been used to remove fibrinogen from blood plasma. Addition of substrates and coenzymes is sometimes helpful in stabilizing enzymes sufficiently to permit a heat treatment for the elimination of other proteins. Variation of pH may be used to precipitate proteins at their isoelectric point, the pH at which electrostatic charge repulsions between molecules is at a minimum. Change of dielectric constant or solution polarity by addition of solvents such as ethanol or acetone is effective in some cases, although one must guard against partial denaturation by organic solvents. The same problem exists in the use of salts of heavy metals such as mercury or lead for selective precipitation, although these compounds have proved effective in some cases.

During the purification of proteins it is often necessary to remove small ions or molecules, or to exchange one ionic environment for another. In many cases this is conveniently accomplished by dialysis. The cellulose tubing commonly used for this purpose permits the passage of salts, small peptides, and the like while retaining macromolecules with molecular weights of the order of 12,000 Daltons or more. Because of the length of time required for this procedure (hours to days, depending on conditions), it is not recommended for unstable proteins. For the handling of many materials, including unstable proteins, dialysis has been largely replaced by the more rapid method of molecular exclusion chromatography. Another technique that depends upon the use of membranes of differential permeability is ultrafiltration. Devices using such membranes in a pressure or vacuum system have proved effective for concentrating protein solutions as well as for buffer exchange. Lyophilization, the drying of samples in the frozen state, is used for removal of solvent and other volatile substances from proteins. Practical information on dialysis, ultrafiltration, and lyophilization is given in Volume 22 of *Methods in Enzymology*.[6]

3 PREPARATIVE PHYSICAL METHODS

The movement of macromolecules in solution under the influence of a centrifugal or electrical field has formed the basis for separation and purification of these substances by a variety of methods. In these physical methods it is possible to maintain careful control over solution conditions (pH, temperature, salts, protein concentration) in order to minimize losses due to total or partial denaturation. The theory of these techniques is presented in Chapters 7 and 10; a brief review of their preparative applications follows.

3-1 Ultracentrifugation

Preparative ultracentrifugation is practically indispensable in the early stages of protein purification, as in the separation of fractions in a tissue homogenate. The most common method, involving sedimentation of materials to form a pellet in tubes of a fixed-angle rotor, is a very effective step in the purification of proteins that are part of high-molecular-weight aggregates or subcellular particles, for example, ribosomal proteins. Centrifugation is also used to remove particulate matter and debris from soluble protein preparations. Separations of mixtures of proteins are not usually possible, however, because the size differences are not sufficient to afford worthwhile fractionation. Certain protein mixtures, however, can be handled effectively by zone centrifugation (see sucrose gradient centrifugation, Chapter 7). Large-scale preparations may be carried out in a zonal ultracentrifuge.[11]

3-2 Electrophoresis

The theoretical basis and analytical application of electrophoretic methods are discussed in Chapter 10. Some zone electrophoretic procedures are useful for preparative purposes and will be briefly mentioned here. As in zone centrifugation, the protein sample is confined to a narrow zone in the apparatus. In this case proteins migrate into the running medium (stabilized against convection) under the influence of the applied electric field, and a series of zones is obtained. The effectiveness of the separation depends upon the nature of the mixture and on the medium employed. One of the most widely used and effective media for preparative zone electrophoresis on a moderate scale is polyacrylamide gel.[12] Other media which have been used in various applications include paper, starch-gel, agar, and granular materials. A related technique, isoelectric focusing (cf. Chapter 10), which may be carried out in polyacrylamide gel or a density gradient, affords extremely high resolution of complex mixtures of peptide and proteins. It can be applied on a moderate preparative scale for the purification of macromolecules that are stable and soluble at their isoelectric point.

4 CHROMATOGRAPHY

Chromatography has played a central role in the study of amino acids, peptides, and proteins. Indeed, its use throughout biochemistry and molecular biology is so widespread that a thorough understanding of the principles and procedures of chromatography is important for all biological scientists. Chromatographic methods may be divided into four major categories based on the type of interactions responsible for separation: partition chromatography, ion-exchange chromatography, adsorption chromatography, and molecular exclusion chromatography. Many procedures, however, involve more than one type of interaction in the chromatographic separation.

4-1 Partition Chromatography

In partition chromatography, separation of materials is achieved on the basis of their differing solubilities in two liquid phases. The simplest example of this technique is the method of countercurrent distribution. Here, the solute is equilibrated between two immiscible solvents (e.g., ethyl acetate and water) in a test tube. The ratio of its concentration in the upper phase C_1 to that in the lower phase C_2 is defined as the partition coefficient K. The upper phase is withdrawn and transferred to a fresh volume of the lower phase solvent in another tube; the original lower phase is added to a tube containing a fresh volume of the upper-phase solvent. After equilibration the process is repeated for each tube. Figure 3-1 shows the concentrations in the upper and lower phases of each tube after various numbers of transfers for a solute with $K = 1$. The resulting distribution curve, as shown for $K = 1$ in Fig. 3-2, is symmetrical about the position $n/2$, where n is the number of transfers. If,

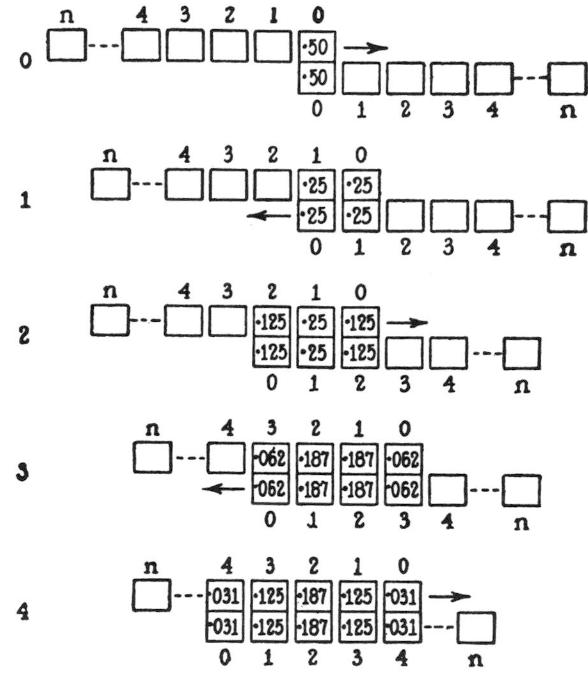

Fig. 3-1 A discontinuous countercurrent fractionation as a function of number of transfers for a solute with partition coefficient $K = 1$. In each diagram the upper set of boxes represents tubes containing pure upper phase solvent while the lower set represents the lower phase solvent. (From L. C. Craig and D. Craig, in *Techniques of Organic Chemistry*, Vol. III, A. Weissberger, ed., Interscience, New York, 1950.)

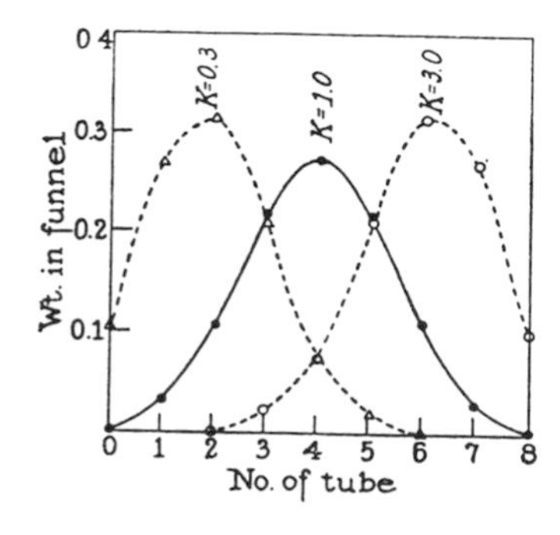

Fig. 3-2 Countercurrent distribution patterns at the eight-tube stage for solutes with partition coefficients of 0.3, 1.0, and 3.0. (From L. C. Craig and D. Craig, in *Techniques in Organic Chemistry*, Vol. III, A. Weissberger, ed., Interscience, New York, 1950.)

however, a solute is present with greater solubility in the upper phase ($K > 1$), it will be concentrated in tubes to the right (in the direction of transfer of the upper phase), whereas one with greater solubility in the lower phase ($K < 1$) will be concentrated in tubes to the left. A countercurrent separation of three components with differing partition coefficients is illustrated by the distribution pattern of Fig. 3-2. After only eight transfers, separation of the solutes with $K = 0.3$ and $K = 3.0$ is almost complete.

Martin and Synge found that it was experimentally more efficient and easier to obtain separation by partition by passing the sample dissolved in a relatively nonpolar solvent through a column containing immobilized water or polar solvent in a supporting medium (e.g., silica gel holding about 50% water). The nonpolar solvent (mobile phase) is passed through the column and the sample distributes itself between the mobile and water phase according to its partition coefficient and rate of movement through the column. Clearly a substance occupying the space offered by the stationary phase to a large extent will be retarded in its passage through the column with respect to a component which is more water insoluble. Column supports for polar stationary phases other than silica gel include starch and cellulose powders.

Partition chromatography using paper as a supporting medium was introduced by Consden, Gordon, and Martin in 1944 and has become a major tool in biochemical research. A simple apparatus for ascending chromatography is shown in Fig. 3-3. The sample to be chromatographed is placed as a spot or thin line (called the origin) a few centimeters from the edge of the paper, and the edge of the paper is allowed to dip into a solvent placed in the bottom of the chromatography jar. The solvent moves up the paper by capillary action (termed development), causing the sample to migrate at a rate depending on its relative distribution between the mobile and stationary phases. The ratio of distances (position of compound − origin position)/(solvent front position − origin position) is the definition of the R_f for that compound under the specified conditions of temperature, solvents, and so forth. The distance separating different components depends

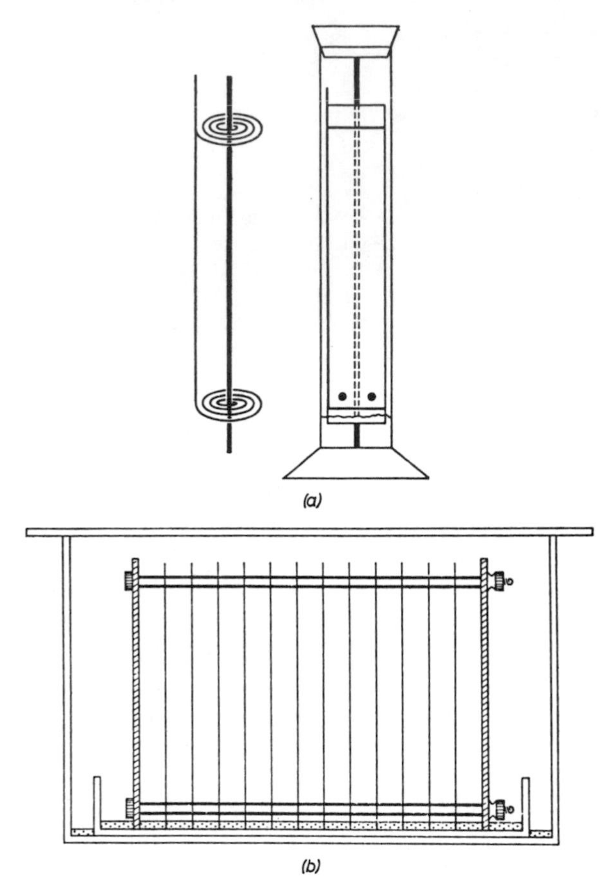

Fig. 3-3 Two types of apparatus for ascending paper chromatography using (*a*) a single sheet on a spiral paper holder or (*b*) multiple sheets. [From R. Ma and T. D. Fontaine, Science **110,** 232 (1949), and S. P. Datta, C. E. Dent and H. Harris, Science **112,** 621 (1950).]

on the distance the solvent front moves; consequently better separation is expected between compounds of similar R_f values by longer development. At some point, however, longer development loses its advantage due to diffusion and adsorptive spreading of the zones. For substances that have low R_f values in a particular solvent, it is often desirable to have the solvent flow a longer distance than allowed by simple ascending technique due to the restrictions on the physical dimensions of the chromatography set-up and the fact that solvent flow rate decreases as the height of the liquid increases. The simple apparatus described may then be modified by placing a pad of folded filter paper at the top to absorb the solvent as it flows off the paper or, as is more

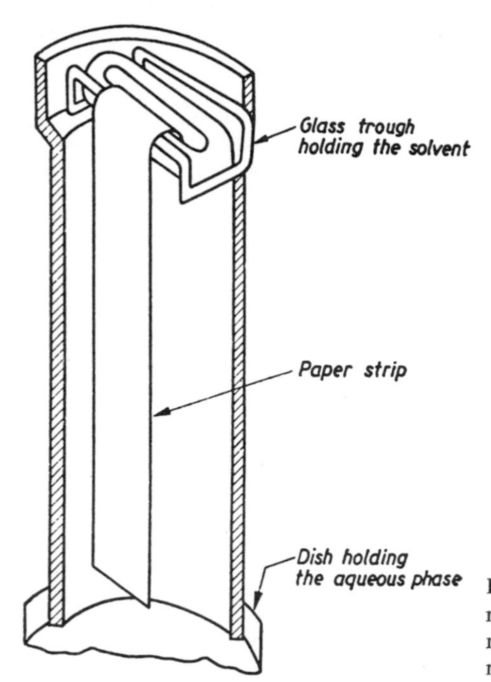

Fig. 3-4 Cutaway diagram of an apparatus for descending paper chromatography in one dimension (container lid not shown).[9]

customary, allow the solvent to flow down the paper and drip off of the bottom continuously (descending chromatography) as shown in Fig. 3-4.

The theory of paper chromatography has not been fully elaborated; however, a reasonable physical picture of the process is afforded by considering it to be one of partition between a mobile phase consisting of the solvent used for development and a more polar immobile phase consisting of a carbohydrate (cellulose)–polar solvent (e.g., H_2O) complex. The stationary phase may arise either by absorption of the polar solvent from the saturated atmosphere or by preferential absorption of the polar component by the cellulose fibers from the leading portion of the moving solvent front. Although true partition is often the major factor in paper chromatographic separation, adsorption and ion exchange also play a role in some systems. The separation of a mixture containing many components is frequently impossible to achieve by development with a single solvent. If the mixture is spotted near one corner of a square sheet of paper and developed with a solvent, it may then be dried (volatile solvents are usually used) and developed with a different solvent at right angles to the first development (2-dimensional paper chromatography). Complete separation of complex mixtures such as amino acids from a protein hydrolyzate may often be obtained by appropriate choice of solvents. A 2-dimensional chromatographic separation of amino acids is shown in Fig. 3-5.

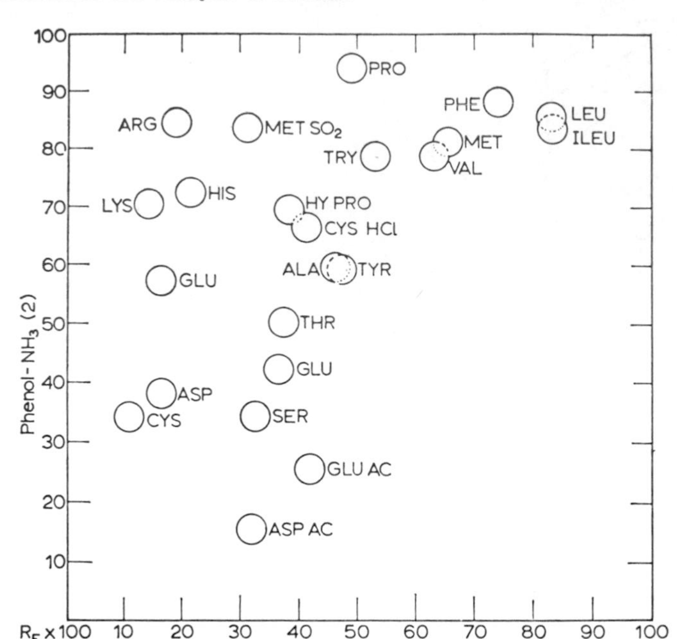

Fig. 3-5 Two-dimensional paper chromatography of amino acids obtained by hydrolysis of lysozyme. Ninhydrin was used for detection.[5]

Reversed phase chromatography on paper, that is, where the more nonpolar phase is the stationary one, may be achieved by chemically reacting the paper to introduce hydrophobic groups onto the hydroxyl functions of the paper or by coating with a water-repellent substance such as silicone. This technique provides high resolution of certain nonpolar compounds.

A technique similar to paper chromatography is that of thin layer chromatography (TLC).[13,14] A slurry of the desired supporting medium and water (or buffer) is spread evenly in a thin layer onto a support such as a glass plate (Mylar or aluminum sheets, etc., may also serve as support). When the solvent has been evaporated, the plate may be developed with appropriate solvents as in paper chromatography. Thin layer chromatography has two principal advantages over paper chromatography. (1) The uniformity and thinness (100–1000 μ) of the chromatographic medium allows the separation and location by appropriate color reaction of extremely small quantities of materials. A ten- to hundredfold greater sensitivity than that possible with paper chromatography is not unusual. (2) A variety of media may be used for the chromatographic support including inorganic material such as silica gel and alumina as well as organic substances such as powdered cellulose. This increases the versatility of the method for chromatographic

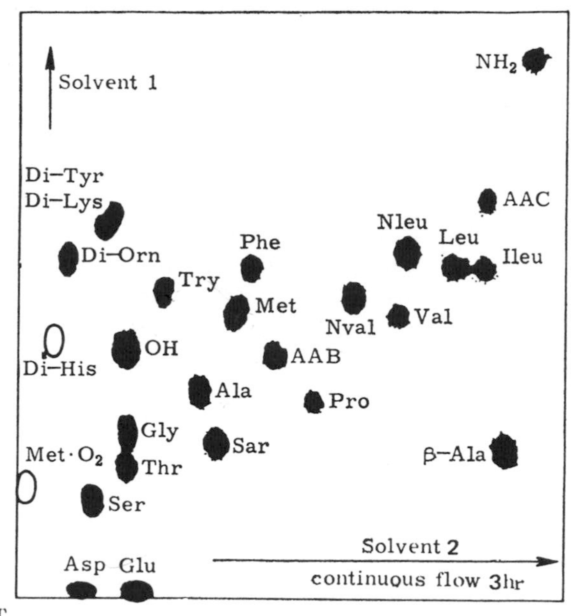

Fig. 3-6 Two-dimensional thin layer chromatography of DNP amino acids (ultraviolet photocopy).[13]

separation and, in the case of inorganic media, allows the use (after development) of chemical reactions to locate compounds that might otherwise prove difficult. Charring at elevated temperature after spraying the plate with concentrated sulfuric acid, for example, will serve to locate any organic compound on silica gel plates. The separation of amino acids in the form of DNP derivatives by 2-dimensional TLC is illustrated in Fig. 3-6.

4-2 Ion-Exchange Chromatography

Large polymers containing charged groups may be made by appropriate polymerization of monomer units or by chemical introduction of such groups into already formed polymers. These polymers (ion-exchange resins) are exceedingly useful in the separation of charged molecules of biochemical interest by ion-exchange chromatography.

The theory of ion-exchange chromatography is quite complex, and the use of the method for large molecules like proteins is for the most part empirical. The following simplified physical picture, however, may be helpful in understanding the nature of the interactions involved. When the insoluble resin is placed in water, it forms a network of charged groups (in the case of molecules which cannot penetrate the resin, the resin particles are seen as a surface of charged groups). If, for example, the Na^+ salt of a strongly acidic resin containing $—SO_3^-$ groups (e.g., Dowex-50) is placed in suspension, some of the Na^+ ions may be displaced upon introduction of other positively charged ions or groups. The relative amounts of the positive ions contributing counterions will depend on both their concentration and affinity for the resin. At a sufficiently low pH and salt concentration, most amino acids can be firmly bound to such a resin. These interactions have considerable strength, since each of many charged groups on the polymer interacts with each cation with a resulting force sufficient to cause strong binding. The Debye-Hückel double layer formed is of very small thickness compared to that formed by interactions of, for example, monovalent salts in solution. Bound ions may be eluted from the resin by reducing this attractive force. This may be accomplished by titrating the bound molecules (or infrequently the resin) to reduce their net charge or by increasing the salt concentration (e.g., Na^+) to a point where the bound molecules are displaced by mass action, even if they are more firmly bound than the salt. The rate of movement of charged species through an ion-exchange column under conditions of elution with a constant composition buffer is therefore determined by the equilibrium constant for binding. A species that spends a large part of the time displaced from the resin will move faster through the column than one more firmly bound. If some components are still bound to the resin after elution with a given solvent, a second solvent of higher ionic strength or different pH may be used to effect their displacement. Such a discontinuous procedure is called stepwise elution. Displacement from ion-exchange columns may also be effected by gradient elution, in which the solvent composition is continuously varied in ionic strength, pH, or both. A simple method for producing a linear gradient between two solvents is illustrated in Fig. 3-7.

Ion-exchange chromatography has become an important tool for separation of proteins. One of the best known media is DEAE–cellulose, in which the diethylaminoethyl group acts as an anion exchanger. The gradient elution pattern of human serum proteins on DEAE–cellulose is illustrated in Fig. 3-8. Other anion exchanging groups include aminoethyl, triethylaminoethyl, and guanidoethyl groups attached to various matrices. Fibrous cellulose is the most commonly used matrix; others used are crosslinked dextran (Sephadex), polyacrylamide gel, and microgranular cellulose. Cation exchanging groups used in protein chromatography include the carboxymethyl group (e.g., CM—cellulose), phosphate (e.g., phosphocellulose), and sulfoethyl group (SE—cellulose). Proteins in solution at pH's above their isoelectric points normally bind to anion exchangers, while below their isoelectric points they will be adsorbed to cation exchangers. The converse may hold either above or below the isoelectric point, however, because of local

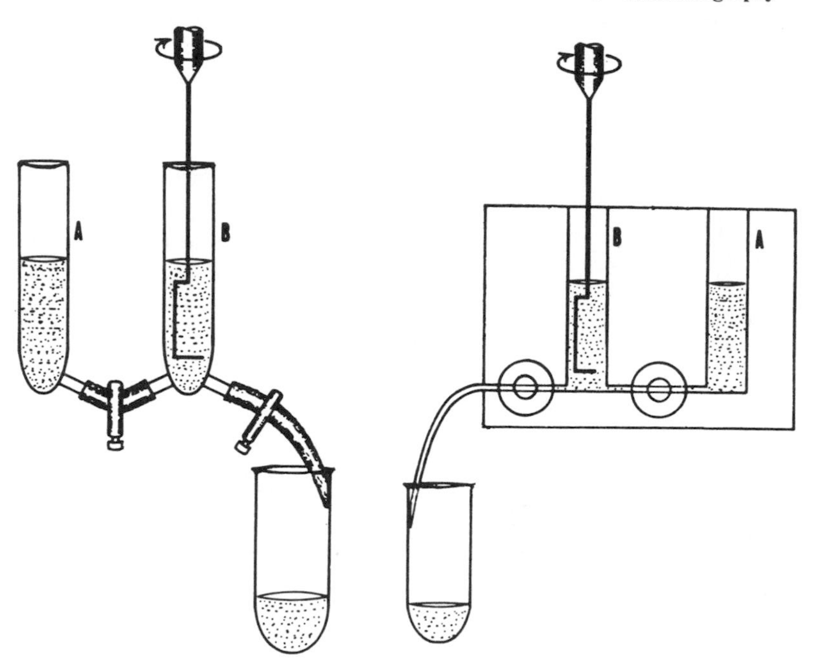

Fig. 3-7 Illustration of simple apparatus for the production of linear gradients for zone centrifugation in density gradients or, on a larger scale, for gradient elution in chromatography. The B reservoir, which is stirred, initially contains the starting elution buffer (or the higher density solution for a density gradient); the A reservoir contains the buffer of the desired final elution concentration (or the low density end of the density gradient). After both reservoirs are filled and leveled, the stopcock closing the connection between them is opened and the outlet from reservoir B is opened. As solution flows from B, buffer moves from reservoir A into B, which is now the mixing chamber, causing a linear change in the concentration of the mixture which flows from the outlet tube. [From J. C. Gerhart, in *Methods in Enzymology*, Vol. XI, C. H. W. Hirs, ed., Academic Press, New York, 1967.]

charge distribution on the surface of the molecules. Elution may be carried out with a salt or pH gradient, which may be linear (obtained as indicated in Fig. 3-7) or of a more complex shape (produced, e.g., with a Varigrad apparatus). Stepwise elution is also useful in large batch purifications or for concentration of a component. Valuable information on types of ion exchangers applicable to protein work and on their preparation for use are given by Himmelhoch in Reference 6.

4-3 Adsorption Chromatography

Solutes may be adsorbed to many kinds of materials: both nonpolar substances, such as charcoal, and polar adsorbents, such as certain oxides and

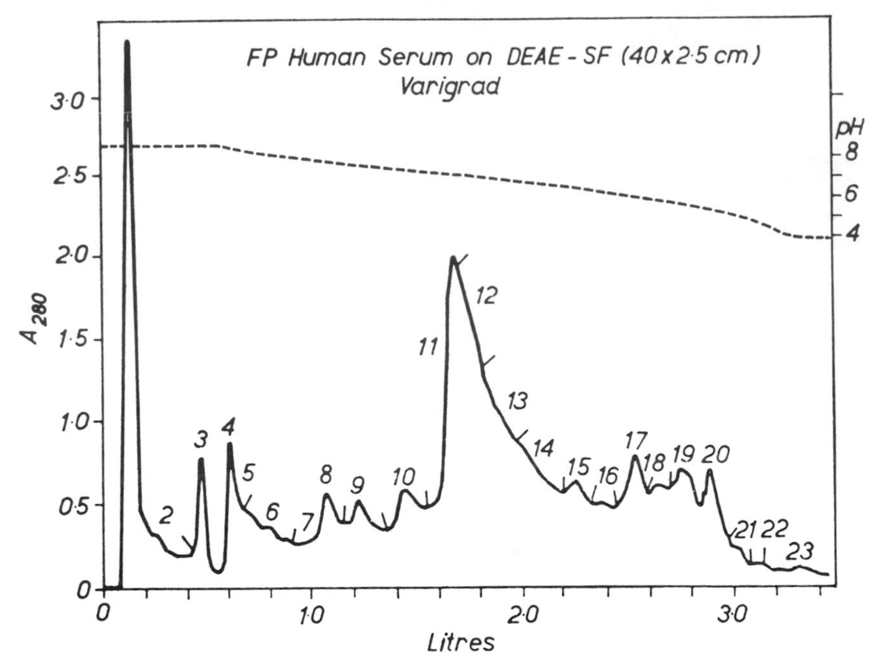

Fig. 3-8 DEAE–cellulose chromatography of human serum proteins with elution by a pH gradient.[8]

salts (e.g., alumina, silica gel, calcium phosphate gel). One of the most important substances for adsorption chromatography today is a form of calcium phosphate, $Ca_{10}(PO_4)_6(OH)_2$, called hydroxyapatite. The preparation of this material and the experimental techniques for its use have been reviewed by Bernardi in Reference 6.

The precise nature of the interactions involved in adsorption chromatography is not known in most cases, particularly those involving nonpolar adsorbents. For different adsorbents the forces involved may include ionic attraction, hydrophobic and van der Waals interactions, or hydrogen bonding. Adsorption effects are frequently noted in systems where ion-exchange or partition is the dominant interaction. In adsorption systems, fractionation is achieved in most cases by sequential elution of the adsorbed substances through displacement by increasing concentrations of another substance that is bound by the adsorbent. In hydroxyapatite chromatography displacement is obtained with sodium or potassium phosphate buffers. Other means for elution of adsorbed materials include alteration of the physical state of the adsorbent or the bound solute (e.g., by titration of charged groups) or a reduction in the net force involved in adsorption (e.g.,

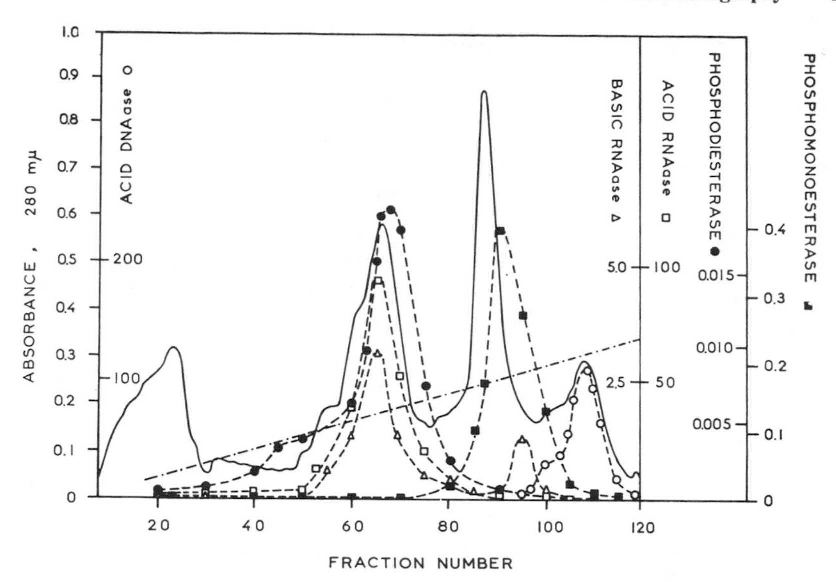

Fig. 3-9 Hydroxyapatite chromatography of a mixture of spleen enzymes. The pattern of elution of five nuclease activities by a phosphate gradient (0.05 to 0.5 *M*), pH 6.8, is shown; the solid line represents absorbance at 280 nm. [From G. Bernardi, A. Bernardi, and A. Chersi, *Biochem. Biophys. Acta*, **129,** 1 (1966).]

lowering the dielectric constant, damping out polar interactions with increased salt).

An example of a protein elution pattern from hydroxyapatite is shown in Fig. 3-9. The adsorption of proteins to hydroxyapatite (prepared so as to have a net positive charge) appears to be primarily dependent on the distribution of negatively charged groups on the surface of the protein molecules. These groups interact with calcium ions at the surface of the hydroxyapatite crystals, the strength of binding being determined by both the number and distribution of negative charges. Polypeptides containing a large number of carboxylate groups show a high affinity for hydroxyapatite; poly-L-glutamate and poly-L-aspartate elute at about 0.25 *M* and 0.35 *M* potassium phosphate, pH 6.8, respectively; highly basic polypeptides and proteins are also strongly adsorbed. Most ordinary proteins elute at lower phosphate molarities.

4-4 Molecular Exclusion Chromatography (MEC)[15,16]

Separation of proteins (and of other molecules) based on molecular weight differences is possible with a technique known variously as molecular exclusion chromatography, gel filtration, or molecular sieve chromatography.

The material used to form the matrix for MEC is a polysaccharide, polyacrylamide, or similar substance and is obtained dehydrated in the form of small uniform particles or beads. It takes up solvent rapidly and reversibly to produce a slurry that may be used to pack a column or to coat plates for TLC. The gel thus produced contains pores of fairly uniform size, and it is the relative ability of solutes to diffuse through these pores into the gel matrix that provides the basis of their separation.

Solutes that are small with respect to the pore size diffuse into the gel particles, as does the solvent, and will pass through the column at the same rate as the solvent. These substances are said to be "totally retarded." On the other hand, solute molecules much larger than the gel pores cannot enter the gel matrix and pass through the column more rapidly, corresponding to "total exclusion." Thus, when a small volume of solution containing such large molecules is layered over a column of an appropriate gel, these molecules will come off the column immediately after the "exclusion volume," that is, the volume of solvent outside the gel particles. In some cases the ratio of the total liquid phase volume in the column to the "excluded" volume is as great as 2, and thus totally retarded molecules, which do not come off the column until solvent equal to the total volume has eluted, are well separated from the excluded molecules. Solutes of intermediate sizes (i.e., of the same order of magnitude as the gel pore size) will elute from the column at times intermediate between the two extremes.

Many gel materials are now available and, by appropriate choice of pore size, excellent separations can be achieved even with molecules of quite

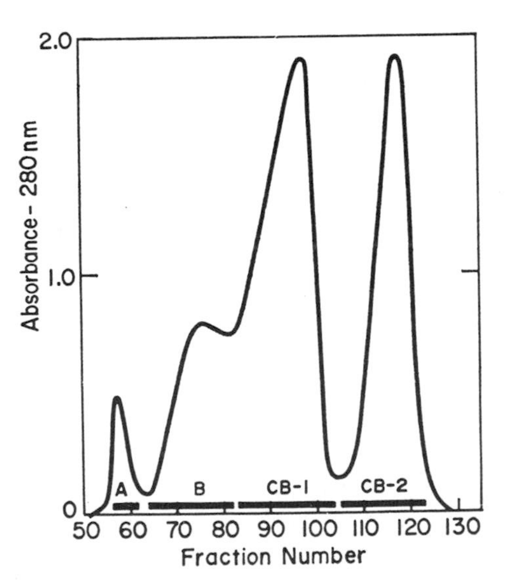

Fig. 3-10 Separation by Sephadex chromatography of peptides of 90 residues (CB-1) and 33 residues (CB-2) obtained by cyanogen bromide cleavage of S-aminoethyl-α-lactalbumin. A is a nonprotein contaminant and B is uncleaved protein. [From K. Brew and R. L. Hill, *J. Biol. Chem.*, **245**, 4559 (1970).]

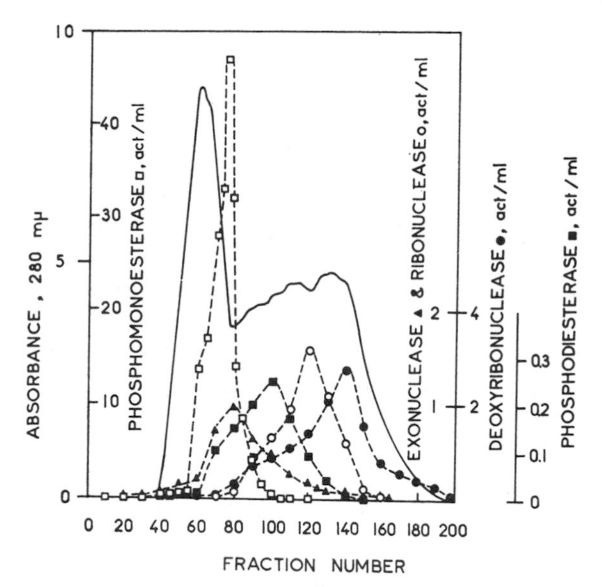

Fig. 3-11 Distribution of nuclease activities from hog spleen extract chromatographed on Sephadex G-100; equilibration and elution with 0.001 *M* potassium phosphate, pH 6.8. The solid line indicates absorbance at 280 nm. [Courtesy of G. Bernardi.]

similar molecular weight (e.g., differing by as little as a factor of 1.3–2). Sephadex and Bio-Gel may be obtained in a range of pore sizes useful for molecular weights in the range of a few hundred to several hundred thousand. Agarose gels have recently been developed for the separation of bacteria and high-molecular-weight cell components (nuclei, ribosomes, etc.).

MEC has proved to be an exceedingly valuable tool in biochemistry, and gives excellent results in many systems. An example of the separation of peptides with MEC is shown in Fig. 3-10. Figure 3-11 shows the fractionation on Sephadex G-100 of the same group of enzymes illustrated in Fig. 3-9 for hydroxyapatite. In addition to protein and peptide purification, MEC is used effectively for desalting (and removal of other small molecules), binding studies, and molecular weight estimation.

4-5 Affinity Chromatography[6]

Selective isolation and purification of certain proteins and other macromolecules has been achieved by taking advantage of their interaction with specific ligands that have been immobilized in a suitable chromatographic matrix. When a mixture of proteins in solution is passed through a column of this type, proteins that do not interact with the immobilized ligand will pass through the column without any retardation. Those that do interact will be retarded to varying extents depending of their affinities for the ligand under the conditions employed. Use of the method depends upon the availability of a suitable specific ligand for the protein to be isolated, which can then be covalently attached to the matrix material. An effective matrix is obtained with beaded derivatives of agarose (a crosslinked polysaccharide); the resulting gel is sufficiently porous to allow noninteracting macromolecules with molecular weights as high as the millions to pass through freely. In cases where the desired protein is strongly bound to the immobilized ligand, elution is achieved by a change in solution conditions, (e.g., pH, temperature, salts), addition of a more strongly interacting ligand to the eluting solution, or cleavage of the matrix–ligand bond. Further information and experimental procedures for this important new chromatographic technique are given by Cuatrecasas and Anfinsen in Reference 6.

A related chromatographic technique based on protein–ligand affinities involves specific elution of proteins bound to a nonspecific matrix by addition of a substrate or other ligand of the desired protein to the eluting solution. Any protein that interacts with the added ligand is likely to undergo a change in its affinity for the matrix, with the result that a selective elution occurs (see Pogell and Sarngadharan in Reference 6 for details).

5 QUANTITATION OF PROTEINS AND AMINO ACIDS

A great variety of highly reliable detection and quantitation methods for use with purified preparations of proteins are available. Difficulties, of course, arise in the presence of other materials, and the choice of method may depend upon the nature of these other components. A table of methods assembled by Sober and his associates[1] is reproduced here (Table 3-1).

Direct dry weight determination is, in principle, free from spurious contributions, provided a sufficient amount of sample is available and other substances have been removed. Approximate protein concentrations for tissue homogenates, for example, may be obtained in this way following a fractionation scheme such as that of Schneider.[17] Here, protein is obtained as a precipitate in hot trichloroacetic acid; final weights are corrected for losses in purification on the basis of a suitable standard (checked, for example, by nitrogen analysis). Other general methods listed in Table 3-1 have various applications and the original references should be consulted. The measurement of refractive index is particularly useful in purified systems; it has been developed as a means of monitoring column effluents and is also the basis for two optical systems for concentration determination in the ultracentrifuge.

The biuret reaction is the most specific of the protein methods. The assay is based on the observation that compounds containing peptide bonds

Table 3-1

Methods	Approximate sample required (mg.)	Destructive (D) or conservative (C)	Automation available
A. General methods			
Dry weight			
Standard	10	D	−
Micro	1	D	−
Refractive index	2	C	+
Dichromate reduction	10^{-1}	D	−
Flame ionization	10^{-6}	D	+
B. Protein methods			
Biuret	1	D	+
Micro-Kjeldahl	1	D	−
Dumas N	1	D	−
Nessler NH_3	1	D	−
Absorption, 280 mμ	10^{-1}	C	+
Absorption, 210 mμ	10^{-2}	C	+
Lowry-Folin	10^{-2}	D	+
Ninhydrin (with hydrolysis)	10^{-2}	D	+
Fluorescence	10^{-3}	C	+
^{64}Cu binding	10^{-4}	C(?)	+

produce a characteristic blue-violet color when treated with copper sulfate in alkaline solution. The substance known as "biuret," $NH_2CONHCONH_2$, gives this reaction, which is apparently due to formation of a coordination complex between the cupric ion and four nitrogen atoms. The biuret reaction forms part of the more sensitive procedure developed by Lowry and associates, utilizing the Folin-Ciocalteau reagent. Additional color development here is due to the presence of tyrosine and tryptophan in proteins which react with phosphomolybdate–phosphotungstate salts in the reagent. Certain amino acid sequences are more chromogenic than others are, and standardization of the procedure must take into account the widely varying response to be expected for different proteins. A number of substances interfere with the assay,[18] and thus it is not advisable to use it in tissue homogenates.

Nitrogen analysis yields an accurate measure of protein concentration provided other nitrogen-containing compounds (e.g., nucleic acid) are absent and that the amino acid composition, including amide nitrogen, is known (see Table 3-1 and Reference 19). For proteins of unknown amino acid composition, quantitation is based on an approximate nitrogen content in proteins of 16%. The typical range for nitrogen content is about 12 to 18%, although a few proteins fall well outside that range (e.g., protamine has 30% nitrogen). The Kjeldahl method is based on titration of ammonium ion produced by digestion of the protein with concentrated sulfuric acid. The Nessler procedure utilizes a colorimetric assay for quantitation of ammonium ion. A useful micro-Nessler procedure is available.[20]

Ultraviolet absorption procedures are convenient, rapid, and particularly recommended for purified proteins whose extinction coefficients have been determined by other means (e.g., weight or nitrogen analysis). In other cases, absorption at 280 nm (due primarily to tyrosine and tryptophan content) can be used only as a very approximate measure of protein content. This is a convenient routine means for locating protein in column chromatographic procedures. Use of lower wavelengths for quantitation of peptide bond absorption is complicated by contributions from other substances, including some common buffers. Fluorescence (primarily of tryptophan, see Chapter 9) provides a sensitive and specific method for detection of proteins and for quantitation if suitably standardized. The variability in tryptophan content of proteins, however, precludes its more general use for protein quantitation.

The reaction of amino acids with ninhydrin to produce a colored product has already been described. Ninhydrin reagent sprays are particularly useful in locating amino acids on chromatograms. Depending on the solvent used, the amino acids appear as blue, purple, or reddish spots. Proline (and hydroxyproline) give a yellow color. A valuable micromethod for amino acid determination depends upon the absorption produced at 230 nm by a complex with Cu^{++}.[21] This method may be used for estimation of protein content after hydrolysis or of free amino acids in extracts of tissue homogenates.

REFERENCES

1. H. A. Sober, R. W. Hartley, Jr., W. R. Carroll, and E. A. Peterson, "Fractionation of proteins," in *The Proteins*, 2nd ed., Vol. 3, H. Neurath, Ed., Academic Press, New York, 1965, p. 2.
2. R. A. Keckwick, Ed., "The separation of biological materials," *Brit. Med. Bull.*, **22,** 103 (1966).
3. P. Alexander and H. P. Lundgren, Eds., *A Laboratory Manual of Analytical Methods in Protein Chemistry Including Polypeptides*, Pergamon Press, New York, 1966.
4. C. H. W. Hirs, Ed., *Enzyme Structure*, Vol. 11 of *Methods in Enzymology*, Academic Press, New York, 1967; C. H. W. Hirs and S. N. Timasheff, Eds., *Enzyme Structure*, Parts B and C, Vols. 25 and 26 of *Methods in Enzymology*, Academic Press, New York, 1972.
5. J. L. Bailey, *Techniques of Protein Chemistry*, 2nd ed., American Elsevier, New York, 1967.
6. W. Jakobi, Ed., *Enzyme Purification and Related Techniques*, Vol. 22 of *Methods in Enzymology*, Academic Press, New York, 1971.
7. E. Heftman, Ed., *Chromatography*, 2nd ed., Reinhold, New York, 1967.
8. C. J. O. R. Morris and P. Morris, *Separation Methods in Biochemistry*, Pitman Publishing, London, 1963.
9. E. Lederer and M. Lederer, *Chromatography*, 2nd ed., Elsevier, New York, 1957.

10. T. S. Work and E. Work, *Laboratory Techniques in Biochemistry and Molecular Biology*, Vol. I, North-Holland Publishing Co., London, 1969. (Covers gel electrophoresis, chromatography, and immunoelectrophoresis.)
11. N. G. Anderson, "Preparative particle separation in density gradients," *Quart. Rev. Biophys.*, **1**, 217 (1968).
12. A. Chrambach and D. Rodbard, "Polyacrylamide gel electrophoresis," *Science*, **172**, 440 (1971).
13. K. Randerath, *Thin-Layer Chromatography*, Verlag-Chemie, Weinheim-Bergstr., Germany 1963.
14. E. Stahl, Ed., *Thin-Layer Chromatography; A Laboratory Handbook*, 2nd ed., trans. M. R. F. Ashworth, Springer-Verlag, New York, 1969.
15. G. K. Ackers, "Analytical gel chromatography of proteins," *Adv. Protein Chem.*, **24**, 343 (1970).
16. H. Determann, *Gel Chromatography*, trans. E. Gross, Springer-Verlag, Berlin, 1968; 2nd ed., Springer-Verlag, New York, 1969.
17. W. C. Schneider, "Determination of nucleic acids in tissues by pentose analysis," *Methods Enzymol.*, **3**, 680 (1957).
18. J. Bonitati, W. B. Elliott, and P. G. Miles, "Interference by carbohydrate and other substances in the estimation of protein with the Folin-Ciocalteu reagent," *Anal. Biochem.*, **31**, 399 (1969).
19. S. Jacobs, "The determination of nitrogen in biological materials," *Methods Biochem. Anal.*, **13**, 241 (1965).
20. F. L. Schaffer and J. C. Sprecher, "Routine determination of nitrogen in the microgram range with sealed tube digestion and direct Nesslerization," *Anal. Chem.*, **29**, 437 (1957).
21. J. R. Spies, "An ultraviolet spectrophotometric micromethod for studying protein hydrolysis," *J. Biol. Chem.*, **195**, 65 (1952).

IV Primary Structure of Proteins

1 INTRODUCTION

The primary structure of a protein has been defined as the linear sequence of amino acid residues making up the polypeptide chains of the molecule. It is, however, convenient to include in the primary structure other aspects of the chemical structure such as disulfide bonds and other covalent cross-linking bonds and any groups or molecules of other types that are bonded covalently to the polypeptide chains. The latter include groups responsible for modification of amino acid residues, as in sulfonated tyrosine or phosphorylated serine, and covalently-linked carbohydrate components in glycoproteins. It may be helpful to distinguish three components of primary structure. (1) The linear sequence of amino acids as laid down by direct translation of the genetic information during protein synthesis. As far as we know, only the 20 amino acids given in Table 2-1 can be incorporated into the sequence in this way, with the exception of those that have only a transitory existence in the primary structure, such as the *N*-formylmethionine which functions in chain initiation. (2) Modified amino acids found to be part of the sequence of native proteins as isolated from living systems but produced by separate reactions during or after polypeptide chain assembly (e.g., cystine, desmosine, hydroxyproline, thyroxine). (3) Finally other substances such as polysaccharide chains and covalently attached prosthetic groups such as pyridoxal. For some cases of modified amino acids it is not yet certain whether they belong to category (1) or (2), although the latter seems more probable. An exciting area of current research is the elucidation of mechanisms by which modifications of primary structure are achieved *in vivo*.

Although knowledge of the primary structure alone does not tell us all we want to know about the nature and function of a protein, this information has varied applications. For example: (1) The position of reactive groups (e.g., SH groups) can be determined and related to protein folding. (2) Knowledge of the primary structure is helpful in understanding the role of particular amino acid residues in relation to enzyme activity. (3) On the basis of sequence information, polypeptides can be synthesized with altered sequence in order to determine which amino acid residues are necessary for biological activity (e.g., peptide hormones). Rapid advances in organic chemical methods for peptide synthesis may soon permit systematic alterations in protein primary structures for the study of the active sites of enzymes and of protein folding mechanisms. (4) Determination of amino acid replacements caused by natural or induced mutations has been of value in elucidation of the genetic code and in understanding of congenital diseases such as sickle cell anemia.[1] (5) Comparison of amino acid sequences in proteins with similar function from different species (e.g., hemoglobins) contributes to our understanding of evolution. (6) Primary structure information is used in the solution of three-dimensional structure by X-ray crystallography. (7) Even where complete "sequencing" has not yet been possible, the chemical and enzymatic methods discussed here are valuable in identification and characterization of proteins, for example, determination of the number of polypeptide chains in the protein, the nature of the *N*-terminal and *C*-terminal amino acid residues, or the characteristic "fingerprint" of peptides produced by tryptic digestion.

The list of proteins whose sequence has been determined is growing rapidly. New results from sequence analyses are published annually in review form.[2]

2 THE PEPTIDE BOND

2-1 Structural Properties[3,4]

The bond distances and angles for a peptide bond are shown in Fig. 4-1. The C—N bond length (1.32 Å) is shorter than that of a normal C—N single bond (1.47 Å) while the C═O distance (1.24 Å) is slightly longer than a typical carbonyl double bond (1.215 Å). This suggests that the peptide bond has partial double bond character resulting from contributions from two resonating structures:

$$\begin{array}{c} \text{O} \\ \| \\ -\text{C}-\underset{\text{H}}{\text{N}}- \end{array} \rightleftharpoons \begin{array}{c} \bar{\text{O}} \\ | \\ -\text{C}=\underset{\text{H}}{\overset{+}{\text{N}}}- \end{array}$$

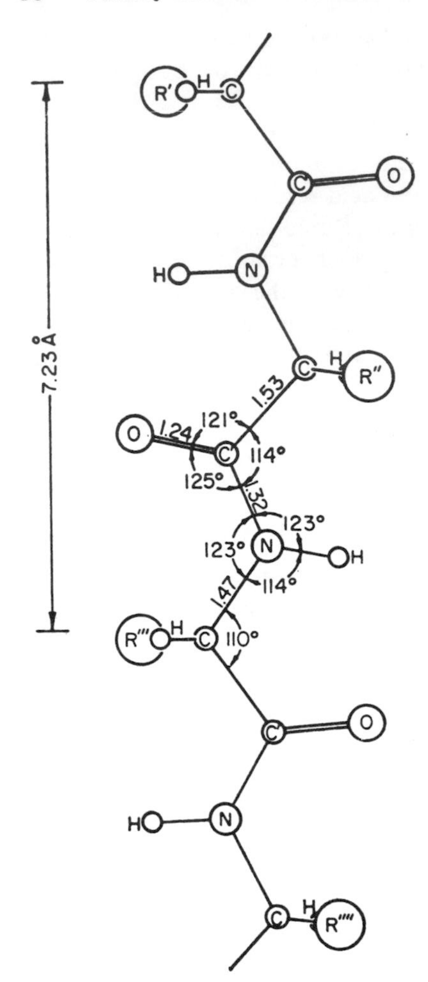

Fig. 4-1 Bond lengths and angles in a fully extended polypeptide chain with amino acids in the L-configuration. [From R. B. Corey and L. Pauling, *Proc. Roy. Soc.*, **B141,** 10 (1953).]

The partial double bond character of the peptide C—N bond results in restricted rotation about that bond and thus in the possibility of *cis* and *trans* isomers. In both cases the four atoms of the peptide bond (C, O, N, H) plus the two adjacent α-carbon atoms lie in the same plane. The α-carbon peptide backbone bonds have the dimensions of single bonds (1.53 Å for the C—C bond and 1.47 Å for the C—N bond) and rotation about both of these bonds is possible. This rotation can give rise to a myriad of possibilities for the 3-dimensional configuration of a protein. Of course, even without the constraints produced by noncovalent interactions in the native protein (cf. Chapter 6), there is some restriction on orientation about the α-carbon bonds because of steric hindrance between the amino acid side chains. The subject of steric factors and their possible effect on protein conformation has been reviewed by Schellman and Schellman[3] and by Ramachandran and Sasisekheran.[4]

Although the peptide bond may conceivably exist in either the *trans* form (as shown in Fig. 4-1) or the *cis* form, crystal structure analyses of small peptides and proteins have indicated that the *trans* isomer is by far the prevalent form (cf. Chapter 14). A variety of physical measurements also indicate that in general the *trans* conformation of the peptide bond is energetically favored in peptides formed from α-amino acids. In proline and hydroxyproline, however, the *cis* and *trans* isomers possess similar energies. Thus the occurrence of *cis* configurations in proteins is more likely where these amino acids are located. The synthetic polypeptide poly-L-proline has been found to occur with all residues in the *cis* form as well as with all residues in the *trans* form (cf. Chapter 14). These structures may be interconverted by temperature or solvent changes.

2-2 Chemical Synthesis of Peptides[5,6]

Chemical synthesis of polypeptides has contributed much to biochemical knowledge. Small peptide antibiotics and hormones have been prepared with amino acid deletions or substitutions to determine which components of these molecules are essential for biological activity. Synthesis of large protein polypeptide chains *de novo* (or partial synthesis using peptides isolated from enzymatic digestion) is now feasible and will eventually permit direct correlation of variations in amino acid sequence with enzymatic function and three-dimensional structure.

The history of peptide synthesis goes back to the preparation of benzoylglycylglycine by Curtius in 1882. In 1903 Fischer and Otto developed the α-halogeno-acylamino acid chloride method and 4 years later Fischer synthesized an octadecapeptide of leucine and glycine. In 1939 the procedures for synthesizing peptides containing polyfunctional amino acids were applied to glutamic acid, lysine, and arginine by Fruton and Bergmann and co-workers. The methodology for peptide synthesis has progressed with remarkable speed, and an automated solid-state method is available. A detailed account of peptide synthesis and appropriate references to the original literature may be found in the excellent review by Hofmann and Katsoyannis.[5] The development of automated techniques is reviewed and referenced by Merrifield and his associates.[6]

The general method employed for peptide synthesis may be summarized as follows: The "carboxyl component" (the amino acid or peptide that is to contribute the carboxyl group to the peptide bond) is prepared by blocking its α-amino group with a protective group Y. The "amino component" is prepared by adding a protective group Z to the α-carboxyl group of the amino acid or peptide that is to donate its amino group to the peptide bond. If either of these components contains a reactive side chain, it must be protected with a group W (this is frequently done before amino or carboxyl protection). The carboxyl component is then activated with a group X; and condensed with the amino component to yield the protected peptide according to the reaction

$$\mathrm{Y{-}N(H){-}C(H)(R_1{-}W){-}C(=O){-}X + NH_2{-}C(H)(R_2){-}C(=O){-}OZ \longrightarrow}$$

$$\mathrm{Y{-}N(H){-}C(H)(R_1{-}W){-}C(=O){-}NH{-}C(R_2){-}C(=O){-}OZ + HX}$$

If further chain elongation is desired, Y or Z must be removed (and the carboxyl group activated if necessary), and the peptide is further reacted with another appropriately protected amino acid or peptide. The final product is obtained after removal of all blocking groups. Ideally, one wishes to have blocking agents that can be put on and taken off without affecting other blocking groups, the peptide bond, or the amino acid side chains. No such ideal reagents exist, and the choice of W, X, Y, and Z depends on the nature of the polypeptide being synthesized. Even with the best choice of blocking groups, some side reactions occur, and it is frequently necessary to purify the desired product to eliminate by-products at each step of the synthesis. A list of blocking and activation groups used in peptide synthesis and details of the conditions for reaction and preferential deblocking are to be found in Reference 5.

The synthesis of larger peptides may be accomplished by first preparing suitable small blocked peptides of 2 to 4 amino acid residues, which are then condensed to larger products. This method was used in the synthesis of horse renin substrate (Fig. 4-2), a polypeptide derived from the plasma protein angiotensinogen by the action of trypsin. The octapeptide from the carboxyl end is II-leu^5-angiotensin, highly active as a hypertensive agent. (The II-leu^5 designation is derived from the fact that a number of angiotensins from different species differ only in the fifth amino acid; the II indicates that it is the second product of enzyme action on polypeptide renin substrate.)

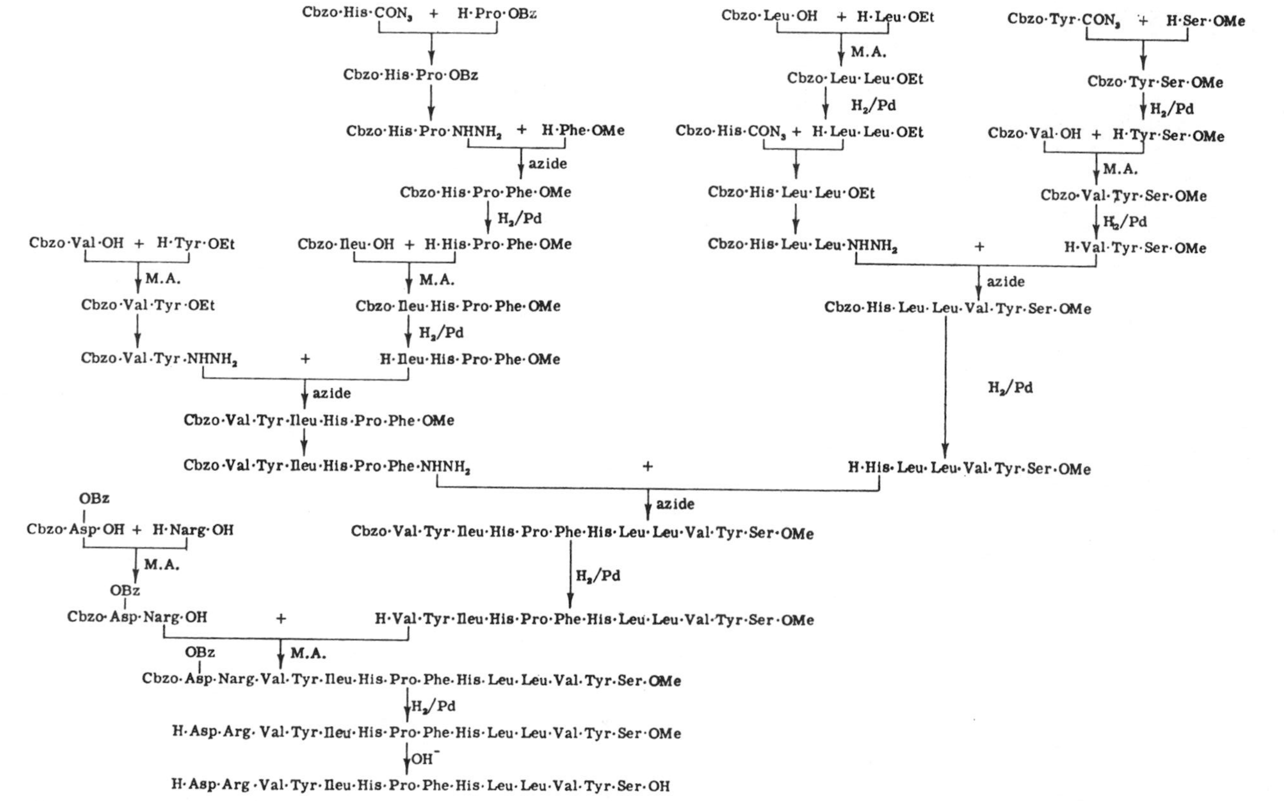

Fig. 4-2 Synthesis of horse renin substrate. Abbreviations are: Cbzo, carbobenzoxy; Bz, benzyl; Et, ethyl; Me, methyl.[5]

Extensive application of peptide synthesis has been made in the study of structure–function relationships of peptide hormones. The structure of corticotropin and its derivatives, the adrenocorticotropic hormone (ACTH) and melanocyte stimulating hormone (MSH), are shown in Fig. 4-3. Relative activities of various synthetic analogs are also given to indicate the direction of these investigations. Similar studies have been made on the neurohypophyseal hormones oxytocin and vasopressin, and their analogs. These compounds were tested for their ability to produce five biological effects: uterine contraction, ejection of milk from a primed mammary gland, fall in avian blood pressure, increase in mammalian blood pressure, and antidiuresis (increased reabsorption of water by kidney). Table 4-1 presents selected data from a more extensive tabulation by Hofmann and Katsoyannis[5] that show the relative biological activities in the five assays for a number of peptides. Although the reasons are not known, it is interesting to note that amino acid substitutions affect different activities to varying extents. The degree of sophistication now possible in the organic synthesis of peptides is exemplified by the *de novo* synthesis of active ribonuclease.[6]

2-3 Reactions of the Peptide Bond[7]

Peptides undergo hydrolysis at elevated temperatures in either acidic or basic solution. This reaction is discussed in more detail later since it forms the basis for amino acid analysis of proteins. A variety of other addition reactions across the peptide bond are possible, such as hydrazinolysis and methanolysis. The reactivity of the peptide bond varies tremendously, depending on the nature of the side chains on either side of the bond and of neighboring amino acid residues in the sequence. For example, acid hydrolysis at aspartic acid residues is unusually rapid while hydrolysis of the peptide bond between leucyl residues is comparatively quite difficult. Serine and threonine residues may undergo $N \rightarrow O$ acyl shift through nucleophilic attack by the side-chain hydroxyl group, with resultant peptide cleavage. The varying reactivity of peptide bonds involving particular amino acid residues and/or sequences is important for specific cleavage of polypeptide chains (cf. sequence determination).

The geometry of neighboring peptide bonds permits the formation of chelate compounds with metals. One of these, formed by reaction with cupric ions, is thought to involve four peptide NH groups, complexed with copper. This is the basis of the biuret reaction discussed earlier. The peptide hydrogen atom is sufficiently reactive to be displaced, upon treatment with Cl_2, by a chlorine atom that can then be detected with starch-iodide solution. The reaction frequently bears the names of Rydon and Smith who first described

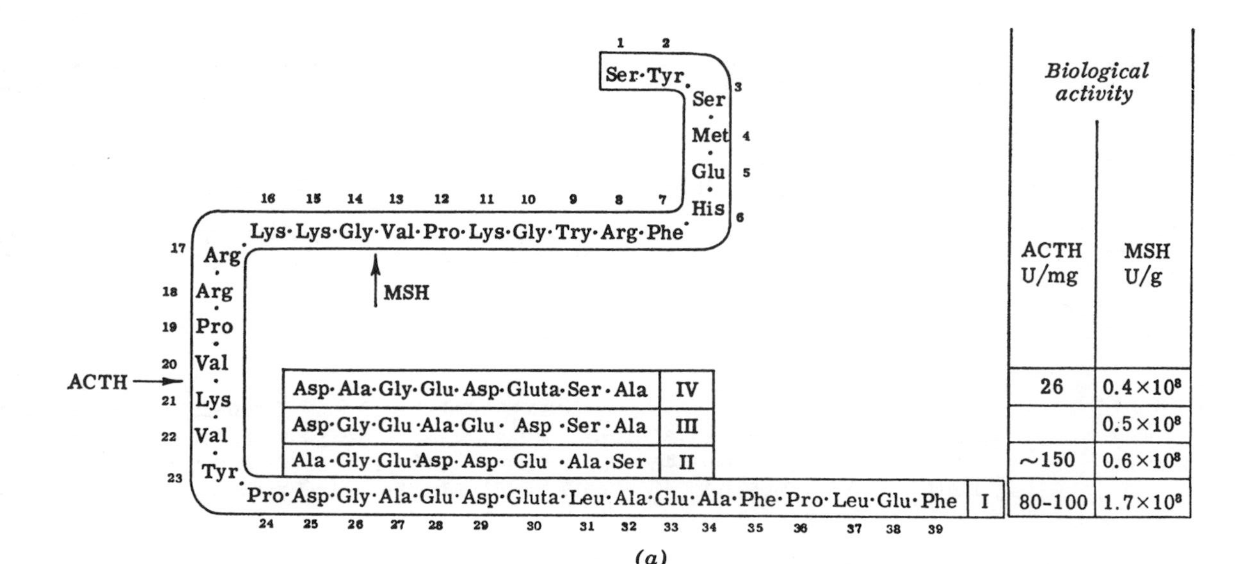

Fig. 4-3 (*a*) Amino acid sequence of pig (I) corticotropin showing the portions of the molecule associated with the activity of adrenocorticotropic hormone (ACTH) and of melanocyte stimulating hormone (MSH). Variations in sequence found in sheep (II), beef (III) and human (IV) corticotropin are shown with the results of hormonal activity assays for the four species.[5]

Analog	MSH (U./gm.)	ACTH (U./mg.)
H·Ser·Tyr·Ser·But·Glu·His·Phe·Arg·Try·Gly·Lys·Pro·Val·Gly·Lys·Lys·Arg·Arg·Pro·Val(NH_2)		~35
NH_2 F F F F		
Ac·Ser·Tyr·Ser·Met·Glu·His·Phe·Arg·Try·Gly·Lys·Pro·Val·Gly·Lys·Lys·Arg·Arg·Pro·Val·Lys·Val·Tyr(NH_2)	2.0×10^8	—
NH_2 F F F F		
H·Ser·Tyr·Ser·Met·Glu·His·Phe·Arg·Try·Gly·Lys·Pro·Val·Gly·Lys·Lys·Arg·Arg·Pro·Val·Lys·Val·Tyr(NH_2)		<1
H·Ser·Tyr·Ser·Met·Glu·His·Phe·Arg·Try·Gly·Lys·Pro·Val·Gly·Lys·Lys·Arg·Arg·Pro·Val·Lys·Val·Tyr·OH	2.0×10^8	103 ± 10.3
H·Ser·Tyr·Ser·Met·Glu·His·Phe·Arg·Try·Gly·Lys·Pro·Val·Gly·Lys·Lys·Arg·Arg·Pro·Val·Lys·Val·Tyr·Pro·OH		106 ± 14
H·Ser·Tyr·Ser·Met·Glu·His·Phe·Arg·Try·Gly·Lys·Pro·Val(NH_2)	1.9×10^9	<0.1
NH_2 F		
Ac·Ser·Tyr·Ser·But·Glu·His·Phe·Arg·Try·Gly·Lys·Pro·Val(NH_2)	1.5×10^8	
NH_2 F F F		
Ac·Ser·Tyr·Ser·Met·Glu·His·Phe·Arg·Try·Gly·Lys·Pro·Val·Gly·Lys·Lys(NH_2)	2.0×10^9	
H·Ser·Tyr·Ser·Met·Glu·His·Phe·Arg·Try·Gly·Lys·Pro·Val·Gly·Lys·Lys·OH	3.7×10^8	<0.1
H·Ser·Tyr·Ser·Met·Glu·His·Phe·Arg·Try·Gly·Lys·Pro·Val·Gly·Lys·Lys·Arg·OH	~10^8	~6
H·Ser·Tyr·Ser·Met·Glu·His·Phe·Arg·Try·Gly·Lys·Pro·Val·Gly·Lys·Lys·Arg·Arg·Pro·OH	1.4×10^7	~35
NH_2		
H·Ser·Tyr·Ser·Met·Glu·His·Phe·Arg·Try·Gly·Lys·Pro·Val·Gly·Lys·Lys·Arg·Arg·Pro·OH		20–30
H·Ser·Tyr·Ser·Met·Glu·His·Phe·Arg·Try·Gly·Lys·Pro·Val·Gly·Lys·Lys·Arg·Arg·Pro·Val(NH_2)	1.1×10^8	111 ± 18

Fig. 4-3 (*b*) Biological activity of synthetic analogs of MSH and ACTH.[5]

Table 4-1 Biological activities of some Synthetic analogs of oxytocin and vasopressin.[5]

Analog	Oxytocic activity (I.U./mg.) Uterus Rat *in vitro*	Avian depressor	Milk ejection	Vasopressin activity (I.U./mg.) Blood pressure Rat	Anti-diuresis Rat
Oxytocin: 1 Cys·2 Tyr·3 Ileu·4 Glu(NH_2)·5 Asp(NH_2)·6 Cys·7 Pro·8 Leu·9 Gly(NH_2) (ring 1–6)					
Cys·Tyr·Ileu·Glu(NH_2)·Asp(NH_2)·Cys (ring)	~3.3	0	~1.1	—	—
Arg-vasopressin (----Tyr·Phe---- ... Arg----)	~20.0	~60	~70	350–440	~400
Lys-vasopressin (Tyr·Phe ... Lys)	~4 5 ± 0.5	~30 40 ± 5	~45 60 ± 10	~280 268 ± 19	180 ~250
Oxypressin (Tyr·Phe ... Leu)	20 20	45 ~30	— ~60	3 ~3	— ~30
Tyr·Phe ... His	~1.5	~4.6	—	1.5	—
Arg-vasotocin (Arg)	~75	~150	~100	~125	74
Lys-vasotocin (Lys)	20 ± 3 —	54 ± 4 190	55 ± 5 —	39 ± 4 130	~13 —
Oxytocin (Ileu ... Leu)	486 ± 5 (42)	507 ± 23 (42)	450 ± 30	3.1 ± 0.1 (42)	2.7 ± 0.2 (42)
Ileu ... Ileu	289 ± 21	498 ± 37	328 ± 21	6 ± 1	1.1 ± 0.1
Ileu ... Val	200 ± 15	280 ± 17	310 ± 20	9 ± 1	0.8 ± 0.1
Val ... Leu	59 ± 8	57 ± 4	207 ± 14	~0.2	~0.8
--Tyr·Leu ... Leu	45 ± 7	42 ± 1	101 ± 13	0.3	~0.2
Phe ... Leu (Oxypressin)	20 20	45 ~30	— ~60	3 ~3	— ~30
Tyr ... Leu	0.1 ± 0.03	~0.03	1.5 ± 0.3	~0.01	—
Try ... Leu	0.04 ± 0.01	~0.1	0.1 ± 0.06	0	—

it as a sensitive method for detecting proteins and peptides in paper chromatography. The characteristic starch-iodine color develops with secondary amines in general as well as with certain other compounds.

3 OTHER COVALENT BONDS OF PROTEINS

3-1 Disulfide Bonds[8,9]

One of the amino acids that occurs in the primary (genetically determined) sequence of proteins is cysteine, which contains a highly reactive thiol (SH) function. The oxidation of two cysteine residues to form a disulfide bond (—S—S—) in a native protein is thought to occur after polypeptide chain folding due to noncovalent interactions has brought certain specific SH groups into close enough contact for reaction. Thus cysteine residues that are quite distant in the sequence may become covalently joined. Disulfide linkages have also been found to occur between cysteine residues belonging to different polypeptide chains in multiple chain proteins. In fibrinogen this situation has been demonstrated by the finding of disulfide bonds between peptides derived from different chains of the parent molecule.[10] Another interesting case is the 7S γ-globulin in which 4 of 16 disulfide bonds appear to be interchain.[11] These bonds impart considerable constraint on the three-dimensional conformation of the protein. The interrelation of disulfide bonds and noncovalent forces with respect to protein conformation is discussed in Chapter 6.

The chemistry of the disulfide (and sulfhydryl) group has been the subject of extensive review.[8,9] Most methods for the determination of SH and SS groups in proteins are based on reactions of the SH groups. The number of cysteines in disulfide linkage is determined from the difference between sulfhydryl yields obtained before and after conversion of all —S—S— to —SH. Difficulties are frequently encountered because the reagent used for SH determination is not entirely specific for this group. Low values may be obtained when SH or S—S groups are buried in the protein and not available for reaction. Undesired oxidation of SH groups or reduction of S—S bridges may occur if appropriate precautions are not taken.

Several methods used for quantitative estimation of —SH groups in proteins involve the formation of highly stable (i.e., undissociated) mercaptides of silver or mercury salts or of organic mercury derivatives such as *p*-chloromercuribenzoate (pCMB). As discussed by Cecil,[9] the silver mercaptides have a tendency to bind additional silver ion, and the mercury mercaptide may form either the single mercaptide protein—S—HgX or a double mercaptide protein—S—Hg—S—protein in which two sulfhydryl groups of the protein are linked through mercury. The occurrence of this form depends upon the steric possibilities. Denatured proteins usually form the latter type of derivative. The quantity of mercury or silver ions bound may be determined by titration, polarography, or elemental analyses of the isolated Hg or Ag protein derivative. The reaction of thiols with organic mercury derivatives is illustrated by the reaction with pCMB:

$$\mathrm{R{-}SH + ClHgC_6H_4COO^- \rightarrow R{-}S{-}HgC_6H_4COO^- + H^+ + Cl^-}$$

Analyses for pCMB are frequently made using the Boyer method, which is based upon an increase in absorption at 250 nm when pCMB is in mercaptide linkage. The spectral change observed is not always proportional to pCMB bound, but in general the method is convenient and reasonably accurate. Titrimetric and amperometric methods may also be used to determine the extent of reaction.

An excellent and very specific reagent for quantitating free sulfhydryl groups in proteins is 5,5′-dithiobis(2-nitrobenzoic acid), commonly referred to as DTNB or Ellman's reagent. The reaction is

$$\mathrm{NO_2{-}C_6H_3(COO^-){-}S{-}S{-}C_6H_3(COO^-){-}NO_2 + RSH \xrightarrow{pH\ 8} H^+ + R{-}S{-}S{-}C_6H_3(COO^-){-}NO_2 + S^-{-}C_6H_3(COO^-){-}NO_2}$$

The free thiophenylate anion can be quantitatively determined by its strong absorbance at 412 nm.

Sulfhydryl groups in proteins can be preferentially alkylated at pH 7 to 8. Iodoacetate and iodoacetamide are most frequently used for the purpose:

$$\mathrm{RSH + ICH_2C({=}O)O^- \longrightarrow R{-}S{-}CH_2C({=}O)O^- + H^+ + I^-}$$

The resulting carboxymethyl derivative (or the corresponding amide) of cysteine is very stable, and the extent of reaction can be quantitated to a high degree of accuracy by determining the yield of carboxymethylcysteine isolated from the acid-hydrolyzed protein. If ^{14}C-iodoacetate is used, the quantitation can be done by radioactive assay of the protein after removal of the reagent. This method is sensitive and usually quite satisfactory, although partial alkylation of other groups (e.g., imidazole or amino groups) or slight adsorption of the reagent will introduce some error. The reaction with iodoacetate has been used extensively to differentiate between "reactive" SH groups of a protein that are readily available to the reagent and those that react only after denaturation of the protein (e.g., after addition of guanidine·HCl). New reagents for SH determinations that contain chromophores with high extinction coefficients have been reported. The location of SH-containing peptides on paper chromatograms or electrophorograms is possible by virtue of the pink to orange color formed in various modifications of the thiol reaction with sodium nitroprusside.

Disulfide bridges are readily formed *in vitro* by oxidation with dissolved molecular oxygen at neutral and alkaline pH:

$$\mathrm{RSH + R'SH \xrightarrow{+\frac{1}{2}O_2} R{-}S{-}S{-}R' + H_2O}$$

The reaction may not occur in this way, of course, *in vivo* (where molecular O_2 may not be available for example); evidence for an enzymatic mechanism has been reported (cf. Chapter 6). Disulfide bonds undergo a variety of reactions which are of importance in protein structural work. Reduction of disulfide bridges to SH groups is possible with a variety of reagents, but the best specificity is obtained by reaction with thiol reagents such as mercaptoethanol, cysteine, or dithiothreitol, in excess. The last-named reagent has the advantage that it is not so prone to oxidation by dissolved oxygen. For example, with mercaptoethanol:

$$\mathrm{R{-}S{-}S{-}R' \xrightarrow[\text{HS-CH}_2\text{-CH}_2\text{OH}]{\text{excess}} RSH + R'SH + HO(CH_2)_2S{-}S(CH_2)_2OH.}$$

Reaction with CN^- also breaks disulfide bonds to yield R—SCN + R′SH This reaction followed by the nitroprusside reaction can be used in the location of disulfide-containing peptides on paper chromatograms.

Oxidation of disulfide bonds (and sulfhydryl groups as well) can be carried out with performic acid to achieve quantitative conversion of cysteine residues to cysteic acid. This method suffers from the disadvantage that oxidative destruction of tryptophan occurs, accompanied by peptide bond rupture. Upon total hydrolysis of a performic-oxidized protein, cysteic acid can be quantitatively determined in an amino acid analyzer. A specific oxidation of cysteine and cystine residues in proteins to their *S*-sulfo derivatives is achieved by reaction with sulfite and cupric ions and is quantitative in the presence of concentrated urea (e.g., 6 *M*). The first step of the reaction is

$$\mathrm{R'{-}S{-}S{-}R + HSO_3^- \rightarrow R'S{-}SO_3^- + HS{-}R}$$

followed by a copper-catalyzed oxidation of the resultant sulfhydryl product to R—S—S—R, which is then further oxidized to the *S*-sulfo derivative.

S-Sulfocysteine is labile to acid hydrolysis, and in amino acid analysis is recovered as cystine.[12]

3-2 Unusual Covalent Linkages

From the point of view of an organic chemist, there are a considerable number of covalent bond possibilities for amino acids besides the peptide bond. Several amino acids contain potentially reactive side chains: the ε-NH_2 group of lysine or the side chain carboxyl groups of glutamic and aspartic acids could participate in peptide-like links with other amino acids or with each other; for example, the dicarboxylic amino acid side chains might form inter- or intrachain crosslinks by reaction with the ε-NH_2 groups of lysine residues. Such linkages are not common, however. The best established cases are in small naturally occurring peptides (γ-glutamyl-amides) and in insoluble fibrin. A great deal of work has been done to establish the presence of γ-glutamyl and β-aspartyl crosslinks in collagen. As in fibrin, such crosslinks may be associated with the formation of highly stable aggregates of these structural proteins (cf. Chapter 17). Evidence that the ε-NH_2 group of lysine is not ordinarily involved in covalent bonding in proteins is based on its availability to react quantitatively with amino reagents such as fluorodinitrobenzene when the protein is completely unfolded by denaturing solvents. Covalent bonds, however, occur with the carboxyl group of biotin or lipoic acid in some enzymes, and an altered form of lysine functions in crosslinking of elastin. Thioether bridges between the iron protoporphyrin group and two cysteine residues occur in cytochrome *c*.

Phosphoproteins contain phosphorylated serine and threonine residues. The formation of phosphodiesters between two serines or threonines is a possible source of crosslinkage, but there is presently little evidence to confirm its existence in proteins. Most phosphate occurring in proteins is released by mild acid hydrolysis (2 N H_2SO_4, 100°C, for 2 hours). The bond is also completely unstable in 0.25 N NaOH (in 24 hours at 37° there is essentially total release of inorganic phosphorus).

The existence of covalent linkage of proteins to carbohydrates via γ-glutamyl, β-asparaginyl, or *O*-seryl glycosidic linkage is well established.[14,15] Covalent linkage to lipid in lipoproteins is also possible, but has not been conclusively demonstrated. The fact that most lipid–protein complexes can be completely disaggregated by detergent, for example, and the lipid components removed by extraction with organic solvents argues against covalent bonding between the components. Among the serum lipoproteins, variation in lipid concentration and composition is sufficient to indicate true molecular heterogeneity, consistent with entirely noncovalent complexing between components.[15] The *N*-terminal residues of proteins are often not reactive as primary amines. Covalent linkage with acetate to give terminal α-*N*-acetyl amino acid residues occurs frequently, and the cyclization of *N*-terminal glutamine to pyrrolidone carboxylic acid residues has also been reported.

4 DETERMINATION OF AMINO ACID COMPOSITION[16,17]

4-1 Hydrolysis

Primary structural analyses generally begin with determination of the amino acid composition. The first step in this procedure is the hydrolysis of the protein into its constituent amino acids. No single hydrolysis method has yet been found that will give quantitative yields of all of the amino acids. Thus more than one hydrolysis procedure must be used.

The most common method of hydrolysis consists of heating the protein in concentrated acid, usually constant boiling (6 *N*) HCl at 105 to 110°C for 20 to 70 hours. The time necessary for peptide bond rupture depends on the nature of the amino acid residues, for example, steric hindrance by the side chains of valine and isoleucine sometimes results in a somewhat slower hydrolysis of peptide bonds involving the carboxyl groups of these amino acids. A disadvantage of acid hydrolysis is that several amino acids are altered in the process: tryptophan is totally destroyed, asparagine and glutamine are converted to aspartic and glutamic acid, respectively, and cysteine is converted to cystine. Serine, threonine, cystine, and tyrosine are also destroyed to some extent. Hydrolyses are therefore carried out for varying times, and the values for those amino acids that are partially destroyed are extrapolated to zero time, while those released slowly are determined from longer times of hydrolysis.

Tryptophan can be determined after alkaline hydrolysis of the protein with 5 *N* sodium hydroxide or barium hydroxide at 110°C for about 20 hours. Alkaline hydrolysis destroys many amino acids and causes racemization. It is therefore most likely to be employed solely for the determination of tryptophan. Tryptophan may also be determined with fair precision simply from the ultraviolet absorbance of the protein after making appropriate corrections for tyrosine content.

The possibility of complete hydrolysis of proteins by enzymatic methods has received considerable attention. This would allow the quantitative determination of tryptophan, glutamine, and asparagine, and would eliminate the necessity of time studies to determine those amino acids that are either partially destroyed or slowly hydrolyzed. In addition, labile linkages to substituent groups on the amino acids, such as tyrosine-*O*-sulfate or carbohydrate moieties on asparagine, would survive and these compounds could then be identified. Certain nonspecific enzymes such as papain or microbial proteases (e.g., from *Streptomyces griseus*) used in conjunction with leucine aminopeptidase and proline iminopeptidase (which will hydrolyze proline peptide bonds resistant to leucine aminopeptidase) show promise for complete hydrolysis of peptides and some proteins. Other proteins, however, are hydrolyzed only 50 to 70% by these methods. Other combinations of known proteolytic enzymes have been tried, and it appears that the most successful combination to be used varies from protein to protein. Because of the possibility of incomplete hydrolysis in enzymatic digestion, results must be correlated with those of acid hydrolysis.

4-2 Amino Acid Analysis

The development of paper chromatography has made possible the separation of most amino acids found in protein hydrolyzates using various solvent systems. Fair quantitative accuracy may be obtained by eluting the amino acid from the paper (after location with ninhydrin) and comparing the ninhydrin color of the eluant with that of known concentrations of amino acids treated the same way. Many workers have preferred to separate the amino acids in one dimension by high-voltage electrophoresis followed by chromatography in a second dimension. An example of such a system is shown in Fig. 4-4. Another method showing great promise in the separation of all the common amino acids involves one-dimensional high-voltage electrophoresis under carefully controlled conditions (Fig. 4-5). Approximate quantitation of the amino acids may be achieved by scanning the paper with a recording densitometer after staining with a suitable ninhydrin solution.

The most accurate method for amino acid analysis uses ion-exchange chromatography for separation of the amino acids and the ninhydrin reaction for quantitation. Routine analyses are made conveniently with an automatically-recording instrument such as that first described by Spackman, Moore, and Stein in 1958. In this procedure amino acids are eluted from a sulfonated polystyrene resin (Amberlite IR-120) with sodium citrate buffer. Ninhydrin reagent is pumped into the effluent stream and the mixture is passed through a heated reaction coil to develop the color. The solution then passes through a photometer recorder system where the absorbance at 570 nm (the wavelength of maximum absorbance of the normal product of the reaction with α-NH_2 groups) and at 440 nm (for the determination of proline and hydroxyproline) is continuously recorded. A separate shorter column is used for the elution of the basic amino acids with citrate buffer at a higher pH. The results of such an analysis are shown in Fig. 4-6. Other columns such as Dowex-50-X12 have also been used with elution afforded by a continuously increasing gradient of salt and pH. The use of automatic amino acid analyzers

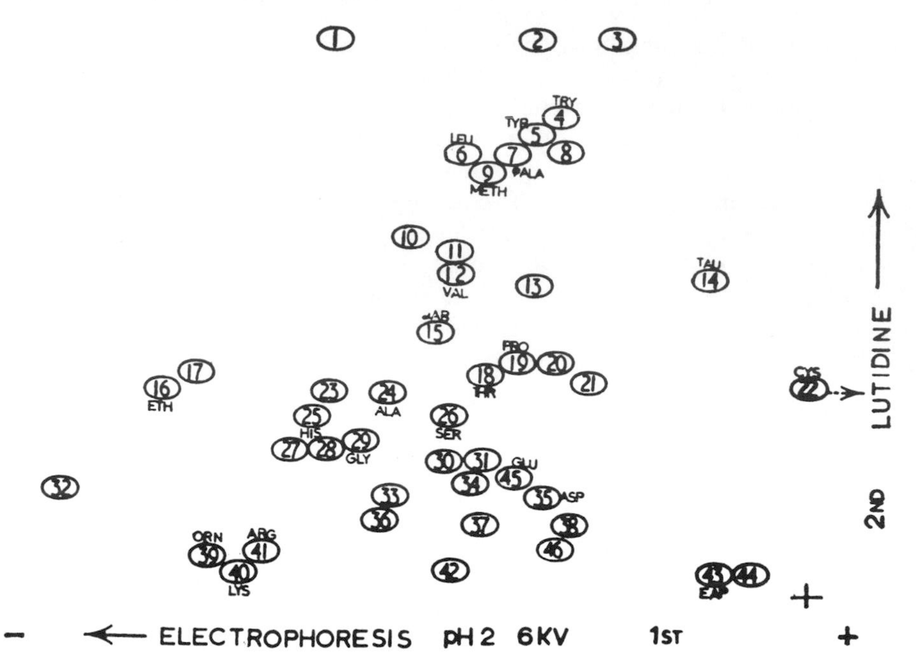

Fig. 4-4 Two-dimensional separation of amino acids using electrophoresis in one direction followed by chromatography at right angles. [From I. Smith, *Chromatographic and Electrophoretic Techniques*, Vol. 2, Wiley, Interscience, N.Y., 1968.]

has greatly facilitated protein chemistry, and they are used routinely in many laboratories. High precision is obtainable (recoveries of 100 ± 3%). Typically, about 2 hours is required for the analysis and about 0.01 μmole of each amino acid. Lesser quantities may be analyzed with only slight loss in accuracy. The methodology is constantly being refined for greater sensitivity and shorter development time.

A very promising method for amino acid analysis is offered by gas–liquid partition chromatography.[18] Very small amounts of amino acids are required (10^{-11} moles) and development time is short (about 30 min). If the difficulties presently involved in quantitative conversion of the amino acid to volatile derivatives can be overcome, gas chromatography may become the method of choice for routine analyses in the future.

5 END-GROUP ANALYSIS[16,17]

Many proteins contain more than one polypeptide chain. Information concerning the number of chains and their molecular weights may be obtained

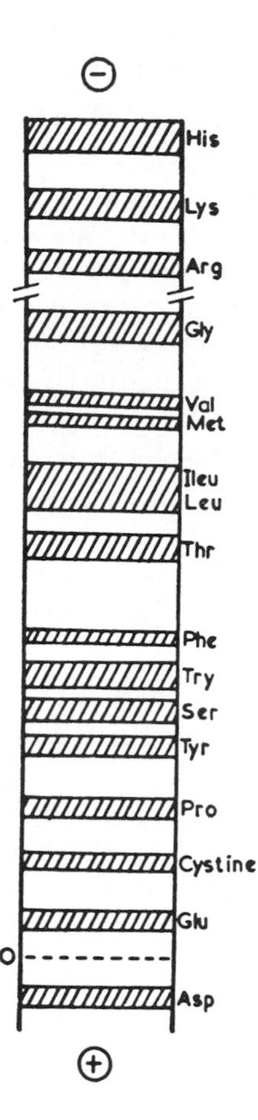

Fig. 4-5 High voltage electrophoretic separation of amino acids at pH 3.6. Overlapping amino acids are separable by pH or temperature alteration.[12]

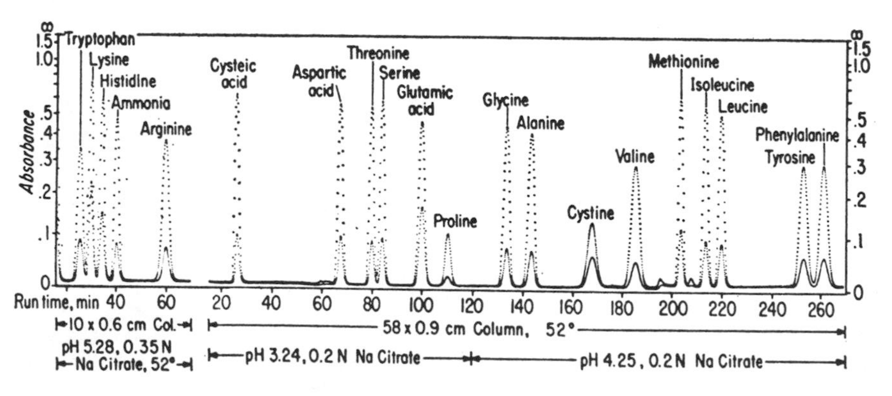

Fig. 4-6 Automatic amino acid analysis of a synthetic hydrolyzate sample; absorbance measured at 570 nm and 440 nm; total time of run = $4\frac{1}{2}$ hr. [From D. H. Spackman, "Accelerated methods" in Reference 17.]

from a quantitative analysis of the amino terminal (*N*-terminal) and carboxyl terminal (*C*-terminal) amino acids. These results may be combined with the molecular weight of the protein obtained by physical methods to determine the minimum number of chains present in the molecule. If more than one kind of *N*-terminal or *C*-terminal amino acid is found in the protein, it demonstrates the existence of at least that number of types of chains in the molecule. When the protein appears to have more than one of a particular end group per molecule (when molecular weight is known), it may signify the existence of identical subunits. In some cases, even very careful end-group analyses (especially *N*-terminal analyses) do not reveal the correct number of polypeptide chains. Both the Sanger and Edman procedures to be described here will fail if the *N*-terminal amino acid is modified in certain ways (e.g., *N*-acetylation or, in the case of glutamine, cyclization). There may be other, as yet undetermined, reasons why proteins do not give quantitative *N*-terminal reactions.

5-1 *N*-Terminal Analysis

The 1-Fluoro-2,4-dinitrobenzene (FDNB) Method. In 1945 Sanger developed a method for *N*-terminal amino acid identification based on the reaction of amino groups of proteins with FDNB. FDNB has many of the properties of an ideal reagent for end-group analysis. It may be reacted with the amino groups of proteins under very mild conditions (e.g., room temperature, pH 8) to form dinitrophenyl (DNP) derivatives according to the reaction

$$R{-}NH_2 + F{-}C_6H_3(NO_2){-}NO_2 \xrightarrow{\text{pH } 8\text{–}9} R{-}NH{-}C_6H_3(NO_2){-}NO_2 + HF$$

Upon complete hydrolysis of the protein (e.g., 6 *N* HCl at 105°C for 16 hours) the *N*-terminal amino acids are recovered as DNP derivatives by extraction with ether (except for arginine). The identity of these derivatives is established by comparison with standards with the use of chromatographic methods (on paper, thin-layer plates, or columns of silica gel or kieselguhr). The yellow DNP derivatives are readily located and can be eluted and analyzed quantitatively by their absorbance at 360 nm (or 385 nm for DNP–proline). It should be noted that one of the prime aims of such methods is to obtain an *N*-terminal derivative that is stable to the acid hydrolysis necessary to cleave the peptide bonds. This is not strictly true for DNP amino acids, and corrections must be applied for losses in order to quantitate the results. The DNP derivatives of glycine, proline, and cystine, for example, are exceptionally labile, and special methods are necessary for their determination as DNP-derivatives. FDNB also reacts with a few groups in proteins other than α-NH_2 groups. Thus *O*-DNP-tyrosine (colorless), im-DNP-histidine (colorless), and ε-DNP-lysine (yellow but ether-insoluble) are formed, but none of these interferes with *N*-terminal analysis. These reactions may themselves be used to advantage in certain studies of protein structure.

Dansyl Chloride Procedure. Several methods of *N*-terminal analysis have been employed that are similar in principle to the FDNB method discussed above. Of these, the dansyl chloride (1-dimethylaminonaphthalene-5-sulfonyl chloride) procedure is extremely useful and has now largely replaced the classical FDNB method. The dansyl derivatives are fluorescent (with 254 nm or 365 nm excitation from a mercury lamp), and as a result, the method is about 100 times more sensitive than the Sanger method, which is based on absorbance. As little as 1 nmole of material may be detected on paper after suitable electrophoretic separation. Another advantage of the procedure is the greater stability of some dansyl amino acids compared with the corresponding DNP derivatives. Only *N*-terminal tyrosine has proved difficult to label and assay by this method. Details of the experimental procedure and references are provided by Gray in Reference 17.

Phenylthiohydantoin (Edman) Method. In 1950 Edman described a relatively simple procedure whereby amino acids could be removed stepwise from the *N*-terminus of proteins and peptides by reaction with phenylisothiocyanate. The method is based on earlier work of Bergmann in 1927 and Abderhalden in 1930, who obtained the first valid end-group data by reaction with phenylisocyanate. These early methods, however, did not prove to be of general usefulness.

The first step of the Edman procedure is an addition reaction of phenylisothiocyanate with the protein or peptide to form the phenylthiocarbamyl peptide (PTC-peptide):

$$\underset{\text{N—C}_6\text{H}_5}{\overset{\text{C=S}}{\|}} + \text{RCH}(\text{NH}_2)\text{CONH—CH}(\text{R}_1)\text{—CO—} \xrightarrow{\text{pH 8–9}} \text{R—CH}(\text{NH—C(=S)—NHC}_6\text{H}_5)\text{—CONH—CH}(\text{R}_1)\text{—CO—}$$

The PTC-peptide is cleaved in anhydrous acid (which favors attack of the sulfur on the carbonyl carbon of the terminal peptide) to give a thiazolinone intermediate and the liberation of the remainder of the peptide intact. The thiazolinone is hydrolyzed to the phenylthiocarbamyl (PTC)–amino acid, which then cyclizes to form the phenylthiohydantoin derivative of the amino acid (PTH–amino acid). This series of reactions may be written as:

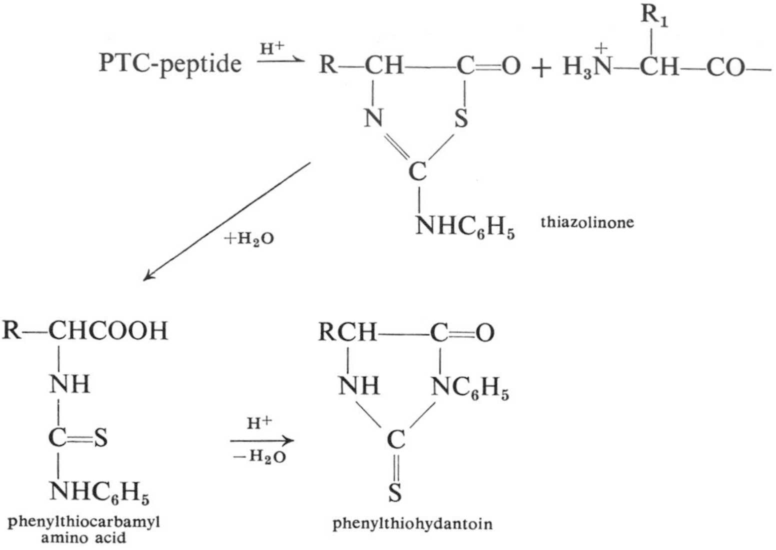

$$\text{PTC-peptide} \xrightarrow{H^+} \text{R—CH—C=O (ring: N, S, C—NHC}_6\text{H}_5\text{)} + H_3\overset{+}{N}\text{—}\underset{}{\overset{R_1}{|}}\text{CH—CO—}$$

thiazolinone

$+H_2O$

$$\text{R—CHCOOH, —NH—C(=S)—NHC}_6\text{H}_5 \xrightarrow[-H_2O]{H^+} \text{RCH—C=O (ring: NH, NC}_6\text{H}_5\text{, C=S)}$$

phenylthiocarbamyl amino acid

phenylthiohydantoin

The PTH–amino acids are extracted with organic solvents and identified by comparison with standards, using paper, column, or thin-layer chromatography. Their strong absorption at about 259 nm permits visualization against a fluorescent screen with an ultraviolet light source. Rapid and sensitive quantitative estimation of PTH–amino acids (except arginine) is afforded by gas–liquid chromatography.

The advantage of the Edman method is its nondestructiveness; the *N*-terminal residue is removed without hydrolysis of other peptide bonds, thereby leaving the rest of the polypeptide chain intact for further cycles of the procedure. This is particularly valuable for peptide sequence analysis, described later. Most amino acid residues react readily and give good quantitative yields. Special problems associated with lysine and histidine are discussed by Schroeder in Reference 17.

Carbamylation by Cyanate. Excellent quantitation of *N*-terminal amino acids is possible in most cases with the cyanate procedure of Stark and Smyth (see discussion by Stark in Reference 17). The carbamylation reaction, conversion to the hydantoin, and subsequent hydrolysis to the free amino acid are as follows:

$$NH_2CHRCONHCHR' \cdots COO^- + {}^-NCO \xrightarrow{pH\,8}$$

$$NH_2CONHCHRCONHCHR' \cdots COO^- \xrightarrow{H^+}$$

$$\begin{array}{c} CHR\text{---}CO \\ | \qquad\quad | \\ HN \qquad\quad NH \\ \diagdown \quad \diagup \\ CO \end{array} + {}^+NH_3CHR' \cdots COO^-$$

$$\text{Hydantoin} \xrightarrow[\text{or HCl 6M}]{\text{NaOH 0·2 M}} {}^+NH_3CHRCOO^- + NH_3 + CO_2$$

The resultant amino acid obtained by hydrolysis of the isolated hydantoin is quantitated by standard amino acid analysis procedures. It should be noted that although this procedure is similar to the Edman method described previously, it cannot be applied as a sequential degradation because of partial peptide bond hydrolysis accompanying the relatively harsh acidic condition required for the cyclization step.

Leucine Aminopeptidase Method. Leucine aminopeptidase is an enzyme that catalyzes the hydrolysis of the peptide bond linking the *N*-terminal amino acid to the polypeptide. It is not actually specific for leucine; however, the action of the enzyme is greatest for amino acids with large side-chains, while it works less rapidly on aromatic residues, and hydrolyzes those residues having polar groups very slowly (proline is not attacked at all). In order to identify the *N*-terminal residue of a polypeptide it is usually necessary to plot the quantitative release of individual amino acids during the reaction as a function of time, since the enzyme will release the second amino acid before all molecules have lost the first, and so on. For proteins containing a single polypeptide chain only, the release of two amino acids at the same rate may result from differences in rates of hydrolysis, for example, for a polypeptide whose sequence begins with the sequence Asp-Leu-. . . , leucine (which is rapidly released as soon as aspartic acid residues are removed) would appear as free amino acid at almost the same rate as aspartic acid, which is released slowly. Rates are also dependent on the neighboring amino acids, thus providing a source of ambiguity. The interpretation of data obtained by this method from proteins containing more than one type of polypeptide chain is difficult.

5-2 *C*-Terminal Analysis

The determination of the *C*-terminal amino acid of a polypeptide chain is most commonly made by hydrazinolysis, by reduction of the terminal carboxyl group to an alcohol, or by release of amino acids from the *C*-terminal end enzymatically with carboxypeptidase A and carboxypeptidase B. In general the chemical methods for *C*-terminal analysis are subject to greater error than those for *N*-terminal analysis, especially for some of the amino acids. A description of the side reactions and the subtleties of the two chemical methods has been given by Bailey.[12]

Hydrazinolysis (Akabori's Method). When a protein is reacted with anhydrous hydrazine at 100°C for 5 to 10 hours, all amino acid residues except the *C*-terminal residue are converted to hydrazides. The *C*-terminal residue is released as the free amino acid:

$$NH_2—\underset{\displaystyle R_1}{\overset{|}{C}}H—CO—NH—\underset{\displaystyle R_2}{\overset{|}{C}}H—CO— \cdots —NH—\underset{\displaystyle R}{\overset{|}{C}}H—COOH \xrightarrow{NH_2NH_2}$$

$$NH_2—CH(R_1)—CO—NH—NH_2$$

$$+ NH_2—CH(R_2)—CO—NH—NH_2 + \cdots + NH_2—CH(R)—COOH$$

Identification of the *C*-terminal amino acid may be made by converting the amino acid hydrazides to their water-insoluble dibenzal derivatives (by reaction with benzaldehyde) and determining the amino acids remaining in the supernatant by methods already discussed. Alternatively, the amino acids may be identified directly in the mixture by chromatography. A good method for identification uses the FDNB reaction to form the DNP *C*-terminal amino acid and the di-DNP derivatives of the hydrazides (or tri-DNP derivatives of lysine, tyrosine, and cysteine). The DNP amino acids are extracted with aqueous bicarbonate from ethyl acetate (contaminated only with the di-DNP hydrazides of aspartic and glutamic acid) and identified by chromatography. Hydrazinolysis cannot be used to determine cystine, cysteine, asparagine, or glutamine; several other amino acids are recovered in quite low yields.

Reduction to Amino Alcohols. Free carboxyl groups in polypeptides are reduced to alcohols by lithium borohydride:

$$R—COOH \xrightarrow{LiBH_4} R—CH_2—OH$$

The amino alcohols can be extracted after hydrolysis of the polypeptide and separated by chromatography on paper or on buffered silica gel columns. This method is complicated by the fact that reductive fission of peptide bonds also occurs to the extent of a few percent producing additional amino alcohols. This fission is selective (e.g., one glycine residue in insulin is particularly sensitive) and leads to erroneous results for large peptides and proteins. The reaction may also be carried out with lithium aluminum hydride, but even more side reactions occur with this harsher reagent.

Carboxypeptidases A and B. The carboxypeptidase method for *C*-terminal analysis is usually the method of choice since the chemical methods are difficult and frequently ambiguous. The specificity of carboxypeptidase A and B for peptide bond hydrolysis requires a free α-carboxyl group. Neither enzyme will hydrolyze a *C*-terminal proline or hydroxyproline residue, nor will either enzyme remove the *C*-terminal amino acid if proline (or hydroxyproline) occupies the penultimate position.

Carboxypeptidase B is specific for the removal of the *C*-terminal basic amino acids lysine, arginine, or ornithine from peptides and proteins. Certain chemically modified amino acids with a positively charged side chain, such as *S*-(β-aminoethyl)cysteine residues are also hydrolyzed if they are carboxyl-terminal, but histidine is not cleaved by this enzyme. The enzyme is influenced by the nature of nearby amino acid residues, and the rate of hydrolysis for susceptible *C*-terminal amino acids is markedly decreased if the next amino acid is charged (either positively or negatively). Carboxypeptidase A catalyzes the hydrolysis of carboxyl-terminal acidic and neutral amino acids (except proline). The rate at which hydrolysis occurs varies greatly among the amino acids and is influenced by the nature of the adjacent amino acids.

It is frequently possible to establish the sequence of several amino acids from the *C*-terminal end of a peptide or protein by removing aliquots from a reaction mixture and quantitatively plotting the release of amino acids as a function of time. However, ambiguities similar to those discussed for leucine amino peptidase will eventually develop. A refinement of the carboxypeptidase method developed by Boyer (see discussion by Ambler in Reference 17) allows positive identification of the *C*-terminal amino acid(s) in proteins even when more than one polypeptide chain is present. In this procedure the hydrolysis is carried out in the presence of $H_2{}^{18}O$:

$$R_1\text{—}\underset{NH_2}{\overset{H}{C}}\text{—}\overset{O}{\overset{\|}{C}}\text{—NH—}\overset{R_2}{\overset{|}{C}}H\text{—}\overset{O}{\overset{\|}{C}}\cdots\text{NH}\overset{R_{n-1}}{\overset{|}{C}}\text{—}\overset{O}{\overset{/\!/}{C}}\text{—NH—}\overset{R_n}{\overset{|}{C}}H\text{—}\overset{O}{\overset{\|}{C}}\text{—OH}\xrightarrow{H_2{}^{18}O}$$

$$R_1\underset{NH_2}{\underset{|}{C}}H_1\text{—}\overset{O}{\overset{\|}{C}}\text{—}^{18}OH + \cdots\cdots + H\text{—}\underset{NH_2}{\underset{|}{\overset{R_{n-1}}{\overset{|}{C}}}}\text{—}\overset{O}{\overset{/\!/}{C}}\text{—}^{18}OH + H\text{—}\underset{NH_2}{\underset{|}{\overset{R_n}{\overset{|}{C}}}}\text{—}\overset{O}{\overset{/\!/}{C}}\text{—OH}$$

As can be seen, only the *C*-terminal amino acid is *not* enriched in ^{18}O during the hydrolysis, thus providing positive identification if a careful isotopic analysis is made.

6 SEQUENCE DETERMINATION[16,17]

No single procedure can be given as the best way to determine the amino acid sequence of a protein. The exact procedure followed varies with both the special problems presented by a particular protein and the preferences of the investigator. The following general outline, however, will serve as a guide to the methods of sequence determination.

1. The protein is separated into its individual polypeptide chains if more than one kind is present.
2. The polypeptide chains are cleaved into smaller units at specific points, usually by the action of cyanogen bromide or trypsin, and the smaller peptides are separated and purified. In the case of large peptides, further enzymatic or chemically specific cleavage followed by product separation may be required.
3. The amino acid sequence of the peptides is determined.
4. The order of the peptides in the original chain is established (frequently called determining the "overlap").
5. The pairing of cysteine residues involved in disulfide bridges is determined.

6-1 Chain Separation

In multichain proteins with either inter- or intrachain nonpeptide covalent bonds, the first step in chain separation is usually the cleavage of such linkages. This must be done by sufficiently gentle means so that peptide bonds are not cleaved. Chemical methods that meet the above criteria are well established in the case of disulfide bridges, but are still the subject of controversy for other possible linkages. The original method of S—S bond cleavage used by Sanger on insulin, performic acid oxidation to cysteic acid, is seldom used today because of side reactions which result in the destruction of tryptophan and subsequent cleavage of the polypeptide chain (no problems arose with insulin since it does not contain tryptophan). Most current investigators prefer either the reduction of —S—S— to SH followed by alkylation (e.g., with iodoacetate or ethyleneimine) to prevent reoxidation to disulfides, or the oxidation of —S—S— to the *S*-sulfo derivatives (cf. Section 3 of this chapter).

The dissociation of a protein into its constituent polypeptide chains also requires disruption of structure maintained by noncovalent forces between chains. The ease with which this can be accomplished varies widely among proteins. Often an acid or alkaline pH will dissociate the chains. In some cases temperature-induced noncovalent bond rupture is irreversible and the chains remain separated upon cooling (e.g., collagen, Chapter 17). A variety of chemical agents, such as detergents, are useful for dissociation of polypeptide chains. Denaturing solvents such as 6 to 8 *M* urea are effective for dissociation and can be employed throughout the chromatographic procedures; the urea is readily removed by dialysis after the chains have been separated.

The isolation of the individual chains is most frequently accomplished by ion-exchange chromatography (carboxymethyl-cellulose, DEAE-cellulose, etc.). Adsorption chromatography (e.g., on calcium phosphate gels), molecular sieve methods (e.g., Sephadex chromatography), or isoelectric focusing may be effective in some cases. All of the above methods can be employed in the presence of a dissociating agent such as 8 *M* urea. It is not unusual to encounter considerable difficulty in chain isolation since in many cases, probably through gene duplication during evolution, proteins have come to contain nonidentical polypeptide chains that have nearly identical molecular weights and charge properties.

6-2 Specific Cleavage to Smaller Peptides

Sequence analysis is not usually directly applied to large polypeptides. The difficulty of a direct approach becomes much greater as the chain length increases to more than 15 to 20 amino acids. However, solid phase and automated Edman degradation may be feasible for some larger peptides or for proteins themselves. The preparation of specifically cleaved fragments of workable size in high yields is generally a desirable second step after chain separation. The larger the peptides are, the less effort is required in overlap determination. If the fragments are too small, and thus numerous, the determination of their order in the original chain will present far more difficulty.

The choice for the initial method of cleavage of a large protein varies among investigators. Although at one time sequence work was usually initiated with trypsin digestion, there seems to be an increasing preference for initial cleavage with cyanogen bromide. The reaction scheme, as discussed by E. Gross in Reference 17, is shown below. This reagent is highly specific for peptide bonds in which the carbonyl group is contributed by methionine. High yields are obtained in most cases. Most proteins contain relatively few methionyl residues. Thus the peptides obtained after cyanogen bromide

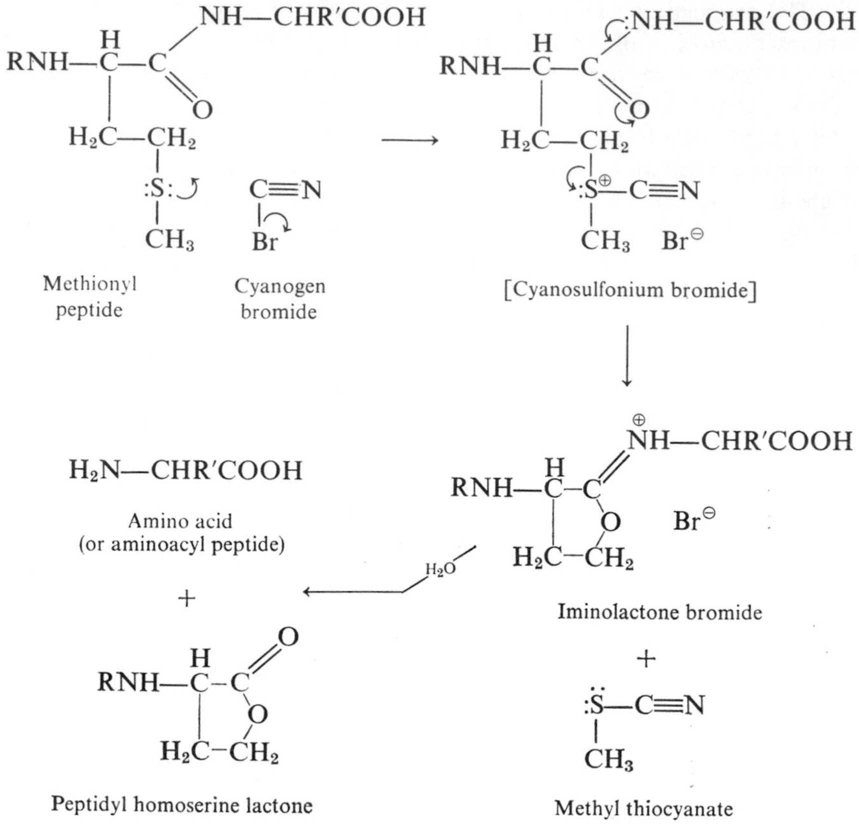

$RNH{-}CH(CH_2CH_2SCH_3){-}C(=O){-}NH{-}CHR'COOH$ + $Br{-}C{\equiv}N$ ⟶ $[RNH{-}CH(CH_2CH_2S^{\oplus}(CH_3){-}C{\equiv}N){-}C(=O){-}NH{-}CHR'COOH]\ Br^{\ominus}$

Methionyl peptide

Cyanogen bromide

[Cyanosulfonium bromide]

↓

Iminolactone bromide ($Br^{\ominus}$) + $CH_3{-}S{-}C{\equiv}N$

Methyl thiocyanate

$\xrightarrow{H_2O}$

$H_2N{-}CHR'COOH$

Amino acid
(or aminoacyl peptide)

+

Peptidyl homoserine lactone

cleavage are likely to be manageable in number, and separation problems are reduced. An example of separation of CNBr peptides by molecular exclusion chromatography has been shown (Fig. 3-10). Other methods are ion-exchange chromatography and isoelectric focusing. After purification the larger CNBr peptides may be further digested with trypsin to produce smaller peptides for sequence analysis.

The high degree of specificity of trypsin for arginine and lysine residues makes it very effective for preparation of reasonably sized peptides from proteins or large peptides. In a typical protein about 10% of the amino acid residues are lysine and arginine, and thus the average tryptic peptide would contain about 10 amino acid residues. Of course the digest will contain a distribution of sizes possibly ranging from free lysine and arginine to very large peptides. (Large insoluble peptides are frequently referred to as the trypsin-resistant "core." TMV protein contains a "core" peptide 40 amino acid residues in length.) The total number of peptides expected is equal to the sum of the lysine and arginine residues present (obtained from amino acid analysis and chain molecular weight) plus one if the carboxyl-terminal amino acid is not lysine or arginine. This number is not always obtained since some lysine or arginine bonds are somewhat resistant to attack (such as lysine or arginine residues adjacent to each other or next to negatively charged amino acids or proline). If the reaction at some bonds is not complete, an excess number of different peptides would be obtained but this would be revealed by irregular yields. The reaction may be followed by removing aliquots for ninhydrin assay or by titration. Tryptic digestion over long periods of time may produce unwanted peptides due to the presence of slight chymotryptic activity in the enzyme preparation, and care must therefore be taken to use trypsin that is free of other proteolytic activity.

Separation of Peptides. This is usually achieved by ion exchange chromatography on Dowex 50 or Dowex 1 (or a combination of the two), eluting with a pH, and/or salt gradient. Other ion exchange resins useful in peptide fractionation are discussed in a number of articles in Reference 17. Figure 4-7 shows the separation of the tryptic peptides of hemoglobin by ion-exchange. Peptides that are eluted at the same position from a particular column often can be separated by changing elution conditions. In some cases a prior separation on the basis of chain length by a molecular sieve column may minimize the overlapping obtained in ion exchange chromatography. The fractionation of ribonuclease tryptic peptides on Sephadex G25 is illustrated in Fig. 4-8. Size fractionation may also be carried out after ion exchange chromatography.

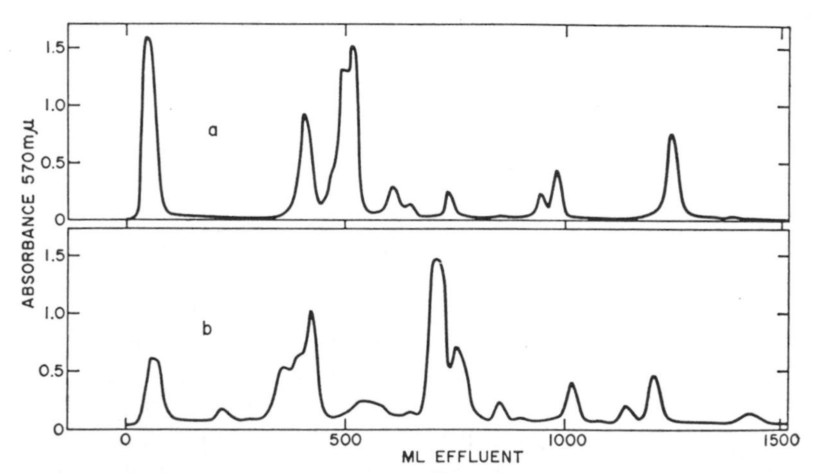

Fig. 4-7 Separation of the soluble tryptic peptides of hemoglobin (*Propithecus verreauxi*) by ion-exchange chromatography: (a) α-chains; (b) β-chains. [From R. L. Hill and R. Delaney, "Peptide mapping with automatic analyzers: use of analyzers and other automatic equipment to monitor peptide separations by column methods" in Reference 17.]

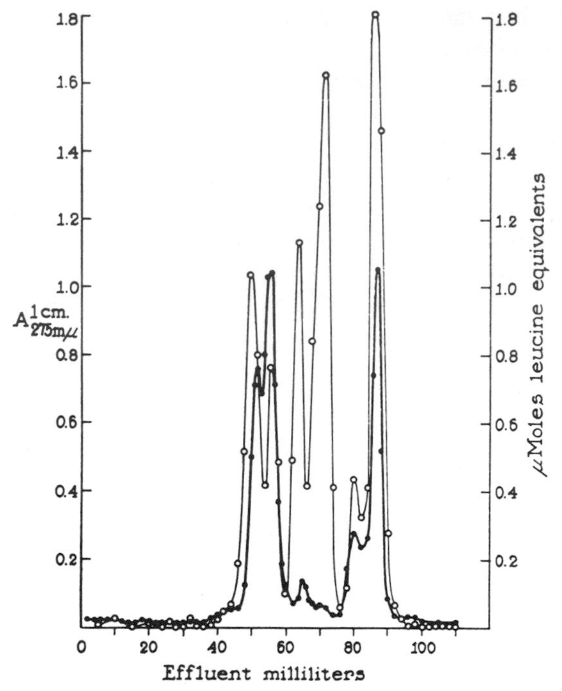

Fig. 4-8 Fractionation of tryptic peptides of oxidized ribonuclease by gel filtration on Sephadex G-25 in 0.2 *M* acetic acid. Open circles represent ninhydrin analysis and solid circles are absorption at 275 nm. [From J. Porath, *Adv. Protein Chem.*, **17**, 209 (1962).]

Some investigators prefer to purify the peptides by paper chromatography and/or paper electrophoresis. Two-dimensional separation is almost always required, for example, electrophoresis in one direction followed by chromatography (or electrophoresis at a different pH) at right angles. The peptides are located with a suitable spray such as ninhydrin or with the Rydon-Smith Cl_2-starch-I_2 method. The peptides are isolated by elution from the appropriately excised area of a companion unstained paper run simultaneously under identical conditions. (If dilute ninhydrin is used to locate the peptides, only a small proportion of the peptide is reacted and good recovery is achieved by direct elution of the stained area.) Overlapping peptides may be reapplied to another paper and subjected to further chromatography or electrophoresis. This procedure is similar to the "peptide mapping" or "fingerprinting" technique discussed in Section 7. Peptides may also be separated on the basis of their differing solubility in various solvents by countercurrent distribution or partition chromatography (see the article by Hill in Reference 17). Although of lower resolution, this method is able to handle larger quantities of material. After fractionation the amino acid composition of each purified tryptic peptide is determined by standard hydrolysis and analytical procedures. The composition is itself a good measure of purity (whole number molar ratios are required) and, from the lowest molar concentration per unit weight, serves to establish the minimum molecular weight (usually equal to the true molecular weight of the peptide since most peptides will contain at least one amino acid at a concentration of 1 mole/mole peptide). The amino acid composition also indicates which method of further fragmentation might profitably be employed in cases where direct sequence determination is not feasible due to size of the peptide or unusual amino acid composition.

Further Specific Enzymatic or Chemical Cleavage.[7,17,19,20] This may be carried out on peptides containing the appropriate amino acid residues (or sequence). Three different approaches are available:

Proteolytic cleavage by another endopeptidase. The specificities of other endopeptidases are listed in Table 4-2. Chymotrypsin preferentially catalyzes the hydrolysis of peptide bonds involving the carboxyl group of the aromatic amino acids tryptophan, tyrosine, and phenylalanine. However, significant rates of hydrolysis of other bonds are also observed and the influence of neighboring amino acids in some cases may cause more rapid reaction of nonaromatic amino acid peptide links than of nearby aromatic residues. Thus chymotryptic digestion of large proteins is likely to give a somewhat confusing picture. The enzyme is more useful for analysis of moderately sized peptides which have been obtained by other means. A few other endopeptidases have been used for selective cleavage, but their broad specificity limits their application.

Chemical modification of certain amino acid residues to render them susceptible to tryptic action. The specificity of trypsin requires a positive charge on the amino acid side chain located on the carboxyl side of the peptide bond to be cleaved, and the charge must be a certain distance from the α-carbon atom. Neither ornithine (one CH_2 group less than lysine) nor homoarginine (one CH_2 group more than arginine), when incorporated into peptides by chemical modification, is susceptible to tryptic action. The most successful

Table 4-2 Activities of proteolytic enzymes

Enzyme	Major sites of action in proteins	Other sites of action
Trypsin	Arg, Lys	
Chymotrypsin	Tyr, Phe, Trp	Asn, Gln, His, Leu, Lys, Met, Ser, Thr
Pepsin	All except Pro and Ile	
Papain	Arg, Lys, His, Gly, Glu, Gln, Leu, Tyr	Various
Subtilisin	Broad specificity	
Pronase	Virtually all peptide bonds	

means for production of artificial trypsin-susceptible sites in proteins is the alkylation of cysteine residues with ethyleneimine or 1-bromo-2-ethylamine to yield *S*-(β-aminoethyl)-cysteine residues (thialysine). The reaction proceeds smoothly under mild conditions and the modified cysteine is readily attacked by trypsin. Potentially useful but less well characterized modifications producing trypsin susceptibility are the conversion of γ-glutamyl carboxyl groups to γ-hydrazides and the modification of serine by introduction of a glycyl group through an ester linkage. Details and references to these methods may be found in the reviews by Witkop and associates.[7,20]

Selected chemical cleavage of the peptide chain at specific amino acid residues. The difficulty of the problem has been matched by the ingenuity of organic chemists in devising useful reactions with promise of more to come. At present the success of some of the reactions is dependent on the skills and experience of the investigator employing them. The reaction of cyanogen bromide to cleave methionyl bonds has already been discussed. The cleavage of tryptophanyl bonds by *N*-bromosuccinimide has also been applied to several structural determinations with good results. Acid hydrolysis for preferential removal of aspartyl residues has been effective in a number of cases. Details of the reactions are given in the reviews cited above.

6-3 Peptide Sequencing

The particular approach used depends greatly on the size of the peptide, its amino acid composition, the skills and experience of the investigator, the available equipment, and the quantity of peptide available. Any of several alternate methods will serve equally well to establish all but very unusual sequences. Clearly the fullest application of each of the several methods of attack to be outlined below would yield redundant information. This, coupled with the possibility that some of the methods listed may not work for a particular sequence, means that the following list should not be construed as a flow sheet for sequencing.

Determination of the *N*-Terminal and *C*-Terminal Amino Acid Residues. Identification of the terminal residues of the peptide can be carried out by any of the methods already discussed for the polypeptide chains of the protein. If no duplication of terminals occurs among the peptides, then the peptide that came from the *N*-terminal of the original chain and that from the *C*-terminal can be immediately identified.

Determination of Sequence at the *N*-Terminal End by Stepwise Application of the Edman Phenylisothiocyanate Method. In favorable cases the sequence of 10 or more amino acids from the *N*-terminal end has been determined by classical Edman procedures.[17] This may be accomplished either by extraction and identification of the phenylthiohydantoin after each step or by the subtractive method whereby an aliquot of the unreacted peptide is hydrolyzed and subjected to amino acid analysis after each step and the missing amino acid is thus identified. Automatic amino acid analysis or paper electrophoresis may be profitably employed in the subtractive method. The former is more accurate and may be required if the terminal amino acid also appears elsewhere in a long peptide. The latter method is more rapid and quite adequate for small peptides. Figure 4-9 shows the use of electrophoretic

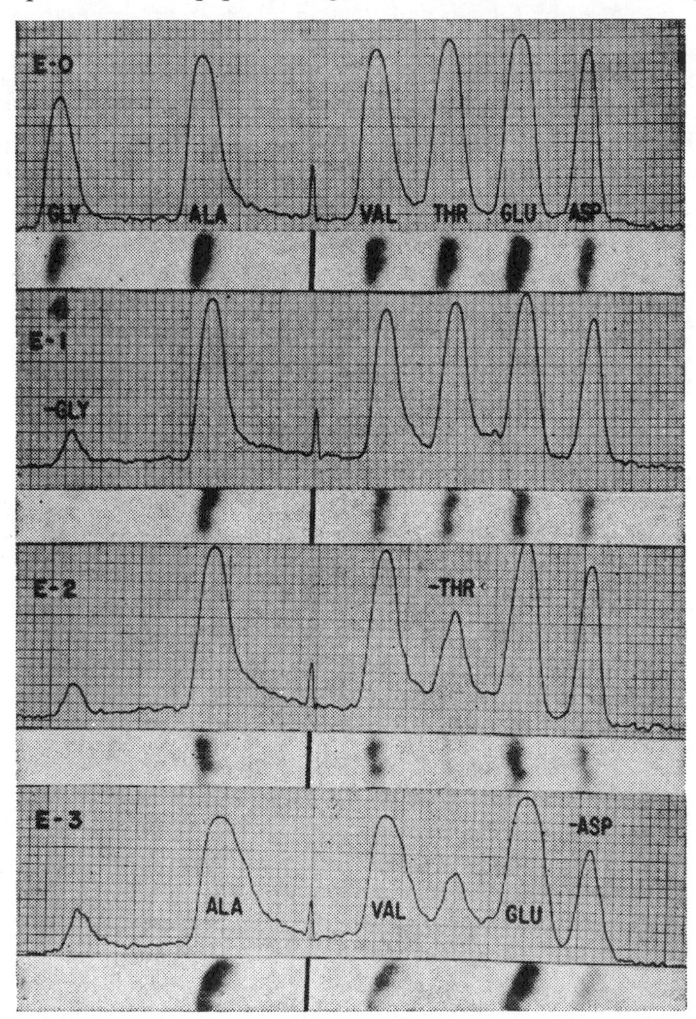

Fig. 4-9 Substractive analysis applied to the Edman degradation of the hexapeptide (Gly-Thr-Asp-Val-Glu-Ala) isolated from lysozyme. The original peptide (E-0) and that remaining after 3 steps of the Edman cycle (E-1, E-2, E-3) were hydrolyzed and subjected to high voltage electrophoresis. Densitometer tracings are shown above each stained strip.[16]

separation in the subtractive method. The success of the Edman procedure depends on the nature of the sequence; unfavorable sequences may result in low yields of the *N*-terminal derivative and some degradation of other peptide bonds. In such cases sequencing must be terminated sooner than would otherwise be necessary. New procedures are, however, available which reduce the level of interfering side reactions and permit much higher yields to be obtained at each step. As a result, it has been possible to sequence polypeptides of up to 60 residues in length. An automated procedure (sequenator) has been developed by Edman and Begg in which environmental conditions and reagent purity are carefully controlled. An alternative procedure termed "solid phase degradation" has been devised, which greatly facilitates handling. In methods developed independently by Laursen and by Dintzis the peptides are attached to an insoluble support through carboxyl groups and the reagents are sequentially passed over the solid phase. Stark and co-workers have described a procedure whereby the reagent itself forms the insoluble phase. Further details and references may be found in the review by Stark.[19]

Stepwise Degradation from the *N*-Terminal End with Leucine Amino Peptidase (LAP). This method has already been discussed with regard to *N*-terminal amino acid identification. The variation in rates of reaction with the different amino acids and the influence of adjacent sequences limit the usefulness of this exopeptidase; however, in favorable cases LAP can be used to obtain the sequence of eight or more amino acids from the *N*-terminal of a polypeptide chain. If proline is present in the sequence, the LAP degradation stops when it is reached. However, this residue can be removed with a specific exopeptidase for proline and the reaction with LAP continued. The proline block may be helpful in the use of LAP since it permits re-establishment of synchrony among chains that are being digested "out of synchrony." Time curves of the amino acids released by LAP are essential and may be obtained by withdrawal of a portion of the digest at various intervals for amino acid analysis. Removal of amino acids, two at a time, by enzymatic means may also be possible with the use of mass spectrometry or other methods for identification of the dipeptides.

Stepwise Digestion from the *C*-Terminal End with Carboxypeptidases A and B. Carboxypeptidase A can be used for stepwise release of amino acids (other than proline, lysine, or arginine) from the *C*-terminal end of the chain; however, as in the case of LAP, the rates for different residues vary greatly and usually only a few residues in the sequence can be determined. The specificity of carboxypeptidase B for the basic amino acids lysine and arginine makes it particularly useful in removing these amino acids from their *C*-terminal position in tryptic peptides. The reaction fails if the adjacent residue is proline or is very slow if the adjacent residue is negatively charged.

Partial Hydrolysis and the Use of Nonspecific Endopeptidases (The Jig Saw Puzzle Method). By partial acid hydrolysis of a peptide a large number of small fragments can be obtained in varying yields. These are separated by paper chromatography or paper electrophoresis (or on columns) and *N*- and *C*-terminal analyses are performed along with amino acid analysis. For the di- and tri-peptides this is sufficient to give the sequence. Larger fragments could be sequenced by the methods discussed previously. However, this is frequently unnecessary since other products of the partial hydrolysis may include this information. If not, some investigators prefer to subject the larger pieces to further hydrolysis and repeat the above steps. After many small fragments have been analyzed the results are correlated with the original peptide of known amino acid composition and known *N*-terminal and *C*-terminal amino acids. The appropriate fit of the pieces in such a "jig-saw puzzle" is demonstrated by Sanger's original work on the A chain of insulin as shown in Fig. 4-10.

For smaller peptides it is particularly advantageous to carry out the partial hydrolysis on the DNP-peptide. The fragments containing the DNP group (which also permits easy visualization) must differ only in the number of amino acids remaining from the *N*-terminal end. Hence in the tetrapeptide DNP-Val-Glu-Tyr-Gly, the identification of three DNP-peptides whose compositions on hydrolysis show the presence of Val alone, Val + Glu, and Val + Glu + Tyr would suffice to determine the sequence. Other possible products that would yield redundant information, such as Glu·Tyr, need not even be isolated and analyzed.

A similar approach involves the action (or partial action) of one or several broad specificity proteases. The relative merits of these methods will depend on the specific problem. Partial acid hydrolysis may result in destruction of some amino acids or even an inversion of sequence via a diketopiperazine intermediate. On the other hand the resistance of certain sequences to proteolytic attack limits the enzymatic method.

```
Ser·Leu          Glu·Leu          Asp·Tyr
    Leu·Tyr          Leu·Glu          Tyr·CySO3H
        Tyr·Glu          Glu·Asp
Ser·Leu·Tyr          Leu·Glu·Asp
    Leu·Tyr·Glu          Glu·Asp·Tyr
        Glu·Leu·Glu
Ser·Leu·Tyr·Glu          Glu·Asp·Tyr·CySO3H

Sequence: Ser·Leu·Tyr·Glu·Leu·Glu·Asp·Tyr·CySO3H
```

Fig. 4-10 The sequence of a nonopeptide from the A chain of insulin as deduced from the "jig saw" method. It was known that there were only two tyrosines in the chain. The various small oligopeptides are shown after proper positioning. [Data based on F. Sanger and E. O. P. Thompson, *Biochem. J.*, **53,** 353 (1953).]

The computerized sequence determination of small peptides, without prior separation, by mass spectrometry[21] is a rapid procedure that seems certain to be applied more widely in the future.

6-4 The Order of the Peptides in the Original Protein ("Overlap Determination")

Once the sequence of each of the peptides obtained from the original digest is known, their order in the original protein must be established. The peptides at the *N*- and *C*-terminals of the chain can usually be identified by comparison with the terminals of the original protein chain. The order of the internal peptides is determined by obtaining composition or sequence data on other peptide fragments that overlap the sites of cleavage of the first set of peptides. Ideally these peptides will contain in an internal position the terminal amino acids of the original peptides and will have enough amino acids on either side to avoid ambiguities. For some time the method of choice in determining the overlap for tryptic peptides, for example, involved digestion of the original chain with chymotrypsin followed by analysis of the purified lysine- and arginine-containing peptides. This is laborious and then may not be effective in establishing the sequence around all of the lysines or arginines (e.g., if an aromatic amino acid is too close to them). Other enzymes or even partial acid hydrolysis has been used to determine the overlap in cases where chymotrypsin failed to do so.

The current trend in overlap determination for tryptic peptides is toward the isolation of selectively cleaved fragments larger than the tryptic peptides followed by *N*- and *C*-terminal analysis, tryptic action on the peptide, and identification of the products, usually by amino acid composition alone. Essentially one-half of the overlap problem for tryptic peptides can be solved by selective blocking of the ε-NH_2 groups of lysine with CS_2 or by trifluoroacetylation before digestion with trypsin. The protein is now cleaved only at the arginine sites, the peptides are separated, and the amino acid composition and *N*-terminals are established. The modifying groups may be removed under mild conditions to regenerate lysyl residues, and each lysine-containing peptide can be resubjected to action by trypsin and the products analyzed to the extent required to equate them to the original tryptic peptides (usually amino acid analysis, elution from an ion-exchange column, or mapping). A reversible blocking of arginine has not yet been accomplished, although the arginine residues can be made trypsin-resistant nonreversibly.[7] The blocked protein is then cleaved at the lysine residues and the products analyzed as above except that further tryptic action is impossible, and the application of chymotrypsin or other methods of further cleavage may be required. The selected cleavage of chains at methionine and tryptophan residues may also be used to great advantage in overlap determination. A novel method for determination of order in a polypeptide chain is based on the specific activity of peptides pulse-labeled with a radioactive amino acid during biosynthesis of the protein.[22]

6-5 Determination of Cysteine Pairing in Disulfide Bridges

The most direct procedure for ascertaining which cysteines are paired in disulfide bridges is as follows:

1. The linear sequence of amino acids is determined as described for the polypeptide chains in which disulfide bonds have been broken by reduction or oxidation.
2. Tryptic digestion of the protein is then performed with the disulfide bridges intact.
3. The disulfide bridge-containing peptides are isolated and are oxidized or reduced according to the procedure used for the intact chains in (1).
4. The two peptides obtained from each disulfide-containing peptide are identified by comparison with the tryptic peptides of the linear sequence. If an S—S bridge is present within a single peptide and if other cysteine residues are also present in the peptide (as SH), the latter can be identified by chemical reaction (e.g., iodoacetate) while the S—S bond is intact. Thus the unreacted cysteine residues are identified as those participating in the disulfide bond. The possibility of two S—S bridges in a single tryptic peptide is remote but pairing could be determined by further cleavage for example, with chymotrypsin.

It should be stressed that the chemistry of disulfide bridges is much easier on paper than it is in the laboratory. Oxidation and disulfide exchange reactions that may occur in disulfide peptides at both acid and alkaline pH preclude the use of some analytical methods acceptable for other peptides. Some of the mandatory precautions are discussed by Bailey[12] and by Canfield and Anfinsen.[16]

7 FINGERPRINTING

A mixture of peptides can be separated by chromatography or electrophoresis in two dimensions as outlined previously. If the procedure is carefully controlled, the pattern obtained is characteristic of the particular protein from which the peptides were derived, that is, it is a "map" or a "fingerprint." In addition to the use of the method for peptide purification and isolation, two other valuable applications are possible. One is the rapid determination of the location of amino acid modifications or substitutions resulting from mutations. The classic determination of the difference between normal hemoglobin and sickle-cell hemoglobin is illustrated in Fig. 4-11. The substitution of a single valine for glutamic acid in the β-chain resulted in the translocation of one peptide in the map. Similar procedures have been used to locate active sites of enzymes (by chemical modification of residues at the site) and in many studies of mutations. The technique is, in general, not sensitive enough to detect sequence inversion (change in order of two amino acids) or substitutions involving no change in peptide charge and little change in solubility (e.g., leucine → valine substitution would be difficult to detect). Although peptide maps may be very complicated for very large proteins, particularly those containing multiple chains of different types, specific chemicals may be used to localize selected peptides on the map. For example,

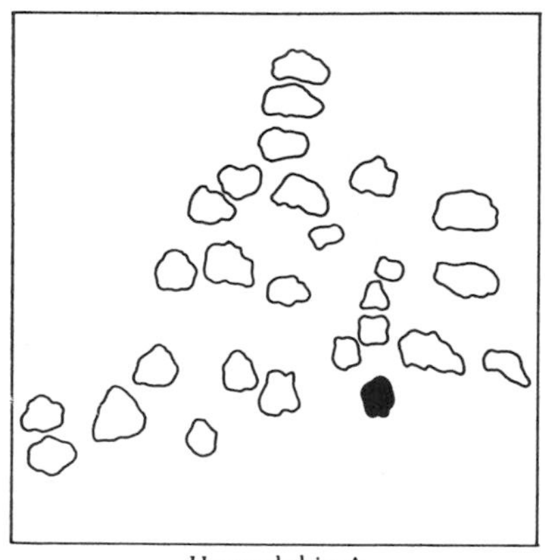

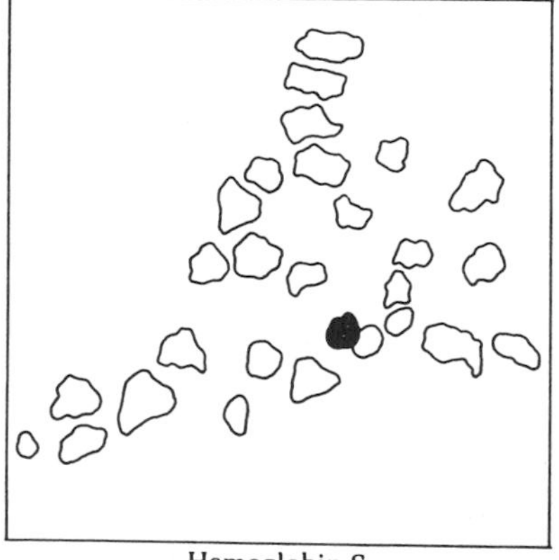

Fig. 4-11 Peptide maps of normal hemoglobin A and sickle-cell hemoglobin. The peptide containing the amino acid substitution (solid spot) is clearly displaced. [From A. L. Lehninger, *Biochemistry*, Worth, New York, 1970, based on data of V. Ingram, *Biochem. Biophys. Acta*, **28,** 539 (1958).]

cysteine-containing peptides are readily located by characteristic sulfhydryl reactions. Fingerprinting is also a valuable adjunct to physical methods for the elucidation of the number of subunits in a protein and the number of different types of chains present. For example, a protein containing 54 lysine and arginine residues shown to have 6 subunits by molecular weight determination would yield 10 tryptic peptides if all 6 subunits were identical, 20 if there were 3 each of 2 different subunits, and so forth.

REFERENCES

1. D. R. Helinski and C. Yanofsky, "Genetic control of protein structure," in *The Proteins*, 2nd ed., Vol. 4, H. Neurath, Ed., Academic Press, New York, 1966, p. 1.
2. *Atlas of Protein Sequence and Structure*, National Biomedical Research Foundation, Silver Spring, Md. An annual series beginning in 1965.
3. J. A. Schellman and C. Schellman, "The conformation of polypeptide chains in proteins," in *The Proteins*, 2nd ed., Vol. 2, H. Neurath, Ed., Academic Press, New York, 1964, p. 1.
4. G. N. Ramachandran and V. Sasisekheran, "Conformation of polypeptides and proteins," *Adv. Protein Chem.*, **23,** 283 (1968).
5. K. Hofmann and P. G. Katsoyannis, "Synthesis and function of peptides of biological interest," in *The Proteins*, 2nd ed., Vol. 1, H. Neurath, Ed., Academic Press, New York, 1963, p. 53.
6. A. Marglin and R. B. Merrifield, "Chemical synthèsis of peptides and proteins," *Ann. Rev. Biochem.*, **39,** 841 (1970).
7. T. F. Spande, B. Witkop, Y. Degani, and A. Patchornik, "Selective cleavage and modification of peptides and proteins," *Adv. Protein Chem.*, **24,** 97 (1970).
8. R. Cecil and J. R. McPhee, "The sulfur chemistry of proteins," *Adv. Protein Chem.*, **14,** 256 (1959).
9. R. Cecil, "Intramolecular bonds in proteins. I. The role of sulfur in proteins" in *The Proteins*, 2nd ed., Vol. 1, H. Neurath, Ed., Academic Press, New York, 1963, p. 379.
10. B. Blomback, M. Blomback, A. Henschen, B. Hessel, S. Iwanaga, and K. R. Woods, "*N*-Terminal knot of human fibrinogen," *Nature* **218,** 130 (1968).
11. G. M. Edelman, B. A. Cunningham, W. E. Gall, P. D. Gottlieb, V. Rutishauser, and M. J. Wardel, "The covalent structure of an entire γ G immunoglobulin molecule," *Proc. Nat. Acad. Sci., U.S.* **63,** 78 (1969). G. M. Edelman and W. E. Gall, "The antibody problem," *Ann. Rev. Biochem.*, **38,** 415 (1969).
12. J. L. Bailey, *Techniques in Protein Chemistry*, 2nd ed., Elsevier, New York, 1967.
13. A. Gottschalk and E. R. B. Graham, "The basic structure of glycopeptides," in *The Proteins*, 2nd ed., Vol. 4, H. Neurath, Ed., Academic Press, New York, 1966, p. 95.
14. A. Gottschalk, *Glycoproteins*, Elsevier, New York, 1966.
15. F. W. Putnam, "Structure and function of the plasma proteins," in *The Proteins*, 2nd ed., Vol. 3, H. Neurath, Ed., Academic Press, New York, 1965, p. 153.

16. R. E. Canfield and C. B. Anfinsen, "Concepts and experimental approaches in the determination of the primary structure of proteins," in *The Proteins*, 2nd ed., Vol. 1, H. Neurath, Ed., Academic Press, New York, 1963, p. 311.

17. C. H. W. Hirs, Ed., *Enzyme Structure*, Vol. 11 of *Methods in Enzymology*, Academic Press, New York, 1967; C. H. W. Hirs and S. N. Timasheff, Eds., *Enzyme Structure*, Part B, Vol. 25 of *Methods in Enzymology*, Academic Press, New York, 1972.

18. E. C. Horning and W. J. A. VandenHeuvel, "Gas chromatography," *Ann. Rev. Biochem.*, **32,** 709 (1963).

19. G. R. Stark, "Recent developments in chemical modification and sequential degradation of proteins," *Adv. Protein Chem.*, **24**, 261 (1970).

20. B. Witkop, "Nonenzymatic methods for the preferential and selective cleavage and modification of proteins," *Adv. Protein Chem.*, **16,** 221 (1961).

21. K. Biemann, "Mass Spectrometry," *Ann. Rev. Biochem.*, **32,** 755 (1963); M. M. Shemyakin, Y. A. Ovchinnikov and A. A. Kiryushkin, "Mass spectrometry in peptide chemistry," in *Mass Spectrometry: Techniques and Applications*, G. W. A. Milne, Ed., Wiley-Interscience, New York, 1971.

22. J. Vuust and K. A. Piez, "Biosynthesis of the α chains of collagen studied by pulse-labeling in culture," *J. Biol. Chem.* **245,** 6201 (1970).

V Chemical Modification of Proteins[1–7]

The side chains of the amino acid residues in proteins are capable of reaction with a variety of small organic reagents and polymeric substances. The discovery of more such reagents can be expected in the future and further applications of these reactions for the study of protein structure and function should be forthcoming. We wish to discuss a few examples in order to indicate the rationale for these methods in relation to particular aspects of protein structure. More detailed information may be obtained in the reviews cited.[1–7] A convenient and extensive summary of reagents used for chemical modification of proteins is given in Table I in the review by Stark.[1]

In considering the choice of a reagent to modify a protein for a specific purpose, several questions need to be considered: (1) What level of completion of the reaction is required (e.g., 95%)? (2) Must it be specific for a particular type of amino acid residue? (3) Are there limitations on the conditions of the reaction (e.g., temperature, pH)? (4) Is it essential that the protein conformation remain unaltered after modification? (5) Will it be necessary to isolate the derivative (e.g., after an acid hydrolysis)? (6) Is reversibility of the reaction a requirement? (7) Is a rapid and convenient assay (e.g., a spectrophotometric assay) required?

The answers to these questions must be weighed to determine if a particular procedure is feasible. Certainly no reagent is yet available that could be judged ideal by the criteria of specificity, yield, reversibility, and maintenance of protein conformation. Some appreciation of the difficulties involved may be obtained from the fact that no reagent has yet been found for specific and reversible modification of carboxyl groups or arginine residues, in spite of extensive efforts. Similarly, the reactivity of many of the best modifying reagents, such as those used for lysyl, histidyl, tyrosyl, and sulfhydryl residues, is not entirely specific because of competing reactions of other residues.

Typical problems encountered may be illustrated with a specific example, namely, the reaction of proteins with iodoacetate. The possible reactions of iodoacetate with amino acid side chains are:

$$ICH_2COO^- + Prot{-}SH \xrightarrow{pH > 7} Prot{-}S{-}CH_2COO^- + H^+ + I^-$$

$$ICH_2COO^- + Prot{-}SCH_3 \xrightarrow{pH > 2} Prot{-}\overset{\overset{CH_3}{|}}{\underset{+}{S}}{-}CH_2COO^- + I^-$$

$$ICH_2COO^- + Prot{-}NH_2 \xrightarrow{pH > 8.5} Prot{-}NH{-}CH_2COO^- + H^+ + I^-$$

$$ICH_2COO^- + Prot{-}\text{(imidazole, NH)} \xrightarrow{pH > 5.5} Prot{-}\text{(imidazole, N)}{-}CH_2COO^- + H^+ + I^-$$

The indicated pH requirement for each reaction reflects the need for the reactive group to be deprotonated. Since methionine is uncharged, it is capable of a specific reaction with iodoacetate at low pH where other side chains are unaffected. The pH given for reaction with the other functional side chains is meant to indicate the acidity level at which a significant rate may be expected. However, some reaction will occur at lower pH values. For example, although most lysyl residues are protonated below pH 8.5 and thus unreactive, a very slow rate of carboxymethylation is possible even at pH 7.

The iodoacetate reaction is most frequently used for modification of sulfhydryl groups. At neutral pH, these are by far the most reactive amino acid residues to iodoacetate, but significant side reactions sometimes occur.

The usefulness of the reagent, however, depends on the information one hopes to get. When carried out with fully denatured polypeptide chains, the reaction is complete for SH groups (disulfides may first be reduced with a thiol reagent) and the modified cysteine residues are quite stable, even to acid hydrolysis. The completeness of the reaction permits its use in quantitation of cysteine content in proteins. The reaction also blocks association of chains through disulfide formation and/or interchange and is often employed prior to physical studies on isolated random coil chains (e.g., for molecular weight determination). Under mild conditions carboxymethylation may cause little disturbance to the conformation of native proteins and thus may be used to compare available surface sulfhydryl groups with total SH groups determined for the denatured protein. Side reactions will not influence the determination of carboxymethylcysteine by amino acid analysis. On the other hand, the side reactions may render the derivative unsuitable for the isolation of specific peptides (e.g., tryptic or cyanogen bromide), since partial reaction with other residues will produce additional peptide species with different charge and chromatographic properties. A considerable degree of experience and judgment is required in order to ensure that reaction is complete and that amounts of additional peptides formed are negligible under the conditions of modification employed. The reagent cannot be used to reversibly modify lysyl residues for use in sequence studies because the product is too stable.

The above discussion does not exhaust the pros and cons of iodoacetate as a reagent for protein structure studies, but is intended to illustrate the kind of considerations that must be made in choosing an appropriate reagent to employ for modification. Several aspects of the study of protein structure in which chemical modification can play a valuable role are discussed below. Only a few of the more important reagents are discussed in each category.

1 MODIFICATION OF THIOL AND DISULFIDE GROUPS

In many cases the dissociation of proteins into their individual polypeptide chains requires the reduction of disulfide bridges followed by some treatment that effectively prevents subsequent air oxidation of SH groups to disulfide bridges and consequent aggregation. One of the ways that this may be done (carboxymethylation) has just been discussed. Other methods (maintenance of a reducing environment, sulfitolysis, and aminoethylation) have been treated in Chapter 4. Reagents that produce charged derivatives are valuable in that the modified protein may show increased solubility.

Quantitative modification of SH and S—S groups usually requires a denaturing solvent such as 8 *M* urea (though care must be taken to avoid unwanted reaction with cyanate, a decomposition product of urea) or 6 *M* guanidine-HCl. The sulfitolysis reaction usually proceeds to completion smoothly, although the method commonly employed is rather time consuming. Some proteins, however, are not suitable for this reaction because of the presence of disulfide bridges that are resistant to the reagent. The modified cysteine residues of the *S*-sulfo protein obtained as the product of sulfitolysis may be converted back to sulfhydryl form with mercaptans. This property is useful if subsequent renaturation is to be attempted, although an alternative procedure, the maintenance of a reducing environment, is often preferred for such studies. A slow reversal of the reaction may occur spontaneously, particularly at extremes of pH, thereby limiting the usefulness of sulfitolysis when such pH conditions are later required, such as in the preparation of cyanogen bromide peptides at low pH.

Carboxymethylation was extensively discussed in the previous section. The side reactions discussed there most often result because of the presence of thiol reducing agents, such as mercaptoethanol, that have been added to reduce disulfide bonds. Sufficient iodoacetate must then be added in order to achieve a slight excess over the total thiol groups present. However, if air oxidation of mercaptoethanol has taken place, too great an excess of reagent will result and lead to significant reaction with histidine and lysine residues, despite the fact that these residues are several hundredfold less reactive than thiol groups.

Ethyleneimine reacts nearly quantitatively and exclusively with free thiol groups in proteins at neutral or basic pH (modification of methionine may occur at low pH) and is the reagent of choice for many studies on sequence and quaternary structure. The modified residue is stable to acid hydrolysis and may be quantitated by amino acid analysis. The aminoethylated residue promotes tryptic hydrolysis at cysteine positions. Although this feature may occasionally be advantageous, it would generally be more desirable to reserve this option for the purpose of further cleavage of normal tryptic peptides.

2 SURFACE CHARGE ALTERATION IN PROTEINS

Increasing the surface charge of proteins, as by conversion of positively charged lysine side chains to a negatively charged derivative, provides a mechanism to introduce repulsive interactions that may promote dissociation of oligomeric proteins. Repulsive expansion sometimes will take place within polypeptide chains as well. By careful control of the extent of reaction, it may be possible to selectively dissociate the oligomer without further conformational changes in the subunits. There is also a good possibility that the dissociated proteins will remain soluble and show little tendency to aggregate in aqueous solvent systems. In some cases modified proteins retain their biological activity. Properties based on charge, such as electrophoretic mobility, are markedly affected by chemical modifications of this type, even when the reaction is only partial. The reversible dissociation–reassociation of a mixture of subunits obtained from partially modified and native proteins will yield mixed hybrids that are likely to be separable on the basis of charge properties. This may be used to great advantage in the study of subunit stoichiometry and symmetry (cf. Chapter 10).

The two reagents most often used to change the positively charged amino groups into anions are succinic and maleic anhydride. The reaction of amino groups with succinic anhydride occurs above pH 7 as follows:

$$\text{Prot—NH}_2 + \begin{array}{l}\text{CH}_2\text{—C(=O)} \\ \quad\quad\quad\quad \text{O} \\ \text{CH}_2\text{—C(=O)}\end{array} \longrightarrow \text{Prot—NH—}\overset{\text{O}}{\overset{\|}{\text{C}}}\text{—(CH}_2)_2\overset{\text{O}}{\overset{\|}{\text{C}}}\text{—O}^- + \text{H}^+$$

This reagent may also attack cysteine, histidine, tyrosine, serine, and threonine residues in proteins; however, the products are readily hydrolyzed with the elimination of the substituent group. The amino derivative is quite stable, and the reaction is not reversible under ordinary conditions.

Maleic anhydride reacts with amino groups in a similar fashion:

$$\text{Prot—NH}_2 + \begin{array}{l}\text{O=C—CH} \\ \text{O} \quad\quad \| \\ \text{O=C—CH}\end{array} \longrightarrow \text{Prot—NH}\overset{\text{O}}{\overset{\|}{\text{C}}}\text{—CH=CH}\overset{\text{O}}{\overset{\|}{\text{C}}}\text{—O}^- + \text{H}^+$$

This reaction is reversible at low pH (half-time at pH 3.7 and 37°C is about 11 hours). Side reactions are spontaneously reversible except with thiol groups where a very stable addition product is formed across the double bond. This side reaction can be largely eliminated by blocking reactive cysteine residues through formation of disulfide bonds with a thiol reagent prior to treatment with maleic anhydride. The sulfhydryl groups of cysteine are later restored by controlled addition of mercaptan. Modification of lysyl residues with maleic anhydride is also very useful for selective tryptic cleavage at arginine residues. The peptides containing the negatively charged modified lysine have improved solubility properties compared to those obtained by other lysine-blocking procedures. The lysyl residues may then be regenerated at low pH, and further tryptic cleavage may be carried out.

The modification of carboxyl residues to increase the overall positive charge of a protein is most conveniently accomplished by titration to low pH (cf. Chapter 10). Carboxyl groups may be esterified by a number of reagents (e.g., see References 1 and 2), but quantitative conversion under mild conditions is difficult to achieve. Both of the above procedures may result in reduced solubility of the protein or peptide.

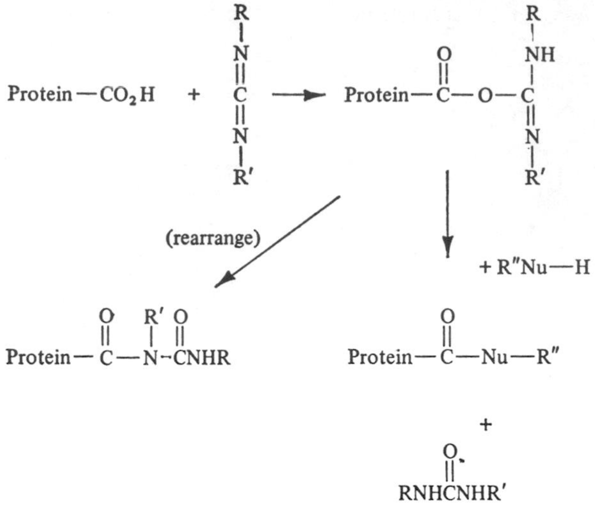

$$\text{Protein—CO}_2\text{H} + \text{RN=C=NR}' \longrightarrow \text{Protein—C(=O)—O—C(=NR}')\text{—NHR}$$

(rearrange) → Protein—C(=O)—N(R')—C(=O)NHR

$+ \text{R}''\text{Nu—H}$ → Protein—C(=O)—Nu—R'' + RNHC(=O)NHR'

Fig. 5-1 Reaction of water-soluble carbodiimides with protein carboxyl groups. The product may undergo further reaction with nucleophilic reagents or rearrangement (usually at a slower rate).[1]

The best method for modification of carboxyl groups involves reaction with water-soluble carbodiimides [e.g., 1-ethyl-3-(3-dimethylaminopropyl)-carbodiimide] at slightly acidic pH.[1] A variety of substituents may then be introduced by carrying out the reaction in the presence of a suitable nucleophilic reagent at high concentration. The reaction by nucleophilic attack is much more rapid than the rearrangement of the activated intermediate. These reactions are illustrated in Fig. 5-1. A positive charge can be introduced by reaction with a nucleophile such as *N*,*N*-dimethylpropylenediamine. This type of modification has not yet been used in gross structural studies but reactions of this type are extensively employed in peptide chemistry.

3 APPLICATIONS OF BIFUNCTIONAL REAGENTS

Reactions of proteins with bifunctional reagents have been reviewed by Wold (in Reference 3). The use of certain bifunctional reagents such as dimethyl-suberimidate for determination of subunit stoichiometry is considered in Chapter 10. Other applications of bifunctional modification that are of interest in structural studies of proteins are noted below.

Proteins may be stabilized against drastic conformational changes by crosslinking surface portions of polypeptide chains. The most extensive application has been in the stabilization of proteins (and tissues or tissue fractions) with glutaraldehyde for electron microscopic investigations. The reaction is thought to involve partial polymerization of the reagent to form small polymers that may react with proteins at lysyl, cysteinyl, and tyrosyl residues.[1] The reaction with amino groups is illustrated in Fig. 5-2. The potential superiority of alternate reagents such as bifunctional imides for preserving specimens for electron microscopy has not yet been extensively investigated.

The reaction of proteins with bifunctional reagents provides a direct chemical approach to the study of protein folding by means of "proximity" determination. The principle of this method is similar to that employed in the determination of naturally occurring disulfide bridges in proteins. When the protein is reacted with the "double-headed" reagent, intrachain cross-linkages are formed between reactive amino acid residues that are sufficiently close together so that the reagent can react with both. Some choice of

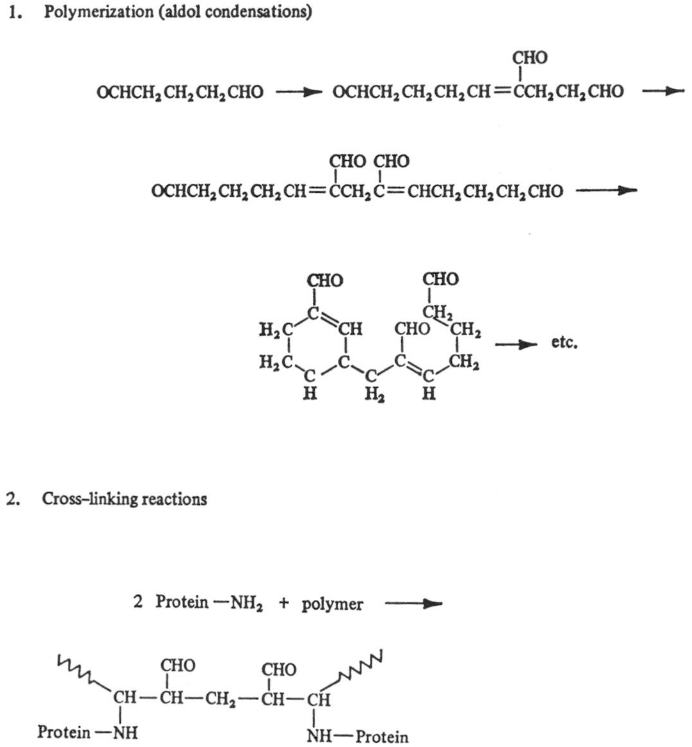

1. Polymerization (aldol condensations)

$$OCHCH_2CH_2CH_2CHO \rightarrow OCHCH_2CH_2CH_2CH{=}C(CHO)CH_2CH_2CHO \rightarrow$$

$$OCHCH_2CH_2CH_2CH{=}C(CHO)CH_2C(CHO){=}CHCH_2CH_2CH_2CHO \rightarrow$$

CHO, CHO, CH_2, H_2C, CH, CHO, CH_2, H_2C, C, C, C, CH_2, C, C, H, H_2, H $\rightarrow$ etc.

2. Cross-linking reactions

$$2\ Protein{-}NH_2 + polymer \rightarrow$$

$$\sim CH(NH{-}Protein){-}CH(CHO){-}CH_2{-}CH(CHO){-}CH(NH{-}Protein)\sim$$

Fig. 5-2 Proposed reactions for the polymerization of gluteraldehyde and reaction of the polymer with protein amino groups.[1]

reagents is available with different lengths between reactive ends. After reaction (often carried out with a radioactively labeled reagent), the cross-linked amino acids are identified, usually by fingerprinting of tryptic peptides and amino acid analysis. Control experiments with analogous monofunctional reagents may provide added confidence that reaction with the reagent itself did not promote conformational changes in the protein.

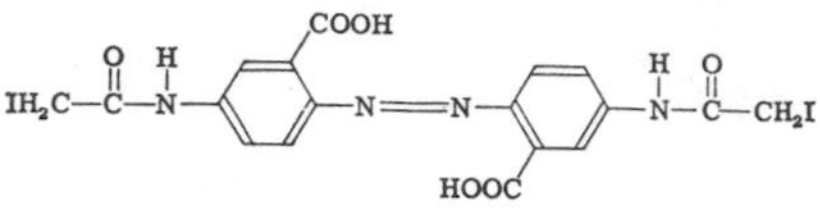

2,2′-Dicarboxy-4,4′-diiodoacetamidoazobenzene

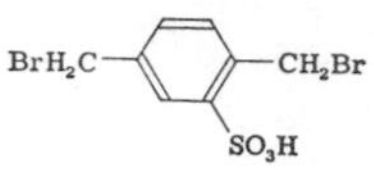

α,α′-Dibromo(or diiodo)p-xylenesulfonic acid

Bifunctional Alkyl Halides

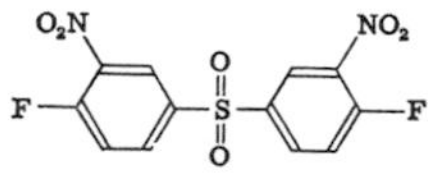

p,p′-Difluoro-m,m′-dinitrodiphenylsulfone

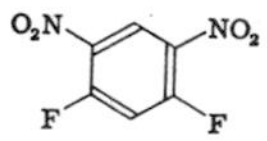

1,5-Difluoro-2,4-dintrobenzene

Bifunctional Aryl Halides

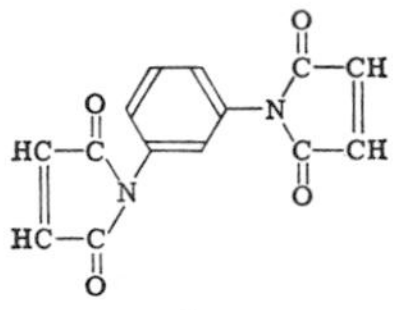

N,N′-(1,3-Phenylene)bismaleimide

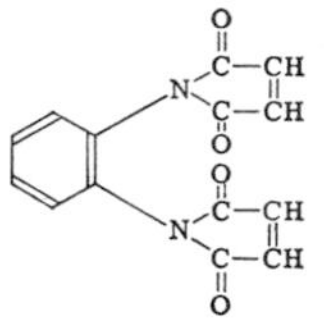

N,N′-(1,2-Phenylene)bismaleimide

Bifunctional Maleimides

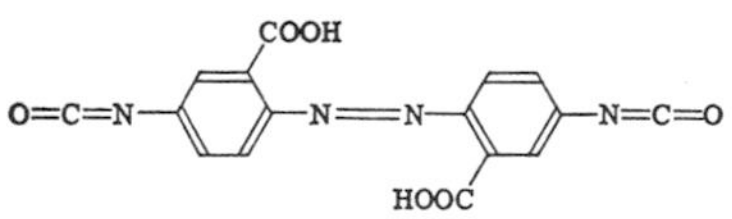

2,2′-Dicarboxy-4,4′-azophenyldiisocyanate

A Bifunctional Isocyanate

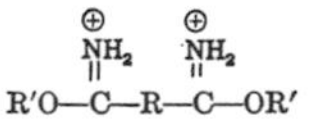

General Structure of a Bifunctional Imidoester

Fig. 5-3 Structures of some representative bifunctional reagents.

Bifunctional reagents are sometimes useful in stabilizing proteins against denaturation. At high protein concentration intermolecular crosslinkages may be formed, which might (for example) serve to render small proteins nondialyzable for enzymatic studies.

One of the most promising applications of bifunctional reagents to the study of biological systems is the crosslinking of adjacent polypeptide chains (and other macromolecules, such as nucleic acids) in heterogeneous and complex systems that are difficult to study by other means. Such studies should give valuable information on geometry in systems such as viruses, ribosomes, ribosome–membrane and ribosome–nucleic acid complexes, antigen–antibody systems, and multienzyme systems.

A brief description of some bifunctional reagents, conditions for reaction, and specificity is given in the review by Wold. The structures of a few representative reagents are shown in Fig. 5-3. The chemistry of the reagents is similar to that of corresponding monofunctional reagents, although other properties such as solubility may differ. Some of these reagents can be cleaved symmetrically in the crosslinked peptides under conditions that leave the peptide bond intact. This permits the crosslinked material to be separated into components for further analysis.

4 DETERMINATION OF "AVAILABLE" AND TOTAL REACTIVE GROUPS

Chemical modification of proteins can be employed in several ways to determine numbers of reactive amino acid residues. If the reaction does not cause any significant conformational changes in the protein, the results will reflect available residues, presumably located at the surface of the molecule. In favorable cases, conformational changes may be monitored by observation of changes in the number of reactive groups. The total number of potentially reactive amino acid residues is usually inferred from amino acid composition but can also be obtained by chemical modification performed on the protein dissolved in a strong denaturing solvent that exposes all residues to the reagent. The latter type of study is often employed to determine relative amounts of cysteine or cystine in proteins, a result that cannot be obtained directly from amino acid analysis.

A variety of strategies are available for the quantitation of reactive groups by chemical modification methods. The extent of reaction of proteins with reagents that are radioactively labeled or which contain a suitable chromophoric group may be measured directly after removal of unreacted reagent (e.g., by dialysis or gel filtration). Counting specific groups by this technique requires a highly specific reagent, or one for which characteristic spectral changes occur upon reaction. Alternatively, if the modified amino acid residue is stable to acid hydrolysis, amino acid analysis to quantitate either the modified residue or the remaining unreacted residues will reveal the extent of reaction. Complete specificity of the reagent for a particular group is not required in this case. A few reagents produce derivatives that are destroyed by acid hydrolysis without regeneration of the original amino acid residue. The extent of modification is then determined from the difference in amino acid composition of unmodified and reacted proteins.

The list of potentially useful reagents for obtaining information on reactive groups is extensive and only a few examples will be given here. A guide to further literature is provided in the summary table given by Stark.[1] The most definitive results are usually obtained with reagents whose products are stable to acid hydrolysis and thus suitable for quantitation by amino acid analysis. A few reagents of this type are listed in Table 5-1. Although the reaction may not be entirely specific for the amino acid residues listed, this method of analysis permits quantitation for the amino acid residue of interest.

The introduction of chromophoric groups into proteins by chemical modification is advantageous in that the extent of reaction may be conveniently estimated spectrophotometrically. Reactions with low-molecular-weight mixed disulfides (e.g., Ellman's reagent, cf. Chapter 4) are specific for cysteine although disulfides will also react if catalytic quantities of mercaptide are included in the reaction. The chromophoric reagent *p*-chloromercuribenzoate (PCMB) is often used for sulfhydryl analyses. It is of interest to note that this reagent is sometimes unreactive with sulfhydryl groups that are readily attacked by other sulfhydryl reagents, such as iodoacetate, most likely because of steric hindrance by the ring structure of PCMB. Factors such as this will influence the interpretation of chemical modification data in terms of "available" groups.

Table 5-1 Reagents for modification of amino acid residues in proteins. Products formed from residues listed can be quantitated by amino acid analysis.

Reagent	Residue(s)[a]
Alkyl halides	Lysine, cysteine, methionine
Ethyleneimine	Cysteine
Cyanates	Lysine
Iodination	Tyrosine
Tetranitromethane	Tyrosine

[a] Other residues reacting with the reagent form a product that is reconverted to the original amino acid or destroyed upon acid hydrolysis.

Dinitrofluorobenzene and dansyl chloride both react with several side chain residues as discussed earlier (cf. Chapter 4). However, only the reaction with amino groups produces a colored derivative; photometric analysis can be used to estimate reaction with lysine residues. The reaction of tyrosine with tetranitromethane (Table 5-1) yields 3-nitrotyrosine which absorbs maximally at 428 nm when ionized. This reaction may therefore be conveniently monitored by absorbance measurements. Tryptophan also produces a colored product, as observed in denatured proteins, but reacts poorly in native proteins. The reaction of sulfhydryl groups with silver or mercury may be monitored polarographically. Titration studies are used for analysis of chemical modifications, such as succinylation, which alter the charge of proteins. Other techniques to determine the extent of reaction in certain cases, such as fluorescence (cf. Chapter 9) and nuclear magnetic resonance (cf. Chapter 12) are likely to be more extensively employed in the future.

5 ACTIVE SITE LABELING[8]

The use of chemical reactions for the introduction of radioisotopes into protein active sites (covalent labeling) has been extensively reviewed by Singer.[8] The results obtained from this type of modification are meaningful for studies of protein structure as well as for elucidation of reaction mechanisms. Singer considered three types of protein active sites: catalytic, regulatory, and antibody-combining sites. This definition of "active site" may reasonably be extended to include the important case of intersubunit binding sets in oligomeric proteins (cf. Chapter 16).

It is clear that the chemical modifications discussed in the previous sections may give rise to a protein whose activity is diminished or abolished. There are at least three possible explanations for such a result: (1) One or more of the residues subject to modification participates directly in the active site; (2) the modification reaction promotes conformational changes that indirectly affect the integrity of the active site; and (3) a modified residue sterically interferes with active site function by reason of geometry. It is often difficult to choose definitively among these alternatives, and thus an element of ambiguity is inherent in the interpretation of such studies.

For some systems specific active site modification can be achieved by the use of limited reactions or because only one or two groups were available for modification (e.g., cysteine). Identification of the particular amino acid residue involved may be carried out, for example, by tryptic mapping after incorporation of a radioactively labeled reagent. However, another more general approach is available, even for cases where the number of reactive residues is substantial. This procedure takes advantage of the fact that residues involved in the active site often are unavailable for modification while the protein is "expressing" its biological activity in some manner. For example, such groups may be unreactive when a substrate or effector is bound to the protein. Residues located at antibody or intersubunit combining sites are unlikely to react with modifying reagents when association is occurring at those sites. Under these conditions all other reactive groups in the protein molecule can be modified using a reagent that is not radioactively labeled. The product should retain substantial biological activity. Then the active site residue is made available by removing the substrate or effector or by dissociation, and reaction with the radioactively labeled reagent is carried out. The isotopically labeled residue is then identified by tryptic mapping (or other suitable technique). This procedure, which involves protection of the critical active site residue, is often referred to as differential labeling.

Another method for covalently labeling amino acid residues at or near the active site is called affinity labeling. The affinity label is designed to fulfill two conditions. First, it should bind specifically to the active site because of its geometrical similarity to the natural active site combinant (e.g., a substrate analog which combines with an enzyme active site). Second, the affinity label must contain a chemical group that is reactive toward some amino acid residue(s). Under suitable conditions, reaction will take place only for those reactive residues that occur at or near the active site. This is because the effective concentration of the reagent there is so much higher than it is in the solution to which other amino acid residues of the protein are exposed.

Less generally applicable to protein structural studies (although of great interest in relation to enzyme mechanisms) is the covalent labeling of enzyme active sites by unique groups such as substrates or coenzymes. Amino acid residues at the active site in many enzymes participate in the catalytic function in a direct way by forming transient covalent crosslinkages with substrates or coenzymes. Unstable intermediates may sometimes be chemically modified so that the substrate or other ligand is covalently trapped (e.g., by borohydride reduction of Schiff base intermediates). Other procedures take advantage of the unusual reactivity of such active site amino acid residues. The interested student will find further examples and references for these techniques in the review by Singer.[8]

REFERENCES

1. G. R. Stark, "Recent developments in chemical modification and sequential degradation of proteins," *Adv. Protein Chem.*, **24**, 261 (1970).

2. L. A. Cohen, "Group-specific reagents in protein chemistry," *Ann. Rev. Biochem.*, **37,** 695 (1968).
3. C. H. W. Hirs, Ed., *Enzyme Structure*, Vol. 11 of *Methods in Enzymology*, Academic Press, New York, 1967; C. H. W. Hirs and S. N. Timasheff, *Enzyme Structure*, Part B, Vol. 25 of *Methods in Enzymology*, Academic Press, New York, 1972.
4. T. F. Spande, B. Witkop, Y. Degani, and A. Patchornik, "Selective cleavage and modification of peptides and proteins," *Adv. Protein Chem.*, **24,** 97 (1970).
5. A. N. Glazer, "Specific chemical modification of proteins," *Ann. Rev. Biochem.*, **39,** 101 (1970).
6. G. E. Means and R. E. Feeney, *Chemical Modification of Proteins*, Holden-Day, San Francisco, 1971.
7. B. L. Vallee and J. F. Riordan, "Chemical approaches to the properties of active sites of enzymes," *Ann. Rev. Biochem.*, **38,** 733 (1969).
8. S. J. Singer, "Covalent labeling of active sites," *Adv. Protein Chem.*, **22,** 1 (1967).

PART TWO

VI Three-Dimensional Structure of Proteins[1–5]

1 INTRODUCTION

Proteins are composed, for the most part, of long chains of amino acid residues linked by peptide bonds, the order and number of residues being determined by the information transmitted from the gene for each particular protein. By means of rotations about single bonds the polypeptide chains that result from this one-dimensional synthetic process have the capability to assume many possible arrangements in three-dimensional space. In solution a newly synthesized polypeptide chain would be subject to continuous bombardment due to the Brownian motion of solvent molecules. In the absence of any form of stabilizing interactions this would tend to produce a tangled string in continuous conformational flux. Such a structure, described by the term *random coil*, is approximately what is found for proteins that have undergone complete denaturation, that is, total loss of the natural or "native" conformation (without damage to the primary structure). But, in the native state, or the functional form of the protein, each polypeptide chain of a genetically distinct protein does not form a family of "random coils"; rather, it exhibits a unique and stable three-dimensional structure. This structure is present, in almost identical form, in every molecule of that protein and permits the protein to be recognized as a definable chemical species as well as a biological entity.

To understand fully the function of a particular protein in relation to biological phenomena, the protein chemist seeks to determine its structure, the forces responsible for the formation and maintenance of the native (functional) state and other conformations which may occur, and the relationship of structure and function. Information must be sought at different levels of refinement or precision, depending on the specific problem at hand. Among proteins one encounters a great diversity of sizes and properties. Individual polypeptide chains in most proteins fall within the molecular weight range of about 10,000 to 100,000 Daltons. Thus roughly 100 to 1000 amino acid residues are contained in each chain. To add to the complexity, most proteins in their functional form contain more than one polypeptide chain, either of the same type or of different types. The protein may consist of dissociable multiple subunits, where each subunit is composed of a single chain or of two or more chains linked through covalent bonds (e.g., disulfide bonds) or through noncovalent interactions. Although some forms of biological activity may be expressed by the separated subunits, full biological function (e.g., sensitivity to the control of activity) may depend on maintenance of this larger organization. These protein aggregates and their symmetry properties will be given special attention in Chapter 16.

To a great extent, increasing size of the component polypeptide chains of a protein means increasing complexity, and a greater effort is required for detailed analysis. The amount of information we now have about different proteins, however, is not correlated with size alone. Other important factors are availability, solubility, ease of purification, and crystallizability. In addition, certain types of structural features are more amenable to study by methods now available than are others. For example, in myoglobin and hemoglobin a high content of a regular structure, the α-helix, facilitated the successful early determination of their crystallographic structures. Relatively few proteins have been analyzed to this level of refinement, however, and in most cases one attempts to learn as much as possible about structure and function by means that are appropriate for each particular protein. For example, the proteins fibrinogen and myosin (cf. Chapter 17) have not been successfully crystallized, yet they have been effectively studied by a variety of other physical techniques.

For convenience in our discussion of protein structure, it is helpful to differentiate certain levels of structure. Primary structure, as indicated earlier, is the sequence of amino acid residues in the polypeptide chains which make up the protein. Linderstrøm-Lang proposed that the three-dimensional structure or conformation of the chains in the native protein be subdivided operationally into two types, secondary structure and tertiary structure. This terminology is still in common usage despite some shortcomings.

Secondary structure refers to structural elements such as the α-helix that involve interactions between amino acid residues fairly close to one another in the sequence. Because of the great diversity in types of near neighbor interactions that occurs in proteins, this definition has proved to be less useful for native proteins than it is for synthetic polypeptides. In a number of special cases (the triple helical structure of collagen or the double helical structure in nucleic acids) the exact meaning of secondary structure depends on the context in which it is used. More generally today, secondary structure in proteins represents arrangements of the polypeptide chain that form more-or-less regular hydrogen-bonded structures, in particular, α-helical and pleated sheet structures.

Tertiary structure, as originally defined, refers to the three-dimensional structure of the polypeptide chain that results from interactions between amino acid residues relatively far apart in the sequence. It may, more generally, be regarded as that arrangement of the chain in three dimensions that is not regular (i.e., not composed of repeating structural elements). In proteins containing more than one polypeptide chain as discussed above, the term is used in reference to the structure within each chain, independently. The three-dimensional structure of the multisubunit protein is then described by the term *quaternary structure*, which refers specifically to that structure resulting from the interactions between polypeptide chains. These definitions will be followed in this book, although, to avoid confusion, we will generally use the term conformation or three-dimensional structure to include all types of structure above the level of primary structure.

The mechanism of protein folding is a difficult problem to approach. After polypeptide biosynthesis (and/or during synthesis), the chain apparently has the capability to spontaneously adopt the specific tertiary structure characteristic of that particular protein. To understand this process of folding, one must first examine the thermodynamics or energy balance of the process. Later attention will be given to the pathway and time course of the process, that is, its kinetics. The random coil or completely disordered state of a polypeptide represents a great family of conformations, depending on the relative freedom of rotation about single bonds, and thus constitutes a state of high entropy (low level of order). From the Second Law of Thermodynamics, one recognizes that energy will be required to transform a system of such molecules from a state of multiplicity of conformations to a single stable conformation, that is, to overcome configurational entropy. This energy cost must be overcome through energetically favorable interactions of the protein chain within itself and with the solvent.

Clearly, the forces involved in protein folding are great enough to overcome configurational entropy since the native state is favored (at least some of the time) over disordered forms *in vivo*. In a number of cases the transition from a fully disordered form (random coil) to a native, biologically active state can be observed *in vitro* as well. However, our understanding of these forces is still of a qualitative nature. Disulfide bonds, which serve to link different segments of polypeptide chains covalently, were once thought to be instrumental in protein folding. Now their role may be regarded as a means for added stability in those proteins in which they occur. Most proteins have no other covalent bonds serving as crosslinks between chains, thus only noncovalent interactions, or "bonds," are left to account for protein folding. These include hydrogen bonds, apolar or hydrophobic interactions (sometimes called hydrophobic bonds), ionic bonds, and van der Waals attractions. Hydrogen bonds were given first attention by physical chemists as the probable major force in structure formation and stabilization. However, proteins are stable (indeed, are formed) in aqueous media, and a number of studies indicated that the strength of protein–protein hydrogen bonds might be comparable to, or even less than, that of protein–water hydrogen bonds. Thus polypeptide chains in water would have little tendency to form internal hydrogen bonds and exclude the solvent from their hydrogen-bonding groups. As a result, scientists came to accord greater importance to apolar interactions in protein folding. At the present time it appears that no one type of bonding can be conclusively demonstrated to be dominant in all proteins but rather that all noncovalent interactions must play a role although their magnitudes are still difficult to evaluate.

In this chapter, following a brief discussion of the role of disulfide bonds, we will review the most pertinent aspects of noncovalent bonding in proteins and the experimental and theoretical methods used to study these interactions. One must keep in mind that knowledge of these interactions to date is based, of course, on solution studies (*in vitro*). We recognize that protein structure will be influenced *in vivo* by the cellular environment, a milieu which at present is not well defined. The following chapters will present the basic principles and some applications of important physical methods currently in use for the study of protein structure. We wish to draw attention to the limitations of these techniques as well as their strengths, and to indicate some successful applications in order that the reader may recognize how to use them correctly and advantageously in any problem involving proteins. These methods are useful not only in the study of the structure of proteins (and other macromolecules) but also in investigations of functional properties. The last part of the book is devoted to a discussion of some examples of particular proteins and to problems of protein dynamics in which physical methods combined with chemistry have been applied with great effectiveness.

2 TYPES OF BONDING IN THREE-DIMENSIONAL STRUCTURE

2-1 Disulfide Bonds[6,7]

The occurrence of disulfide bonds in proteins varies greatly. These bonds appear to be more prevalent among small proteins consisting of a single polypeptide chain (e.g., ribonuclease, lysozyme, trypsin), and particularly in proteins that have a functional existence in extracellular spaces of tissues. A strikingly high content of disulfide bonds is found in bovine serum albumin (molecular weight of 66,500 Daltons, 17 disulfide bonds) and γ-globulin (160,000 Daltons, 16 disulfide bonds). Other examples are given by Schachman.[2] Many large proteins, however, particularly those containing multiple subunits, are completely devoid of disulfide links. Hemoglobin and collagen (for most species) are notable in this group, as well as a number of large enzymes. Where disulfide bonds do occur in multiple-chain proteins, the links are usually intrachain and do not serve to connect different types of chains. However, cases of the latter type have been identified (e.g., in fibrinogen and γ-globulin).

Disulfide bonds clearly contribute greatly to the stabilization of tertiary structure in those proteins which have them. An extreme case is that of keratin (cf. Chapter 17), which is so highly crosslinked through disulfide bridges between polypeptide chains that it is almost impossible to extract it from the tissues in which it occurs. A high content of disulfide bonds relative to its size in ribonuclease (molecular weight of 13,700 Daltons, four disulfide bonds) probably contributes to its resistance to denaturation and the ease with which its structure can be re-formed (*renaturation*) when these bonds remain intact. Thus, in cases of partial denaturation, segments of the chain remain linked, thereby increasing the probability for correct refolding of the denatured portions of the chain. This appears to be a general feature of disulfide proteins, many of which are secreted from the cells in which they are synthesized. The additional stability may be essential for maintenance of function under the more variable extracellular conditions.

The chemistry of disulfide bonds is discussed earlier (cf. Chapter 4); a thorough review on the role of sulfur in proteins has been presented by Cecil.[6] A number of disulfide-containing proteins have been studied extensively in attempts to assess the role of these bonds in conformation and active sites. Pancreatic ribonuclease, which is easily prepared in quantity and has high stability, has received the greatest attention. Reduction of all the disulfide bonds in ribonuclease (or other disulfide proteins) produces a concomitant loss in biological activity. Activity may be recovered by reoxidation in the presence of air at pH 8 [this pH is required to assure sufficient concentration of ionized sulfhydryl groups (cf. Fig. 2-6) for reaction to occur]. The first successful renaturation of a fully reduced protein was that of ribonuclease reported by White in 1961.[7] Recovery of activity, at least in part, after reduction has also been obtained for several other disulfide proteins (e.g., trypsin, molecular weight of 23,800 Daltons, six disulfide bonds).[8] A certain degree of modification of the disulfide bonds may be tolerated without total loss of activity. Reduction of two of the four disulfide bonds does not inactivate ribonuclease, and even after insertion of mercuric ions into all four bonds, some activity is retained.[9]

Studies on ribonuclease have been particularly important in relation to the role of disulfide bonds in protein folding (cf. Chapter 15). When reduced ribonuclease is reoxidized in the presence of an agent (e.g., urea or guanidine hydrochloride at high concentrations) that blocks proper folding by interfering with noncovalent interactions, disulfide bridges form in a random way. The product, after removal of the denaturing agent, is enzymatically inactive. If the order of these steps is reversed, that is, the denaturant is removed first (e.g., by dialysis) and then the reduced protein is reoxidized, the product is active. These experiments demonstrate that noncovalent interactions and not disulfide bond formation direct the proper folding of the polypeptide chain. In the first experiment nonspecific disulfide bonds are formed under conditions where noncovalent forces for folding are blocked. Incorrect pairs result which then prevent proper folding when the denaturing agent is removed. In the second experiment where the denaturing agent is removed prior to disulfide bond formation, the protein is able to partially refold through noncovalent interactions so that the correct pairs of cysteine residues are brought into proximity for reaction. The final oxidized product is then fully active.

Of particular interest with regard to protein folding *in vivo* is the finding that the inactive enzyme described above, with random disulfide bonds, can be rapidly reactivated through sulfhydryl–disulfide interchange catalyzed by a microsomal enzyme.[10] In this case rearrangement of tertiary structure occurs spontaneously once the constraining disulfide bonds have been broken, and the native structure is re-formed. The existence of rapid and reversible disulfide interchange in the cell would thus prevent the production of proteins with incorrect conformations that have been frozen-in by the random formation of crosslinks between available cysteine residues. These findings support the view that disulfide bond formation does not play a significant role in directing the folding of polypeptide chains toward the final stable conformation of the protein. Rather, such bonds confer added stability when they complement the noncovalent interactions that determine the energetically favorable conformation finally achieved.

2-2 Hydrogen Bonds[11,12]

The term hydrogen bond refers to the interaction of a hydrogen atom attached to an electronegative atom such as nitrogen, oxygen, or sulfur with another electronegative atom. Although the bond has primarily an ionic character, it has been known for many years that the force of attraction is strong enough to form aggregates or other relatively stable configurations of molecules.

For small molecules in the vapor state the free energy favoring the formation of hydrogen bonds is considerable, ranging from 2 to 10 kcal/mole, depending on the nature of the donor and acceptor atoms (although one must remember that this is an order of magnitude less than the energies of covalent bonds). Hydrogen bonding is not fully understood theoretically; however, models based on simple electrostatic interaction offer a good approximation for most cases. Hydrogen bonds are important in determining the properties of many liquids, such as water, as well as proteins and other macromolecules. A well-known case of hydrogen bonding is that of the purine–pyrimidine pairs in the nucleic acids.

A great deal of information on hydrogen bonding has come from analysis of crystal structures. Hydrogen bonds are found to have characteristic lengths, depending on the nature of the donor and acceptor atoms. For example, the N—H · · · O hydrogen bond formed by peptide groups has a nitrogen to oxygen distance of about 2.8 Å. The spatial relation of the donor and acceptor atoms is such that the hydrogen atom normally lies very close to the line between them, and bond angles characteristic of the separated atomic groupings are maintained. This linearity restriction has been valuable in the design of protein and polypeptide models based on maximization of the number of hydrogen bonds, although the possibility that nonlinear hydrogen bonds may play a role in proteins cannot be excluded. Other methods for studying hydrogen bonding include measurements of solubility, molar volume, viscosity, hydrogen exchange, infrared spectra, and proton magnetic resonance. The last two are particularly sensitive to hydrogen bond formation and are of great value in small molecule model systems.

Proteins contain a number of groups capable of hydrogen bonding, as shown in Fig. 6-1. The greatest number of possible hydrogen bonds are afforded by the atoms of the peptide linkage. Interpeptide bonding of the atoms of the peptide backbone provides the basis for the Pauling-Corey α-helix and the β-structure (cf. Chapter 14), types of structure now known to occur in segments of the polypeptide chains in a number of proteins. Peptide hydrogen bonds also occur in irregular internal domains. Some of the side chains of the amino acid residues in proteins are capable of hydrogen bonding, as illustrated in Fig. 6-1. These interactions are fewer in number but may be of particular biological importance at specific functional sites on protein molecules.

The existence of hydrogen bonds in proteins is well established although there has been considerable disagreement about the degree of stabilization they provide. In aqueous solution a special situation exists in that the solvent is itself a strong hydrogen bonding agent. Here, both the donor and the acceptor groups of the solute, for example, the N—H and O═C of a peptide bond, are capable of hydrogen bonding with water as well as with (*a*) (Hydrogen bond between peptide groups)

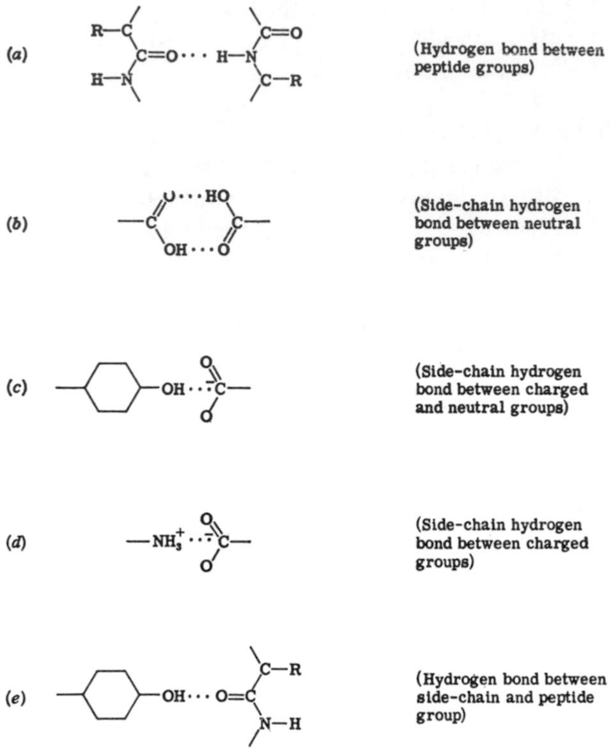

(*b*) (Side-chain hydrogen bond between neutral groups)

(*c*) (Side-chain hydrogen bond between charged and neutral groups)

(*d*) (Side-chain hydrogen bond between charged groups)

(*e*) (Hydrogen bond between side-chain and peptide group)

Fig. 6-1 Possible hydrogen bonds of polypeptides and proteins. (From H. A. Scheraga, "Intramolecular Bonds in Proteins II. Noncovalent Bonds," in *The Proteins*, 2nd ed., Vol. 1, H. Neurath, Ed., Academic Press, New York, 1963.)

each other. The change in free energy (ΔG) for the formation of an internal protein . . . protein hydrogen bond is therefore approximated as that of the reaction:

$$\text{Protein—H} \cdots H_2O + H_2O \cdots \text{Protein} \rightleftharpoons \text{Protein—H} \cdots \text{Protein} + H_2O \cdots H_2O \quad (6\text{-}1)$$

where the protein (or other solute) and the water molecules are acting as both donors and acceptors of hydrogen atoms.

Attempts have been made to determine the thermodynamic parameters for such a process by using model systems in which the degree of association of simple organic substances in water or other solvents is measured as a function of temperature. Because solute–solvent hydrogen bonds must be broken in order to form a hydrogen bond between two solute molecules, as indicated in Eq. 6-1, the change in free energy for the process is much less than that found in gas phase interactions where no solvent is involved. Studies with a model compound, *N*-methylacetamide, have been carried out to determine the thermodynamic parameters of a peptide-like hydrogen bond in solution.[13] The formation of hydrogen-bonded aggregates was followed by measuring the overtone N—H stretching frequency of the amide group in the infrared region. The equilibrium for dimerization is

$$\begin{array}{c} CH_3 \\ | \\ C{=}O \\ | \\ H{-}N \\ | \\ CH_3 \end{array} + \begin{array}{c} CH_3 \\ | \\ C{=}O \\ | \\ H{-}N \\ | \\ CH_3 \end{array} \rightleftharpoons \begin{array}{c} CH_3 \\ | \\ C{=}O\cdots \\ | \\ H{-}N \\ | \\ CH_3 \end{array} \begin{array}{c} CH_3 \\ | \\ C{=}O \\ | \\ H{-}N \\ | \\ CH_3 \end{array} \qquad (6\text{-}2)$$

The shift in the infrared absorption spectrum with increasing solute concentration in the three solvents studied is shown in Fig. 6-2. From these data the degree of association of the solute *N*-methylacetamide can be calculated. The result, as a function of solute molarity, is shown in Fig. 6-3. It was found that aggregation occurred readily in carbon tetrachloride solution, less readily in dioxane, and almost not at all in water except at very high solute

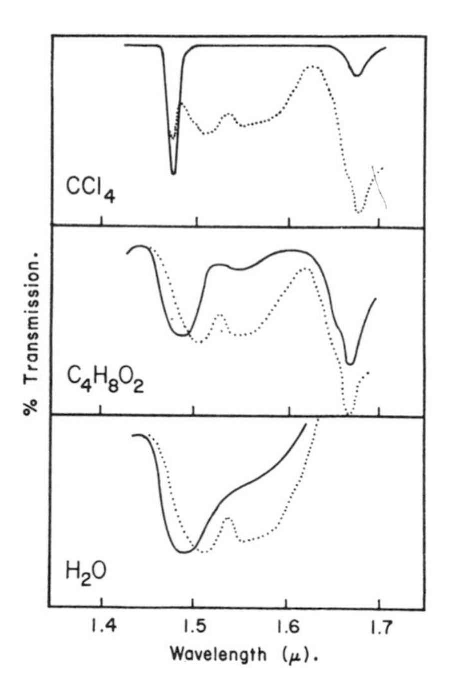

Fig. 6-2 Infrared absorption spectra in the region of the N—H stretching overtone for *N*-methylacetamide as a function of its concentration in three solvents. The solid lines refer to concentrations of 0.01 *M* in CCl_4, 0.2 *M* in dioxane, and 7 *M* in water. The dotted lines represent concentrations of 1 *M* in CCl_4, 3 *M* in dioxane, and 12.5 *M* in water at which spectra characteristic of dimers and higher aggregates of *N*-methylacetamide are observed.[13]

concentrations. Standard free energy changes were determined for association in the various solvents.

In order to apply the thermodynamic parameters derived from model compound studies such as that above, it is necessary to introduce the concept of the unitary free energy.[4] Standard free energies include a so-called *cratic* contribution to the entropy, which is a function of the randomness in location of the molecules in solution. This factor contributes a value of $-R \ln x$ to the partial molal entropy of each substance present at a mole fraction of x in solution, which must be subtracted in order to determine the thermodynamic parameters representing the inherent properties of molecules surrounded by solvent. The desired result, the *unitary free energy change* ΔG_u, and its component unitary entropy change ΔS_u, is obtained by expressing all concentrations in terms of mole fraction units. Thus for the association equilibrium shown in Eq. 6-2:

$$\Delta G_u = -RT \ln K_x \tag{6-3}$$

where the equilibrium constant K_x is based on concentrations in mole fraction units. R is the gas constant, 1.987 cal/mole-degree, and T is the absolute temperature. The association data of Fig. 6-3 yield the following results for ΔG_u of hydrogen-bond formation between N-methylacetamide molecules:

in CCl_4: $\Delta G_u = -2.4$ kcal/mole
in dioxane: $\Delta G_u = -1.1$ kcal/mole
in water: $\Delta G_u = +0.75$ kcal/mole

Thus, under circumstances more comparable to proteins (i.e., with the cratic contribution eliminated), the results for water show the equilibrium to be in

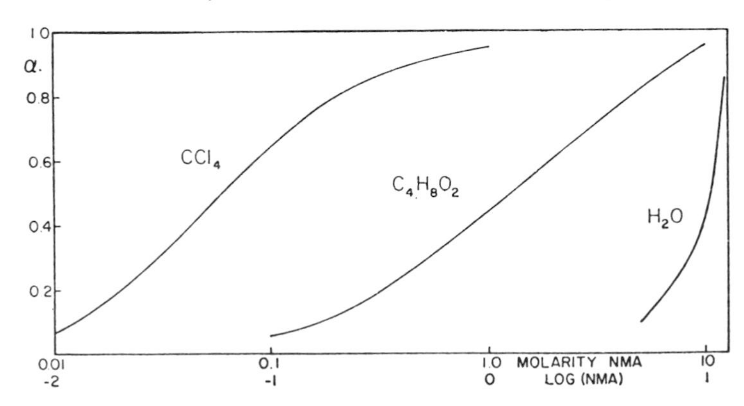

Fig. 6-3 Degree of association (α) of N-methylacetamide as a function of concentration in the solvents carbon tetrachloride, dioxane, and water.[13]

favor of the dissociated molecules. On the other hand, when the solvent is of an organic type, the equilibrium favors the formation of hydrogen bonds of the interpeptide type.

The data described above provide at least a guide for the strength of hydrogen bonds in proteins, although this type of study clearly cannot take into account the many other factors that may come into play in these complex molecules. In particular, the effective concentration of potential hydrogen bonding groups in proteins is difficult to assess. The situation is certainly different from that of independent small molecules in solution, and will influence the frequency of "collision" or close approach leading to bonding. The proximity of polar or charged groups and the presence or absence of water molecules in the bonding region will also affect ΔG. An especially critical factor is the net balance of hydrogen bonds in the equilibrium between random coil and native structure. One would predict that protein–protein hydrogen bonds must be formed when potentially hydrogen bonding groups move into the interior of the structure during folding of the chain. Otherwise the net process would require the expenditure of energy corresponding to the rupture of protein–solvent hydrogen bonds (several kcal/mole), and significant compensation from another source would be required. It follows that even if protein–protein hydrogen bonds do not contribute substantially to the overall free energy change of folding, the conformation finally attained is likely to show a high degree of hydrogen bonding among internal amino acid residues. Thus the final conformation may require a precise "fit" of these groups even if they do not contribute significantly to the net thermodynamic advantage of the native protein over the random coil.

In summary, one is left with a somewhat uncertain picture of the role of hydrogen bonding as a driving force for protein folding. The role of internal hydrogen bonds in the stability of proteins seems assured on the basis of the above argument, if for no other reason. Their part in the net energy balance of protein folding is probably small in relation to the opposing configurational entropy term. Results such as those discussed above for *N*-methylacetamide suggest that ΔG for protein–protein hydrogen bonds may be positive, thus favoring unfolding. In contrast, other investigators have obtained small negative values for ΔG (-150 cal/mole residue) of protein hydrogen bonds based on data for the helix-random coil transition in synthetic polypeptides. In any case the net effect must be viewed as a marginal one, certainly for protein folding in aqueous solutions, albeit the cellular environment may present less "aqueous" conditions that are likely to be more conducive to interpeptide hydrogen bonding. Valuable discussions of the subject are to be found in the articles by Kauzmann, Schachman, Richards, and Tanford.[1–4]

2-3 Apolar (Hydrophobic) Interactions[1]

Proteins contain substantial proportions of amino acid residues that have distinctly nonpolar side chains (alanine, valine, leucine, isoleucine, proline, and phenylalanine) or relatively nonpolar groups (tryptophan, methionine, and cysteine). On the basis of thermodynamic data for solutions of hydrocarbons in polar and nonpolar solvents, Kauzmann predicted that interactions of nonpolar residues in proteins to form intramolecular groupings in order to exclude water would be favored energetically. Thus stable protein conformations might contain a significant proportion of nonpolar side chains located in the interior of the molecule, away from the surrounding solvent. Because these interactions between groups which dislike water lead to their stabilization in spatial relationships involving intimate contact, the term "hydrophobic bond" was introduced to describe them. In contrast to the normal definition of a bond, atoms approach one another only to within their van der Waals radii; nonetheless, the terminology enjoys widespread usage. A more general term, "apolar bond," which refers to the preferential association of macromolecular nonpolar groups in any polar solvent, is also used. Polarity in this regard is a relative term; apolar bonds often involve groups such as the side chains of tryptophan and tyrosine that have a considerable degree of polarity. Figure 6-4 illustrates the molecular close packing that could be expected in the hydrophobic bonding of various amino acid side chains.

The existence of apolar regions in the interior of proteins has been well established by crystal structure analyses. In fact, the three-dimensional structure of several protein molecules is such that the interior contains largely apolar amino acids whereas both apolar and polar residues of the molecule occur on the surface (cf. Chapter 14). Most investigators now agree that the driving force for hydrophobic bonding arises not in the inherent attraction of nonpolar side chains for each other (i.e., through van der Waals attractions), but in the energetically unfavorable effect they have on the structure of water around them. When nonpolar groups associate to form an apolar bond in an aqueous medium, water is transferred from the vicinity of the hydrocarbon to the bulk solvent phase. This process is regarded to be energetically favored at low temperatures, although there is still disagreement about the mechanism.

The difficulties in understanding apolar bonding are apparent when one considers the possible events in the process. When a hydrocarbon side chain is removed from a polar solvent such as water, there may be a change in the overall hydrogen bonding of the water. If more hydrogen bonds are able to form between water molecules in the absence of the side chain, the process would be favored by a negative enthalpy change (ΔH), that is, the evolution of heat. However, in thermodynamic measurements on simple systems at

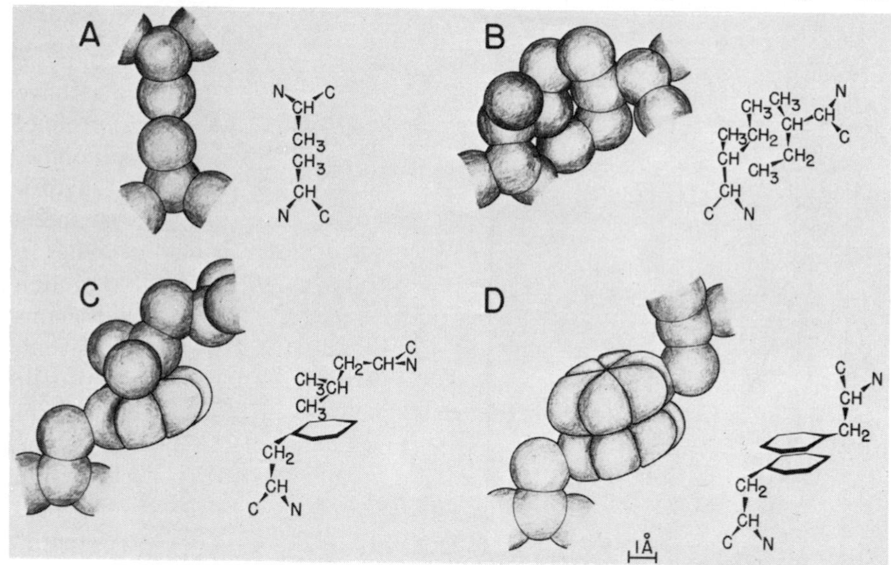

Fig. 6-4 Schematic and molecular model illustration of some possible apolar interactions (hydrophobic bonds) between pairs of amino acid side chains: (A) alanine–alanine; (B) isoleucine–isoleucine; (C) phenylalanine–leucine; (D) phenylalanine–phenylalanine. Van der Waals radii have been reduced 20% for clarity. [From G. Nemethy and H. A. Scheraga, *J. Phys. Chem.*, **66,** 1773 (1962).]

low temperatures it was shown that the transfer of hydrocarbons from water to nonpolar solvents (e.g., benzene, carbon tetrachloride) is not accompanied by a negative ΔH but instead requires heat energy to proceed.[14] On the other hand, the overall free energy change for the process ($\Delta G = \Delta H - T\Delta S$) is favorable ($\Delta G < 0$). The driving force is a large positive entropy change (ΔS), which is due to a change in water structure from a relatively ordered arrangement of molecules about the hydrocarbon to the more disordered nature of liquid water. Thus the thermodynamically favored increase in disorder in the system overrides the enthalpy requirements, and the formation of apolar bonds permits the system to attain a lower energy level.

The relationship of apolar bonding to the term $-T\Delta S$ in the free energy equation indicates a direct temperature dependence; that is, as temperature increases, negative free energy increases and the equilibrium shifts further toward bonding. However, it is experimentally observed that the ordered structure of water molecules around an apolar group is thermally very labile; that is, there is an anomalously high value of ΔC_P, where $\Delta C_P = (\partial\,\Delta H/\partial T)_P$, associated with the transfer of apolar groups to an aqueous medium. Equivalently, an unusually large temperature dependence is introduced into ΔH and ΔS, as is clear in the solubility data for aromatic

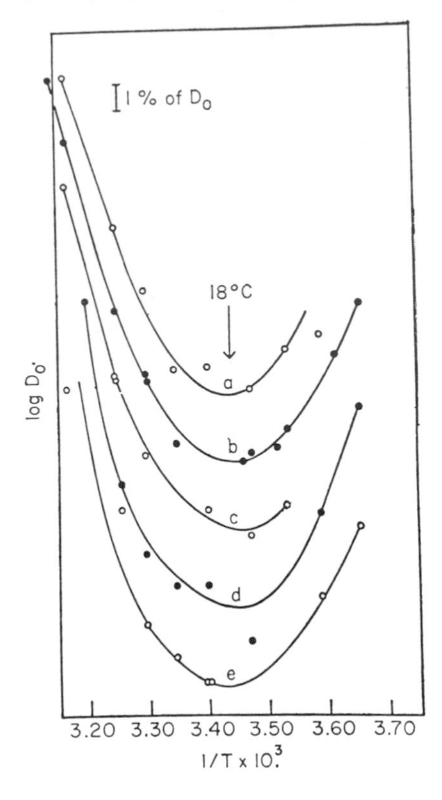

Fig. 6-5 Logarithmic plot of solubilities of aromatic hydrocarbons in water (D_0) as a function of temperature. From top to bottom: benzene, toluene, *p*-xylene, *m*-xylene, ethylbenzene. [Data of R. L. Bohon and W. F. Claussen, *J. Amer. Chem. Soc.*, **73,** 1571 (1951) cited by Tanford.[4]]

hydrocarbons in water shown in Fig. 6-5. The solubility curves pass through a minimum at about 18°C at which ΔH is equal to zero. This suggests that the strength of apolar interactions between these model compounds is greatest at this temperature. The results for ΔH obtained from the slopes of the curves show a linear dependence on temperature, indicating that ΔC_P is approximately constant with temperature. The value of ΔC_P (108 cal/deg-mole) for these organic substances added to water may be compared with their molar heat capacity in the liquid state ($\Delta C_P = 40$ cal/deg-mole). The difference represents the effect of these substances on the structure of water.

A logical approach to determine the free energy contributed to protein folding by apolar side chain residues is the measurement of the energy involved in transfer of these residues from water to a nonpolar solvent. Such data may be obtained (for example) from amino acid solubility data after correction for contributions of groups not in the side chain using results from identical measurements of glycine (e.g., see the review by Tanford[4]). The free energies of transfer for some amino acid side chain residues from water to 100% ethanol are shown in Table 6-1. Tanford has taken the latter to represent a reasonable facsimile of the interior of a protein molecule. The

Table 6-1 Free energy change in kilocalories per mole for transfer from water to 100% ethanol at 25°.[a]

	$\Delta G_{\mathrm{transfer}}$	
	Whole molecule	Side chain only
Glycine	+4.63	0
Alanine	+3.9	−0.73
Valine	+2.94	−1.69
Leucine	+2.21	−2.42
Isoleucine	+1.69	−2.97
Phenylalanine	+1.98	−2.65
Proline	+2.06	−2.60
Methionine	+3.33	−1.30
Tyrosine	+1.76	−2.87
Threonine	+4.19	−0.44
Serine	+4.59	−0.04
Asparagine	+4.64	+0.01
Glutamine	+4.73	+0.10

[a] From C. Tanford, *J. Amer. Chem. Soc.*, **84,** 4240 (1962).

more hydrophobic side chains show a large negative free energy change upon transfer to ethanol, thus providing an energetic basis for the internalization of apolar residues in proteins. One might expect that the cooperative action of many such residues should have a substantial effect on protein folding and stabilization. Attempts to quantitate these factors for proteins will be considered in Section 3 of this chapter and further in Chapter 15.

2-4 Other Types of Noncovalent Interactions

The term *ionic bond* refers to the electrostatic attraction between oppositely charged groups, as in the interaction of sodium ions and chloride ions in the sodium chloride crystal. In proteins under normal conditions of pH such attractions may occur between the negatively-charged side chains of aspartic or glutamic acid (or of sulfhydryl and tyrosyl groups at higher pH) and the positively-charged groups of histidine, lysine, and arginine. Terminal carboxylate and protonated amino groups may also participate in such interactions. All of these groups are also capable of hydrogen bonding, and it is more likely that this is their principal form of interaction when they occur in the interior of protein structures. Little stabilization energy would be expected from ionic bonds in proteins since most charged groups are in contact with solvent both in the native state and in the random coil. In the cellular environment the ionic strength is sufficiently high so that electrostatic attractions between groups on the protein will be damped out by other ions. On the other hand, a concentration of charged groups at a particular surface location or the presence of an ionized group at a ligand binding site could be important in determining the configuration and properties at that site.

Van der Waals forces or London dispersion forces represent electronic interactions of permanent and induced dipoles in atoms. These forces are responsible for the close packing of atoms observed in the crystalline state, enabling the assignment of an approximate van der Waals radius to each type of atom (cf. Chapter 14). The crystallographic structure of the myoglobin molecule gives an impressive demonstration of van der Waals close packing, and there is no question that proteins are subject to these forces. In addition, some aspects of the phenomenon of apolar interactions (i.e., the ordering of water around apolar groups) may perhaps be best explained theoretically in terms of interactions of the van der Waals type. Additional direct involvement of van der Waals forces in the stabilization of native protein structures would require a difference between protein–protein and solvent–solvent versus protein–solvent interactions. Current understanding of these forces, both from an experimental and theoretical point of view, is still far from adequate for a meaningful evaluation of the magnitude of such differences.

3 THE STABILITY OF NATIVE PROTEIN STRUCTURES[1,4]

From the preceding consideration of the factors that are likely to play a role in protein folding and stabilization, we would like to be able to calculate the total free energy change ΔG for the formation of a native structure from the random coil form of the constituent polypeptide chains. The equilibrium of interest is

$$\underset{\text{(denatured or random coil)}}{\text{Polypeptide chains}} \overset{K}{\rightleftharpoons} \text{Native protein} \qquad (6\text{-}4)$$

where ΔG is given in terms of the equilibrium constant K by the usual relationship:

$$\Delta G = -RT \ln K \qquad (6\text{-}5)$$

The free energy change includes the heat energy (ΔH) and entropy (ΔS) terms

$$\Delta G = \Delta H - T\,\Delta S \qquad (6\text{-}6)$$

which, however, are difficult to evaluate theoretically for a process of the complexity of that given in Eq. 6-4. As indicated in the introduction to this chapter, a large factor opposing the transition from the multiplicity of random coil conformations to the single native state is the entropic restriction, which makes an expected large positive contribution to ΔG. This term may be separated out and, following Tanford's notation,[4] one may express ΔG as the sum of a conformational term plus terms for short-range interactions within ordered regions of the molecule ($\Delta g_{i,\mathrm{int}}$), for solvent contacts ($\Delta g_{i,\mathrm{s}}$), and for long-range electrostatic interactions (ΔW_{el}). Thus ΔG may be written

$$\Delta G = \Delta G_{\mathrm{conf}} + \sum_i \Delta g_{i,\mathrm{int}} + \sum_i \Delta g_{i,\mathrm{s}} + \Delta W_{\mathrm{el}} \tag{6-7}$$

where the summations are carried out over all i parts of the molecule.

The approach described above permits utilization of data from model compound studies as presented in the previous section. For proteins whose crystal structures have been solved, the number of residues participating in each kind of interaction (hydrogen bonds, apolar contacts, etc.) can be reasonably assessed. For others an estimate is made based upon amino acid composition and the distribution of residues in known structures. The conformational entropy term is calculated from estimates of the multiplicity of configurations possible for a peptide unit in the random coil state or from conformational potential energy functions based on bond torsional strain, van der Waals repulsions, London attractions, and electrostatic interactions. Varying estimates are used for the accessibility of solvent to groups in the native protein and in the random coil. Calculations of this type, as discussed in detail by Tanford,[4] have indicated that the random coil is energetically favored over the native conformation under conditions comparable to those of the physiological state, a result at obvious variance with observation. Other calculations on solvent accessibility to atoms in the random coil and in the native state of several proteins[15] suggest that the difference between the two states is less than that ordinarily assumed, with the result that the energetic advantage of the random coil is increased.

To illustrate how calculations of ΔG for the random coil–native protein transition are carried out, we follow a model suggested by Schachman[2] modified by the use of more recent data for interaction energies. Figure 6-6 illustrates the various types of noncovalent bonds occurring in proteins and includes estimates of their contribution to ΔG for the process in which 1 mole of a native protein possessing a single polypeptide chain per molecule is formed from the unfolded or random coil chains in aqueous solution at 37°C. The first term, the conformational free energy change, which deals with the change in order in the system, is based upon the estimate by Brant, Miller, and Flory of a value of 10 cal/deg/mole residue for the conformational entropy of a randomly coiled polypeptide backbone.[16] An amount of 1 cal/deg-mole residue is added to represent, on the average, the conformational entropy of the side chains in the random coil. The total conformational entropy for the random coil of 100 residues is then calculated to be 1.1 kcal/deg-mole. If we assume that the native state is completely fixed

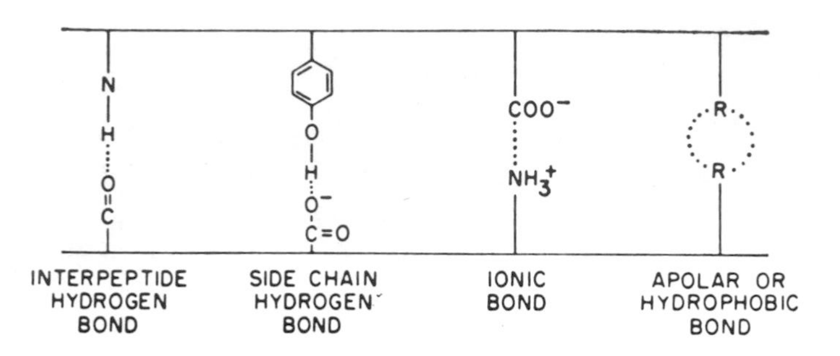

UNFOLDED CHAINS ⇌ NATIVE STRUCTURE

	ΔG (Kcal)
CONFORMATIONAL ENTROPY	+ 340
CONFORMATIONAL ENTHALPY	− 100
APOLAR BONDS	− 130
HYDROGEN BONDS	− 10
IONIC BONDS	− 10
TOTAL	+ 90

Fig. 6-6 Schematic diagram illustrating the types of noncovalent bonds that may contribute to folding and stabilization of a native protein. Estimates for the free energy terms involved in the transition to a native protein are given for a polypeptide chain of 100 residues.[2]

and no free rotation occurs, then the term $-T\,\Delta S$ for the transition to the native state may be calculated directly from the conformational entropy of the random coil. The result, at a temperature of 310°K, is +340 kcal/mole. Although the conformational enthalpy term (ΔH_{conf}) cannot be readily assessed, the data of Brant, Miller, and Flory suggest that a maximum value of −1 kcal/mole residue might be placed on it.[4] This contributes a total of −100 kcal/mole for our example. The next term refers to hydrophobic bonding of the apolar side chains of the protein and represents that part of both of the summations in Eq. 6-7. If we assume that about 65 of the residues of the chain of 100 residues fall in the apolar category, and taking an average value of −2 kcal/mole residue for ΔG of apolar bond formation based on Table 6-1, we obtain a total of −130 kcal/mole for apolar bonding. Here it is assumed that all apolar residues are sequestered in the native structure (in an environment like that of 100% ethanol) and that all are exposed to water in the random coil state. A more reasonable assumption on solvent exposure (e.g., 60% exposure to solvent in the random coil and 25% in the native state) would greatly reduce the total apolar stabilization.

The next term in the calculation represents the hydrogen bonding contribution to ΔG of the transition. We will assume that no net change in hydrogen bonding occurs, that is, that all groups in the polypeptide chain capable of hydrogen bonding are participating in interpeptide or other hydrogen bonds in the native structure. According to the data obtained for *N*-methylacetamide, as discussed earlier, no energetic advantage is obtained from hydrogen bonding within the protein; rather, a small positive ΔG is obtained per residue. However, if the value cited earlier for hydrogen bonding in the random coil–helix transition is used (−150 cal/mole residue) and if we assume the native structure to contain 60 interpeptide hydrogen bonds and five side chain hydrogen bonds, we find a total ΔG contribution of −10 kcal/mole for hydrogen bonding. Finally, we add a term of −10 kcal/mole, as estimated by Schachman, for ionic bonding, although this may be too generous. Long-range electrostatic interactions, as represented by ΔW_{el} in Eq. 6-7, are assumed to have a negligible effect on the transition. The result in our calculation is that ΔG for the transition to the native state is decidedly positive, thus the equilibrium would be in the direction of the random coil. This result is comparable to the calculations by Tanford[4] on ribonuclease and β-lactoglobulin, on which our treatment is based.

In all of these types of calculations, apolar bonding clearly figures as the prime stabilizing influence in native proteins, even though its contribution based on model compound data is not adequate to overcome conformational entropy. However, there are even some reasons for uncertainty about this conclusion. One of these concerns the temperature dependence of the thermodynamic parameters associated with apolar interactions in model compounds (Fig. 6-4). It was shown there that the temperature of maximum stability occurs at about 18°C. Although several proteins have been found to have a temperature optimum in stability in this range, it may vary from less than 0°C (ribonuclease) to as high as 35°C (β-lactoglobulin).[5] The latter, however, was determined in the presence of urea, which may have influenced the result. Another argument against apolar bonding as the primary driving force for protein folding is the finding that only a small change in volume occurs in the transition of native protein to random coil (N → R) compared to that predicted from model compound studies on apolar interactions.

An obvious conclusion is that the quantitative and, in part, the qualitative aspects of the forces responsible for stabilizing native proteins are far from being understood. Evidently, protein interactions include factors of greater complexity than can be adequately accounted for in terms of the interactions of model compounds. Before continuing to the discussion of physical methods used as tools for the study of protein conformation, we summarize a few experimental observations on the stability of proteins. A number of proteins can be made to undergo a reversible $N \rightleftharpoons R$ transition *in vitro*. The native conformation of these proteins therefore clearly represents a minimum in total free energy compared to other conformations. The reservation must be made, however, that the native conformation may not necessarily be the lowest free energy state. Kinetic restrictions in the form of relatively high potential energy barriers may in some cases prevent attainment of a lower energy conformation. Studies on reversible denaturation have clearly shown that the free energy of stabilization of the native structure is, in fact, very little different from that of the random coil. Observed ΔG values for formation of the native structure are in the range of -10 to -50 kcal/mole. Similarly, in studies of association-dissociation equilibria of oligomeric proteins, one finds the interactions between subunits to be energetically weak. The negative ΔG for subunit association in several cases is only a few kilocalories per mole of subunit. Consequently, relatively small changes in amino acid residue interactions could result in substantial conformational changes, a situation that undoubtedly has provided for considerable flexibility in the evolution of functional properties of proteins.

REFERENCES

1. W. Kauzmann, "Denaturation of proteins and enzymes," in *The Mechanism of Enzyme Action*, W. D. McElroy and B. Glass, Eds., The Johns Hopkins Press, Baltimore, 1954. W. Kauzmann, "Some factors in the interpretation of protein denaturation," *Adv. Protein Chem.*, **14,** 1 (1959).
2. H. K. Schachman, "Considerations on the tertiary structure of proteins," *Cold Spring Harbor Symp. Quant. Biol.*, **28,** 409 (1963).
3. F. M. Richards, "Structure of proteins," *Ann. Rev. Biochem.*, **32,** 269 (1963).
4. C. Tanford, "Protein denaturation. Part C," *Adv. Protein Chem.*, **24,** 1 (1970).
5. C. Tanford, "Protein denaturation," *Adv. Protein Chem.*, **23,** 121 (1968).
6. R. Cecil, "Intramolecular bonds in proteins. I. The role of sulfur in proteins," in *The Proteins*, Vol. 1, H. Neurath, Ed., Academic Press, New York, 1963, p. 380.
7. F. H. White, "Regeneration of native secondary and tertiary structures by air oxidation of reduced ribonuclease," *J. Biol. Chem.*, **236,** 1353 (1961).
8. C. J. Epstein and C. B. Anfinsen, "The reversible reduction of disulfide bonds in trypsin and ribonuclease coupled to carboxymethyl cellulose," *J. Biol. Chem.*, **237,** 2175 (1962).

9. R. Sperling and I. Z. Steinberg, "Elongation of the disulfide bonds of bovine pancreatic ribonuclease and the effect of the modification on the properties of the enzyme," *J. Biol. Chem.*, **246,** 715 (1971).
10. D. Givol, F. deLorenzo, R. F. Goldberger, and C. B. Anfinsen, "Disulfide interchange and the three-dimensional structure of proteins," *Proc. Nat. Acad. Sci., U.S.*, **53,** 676 (1965).
11. G. C. Pimentel and A. L. McClellan, *The Hydrogen Bond*, Freeman, San Francisco, 1960.
12. L. Pauling, *The Nature of the Chemical Bond*, Cornell University Press, Ithaca, New York, 1960.
13. I. M. Klotz and J. S. Franzen, "Hydrogen bonds between model peptide groups in solution," *J. Amer. Chem. Soc.*, **84,** 3461 (1962).
14. H. S. Frank and M. W. Evans, "Free volume and entropy in condensed systems. III. Entropy in binary liquid mixtures; partial molal entropy in dilute solutions; structure and thermodynamics in aqueous electrolytes," *J. Chem. Phys.*, **13,** 507 (1945).
15. B. Lee and F. M. Richards, "The interpretation of protein structures: estimation of static accessibility," *J. Mol. Biol.*, **55,** 379 (1971).
16. D. A. Brant, W. G. Miller, and P. J. Flory, "Conformational energy estimates for statistically coiling polypeptide chains," *J. Mol. Biol.*, **23,** 47 (1967).

VII Hydrodynamic Methods

The properties of protein molecules as "hydrodynamic" particles are important both for evaluation of structural parameters and for understanding the nature of macromolecular interactions in the living state. A variety of methods have been developed for the study of proteins in motion in solutions (primarily aqueous media). The rate of movement of molecules from a higher to a lower concentration or from an oriented to a random state (diffusion), the rate of migration in a centrifugal field (sedimentation), or the effect of solute particles on liquid flow (viscosity) are all measurable phenomena that reflect structural properties of the molecules. The following discussion will emphasize the use of hydrodynamic methods for the elucidation of these properties. The methods, however, also have many applications for purposes of identification or purification of macromolecules.

1 TRANSLATIONAL DIFFUSION[1–5]

When a solute is added to a liquid in which it is soluble, it may be observed that the molecules move in all directions and, in time, a uniform distribution is achieved throughout the solution. The same effect is found whenever a concentration gradient exists in a solution, provided no barrier to molecular motion exists. This movement is translational diffusion, and its rate depends upon the size and shape of the particles. The driving force for diffusion is the random Brownian motion of the solvent which causes the solute particles to be continuously bombarded by molecules of the solvent. The net momentum transfer has a Boltzmann distribution of magnitudes and is random in direction. Thermodynamically, the driving force for diffusion may be understood by the concept of entropy, the tendency to randomization.

Although the solute is also acted upon by the force of gravity, this effect is not sufficient, in the case of most proteins, to cause redistribution. For example, at room temperature the thermal energy, kT (k = Boltzmann constant, T = absolute temperature), responsible for Brownian motion is more than 200 times the energy of the gravitational pull on a typical protein of molecular weight 50,000.[6] Hence proteins in solution assume a nearly random distribution through free diffusion.

1-1 The Equations of Diffusion

For one-dimensional diffusion, the concentration is uniform in planes perpendicular to the direction (x) of mass transport. For a single solute component diffusing in a system of constant temperature and pressure, the rate of mass transfer, $\partial m/\partial t$, past a cross-sectional area A, is proportional to A and to the concentration gradient at x, $\partial c/\partial x$. This may be expressed mathematically as

$$\frac{\partial m}{\partial t} = -AD\frac{\partial c}{\partial x} \tag{7-1}$$

The proportionality factor D is the diffusion coefficient and has units of area divided by time (e.g., cm^2/sec); the minus sign is introduced so that D will be positive. Equation 7-1 is known as Fick's first law and describes the physical process of solute diffusion where D is a constant, characteristic of the solute.

The rate of mass transfer at $(x + dx)$ is equal to that at x (Eq. 7-1) plus the change in $\partial m/\partial t$ with x over the interval dx, namely

$$\frac{\partial m}{\partial t} + \frac{\partial^2 m}{\partial t\,\partial x}\,dx = -A\left[D\frac{\partial c}{\partial x} + \frac{\partial}{\partial x}\left(D\frac{\partial c}{\partial x}\right)dx\right] \tag{7-2}$$

The accumulation within the volume $A\,dx$ is obtained by subtracting Eq. 7-1 from Eq. 7-2:

$$\frac{\partial^2 m}{\partial t\,\partial x}\,dx = A\frac{\partial}{\partial x}\left(D\frac{\partial c}{\partial x}\right)dx \tag{7-3}$$

The rate of concentration increase, $\partial c/\partial t$, is equal to the rate of mass increase divided by the volume, $A\,dx$. Therefore, dividing both sides of Eq. 7-3 by $A\,dx$ gives the desired result

$$\frac{\partial c}{\partial t} = \frac{\partial}{\partial x}\left(D\frac{\partial c}{\partial x}\right) \tag{7-4}$$

For systems where D may be considered independent of concentration, Eq. 7-4 may be rewritten as Fick's second law:

$$\frac{\partial c}{\partial t} = D\frac{\partial^2 c}{\partial x^2} \tag{7-5}$$

1-2 The Porous Diaphragm Cell

The measurement of diffusion is illustrated in Fig. 7-1 by one of the earliest and simplest types of apparatus used. It consists of two chambers separated by a porous membrane composed of nearly parallel channels that must be large compared to the macromolecules, but sufficiently small so that bulk flow of liquid through them is negligibly small (cf. viscosity). The solution containing the molecules of interest at a known concentration c_{01} is placed in the upper chamber; the lower chamber contains solution at concentration c_{02} or buffer only ($c_{02} = 0$). As solute diffuses through the porous membrane, the increase of mass Δm in the lower chamber is determined by measuring the concentration as a function of time.

For short-time experiments in which the concentrations in the two compartments are not significantly altered, the concentration gradient may be considered constant over the time interval and is equal to $(c_{02} - c_{01})/l$, where l is the "effective" pore length. Substitution into Eq. 7-1 gives

$$D = \frac{\Delta m}{\Delta t} \frac{l}{A} \frac{1}{(c_{01} - c_{02})} \tag{7-6}$$

The mass increase in the lower compartment, Δm, is equal to $V_L(c_2 - c_{02})$ where V_L is the volume of the lower compartment and $(c_2 - c_{02})$ is the concentration change in that compartment in the time interval $t - t_0$. The diffusion coefficient may then be calculated from

$$D = \left(\frac{l}{A}\right) V_L \left(\frac{c_2 - c_{02}}{c_{01} - c_{02}}\right) \frac{1}{t - t_0} \tag{7-7}$$

The apparatus constant, l/A, is determined by calibrating the instrument with a standard of known D such as KCl. Equation 7-7 can only be used for short time intervals and thus requires high accuracy in concentration determination. In order to take advantage of longer time periods for diffusion, the equation

$$-\left(\frac{A}{l}\right)\left(\frac{1}{V_U} + \frac{1}{V_L}\right) Dt = \ln\left[(c_2 - c_1)/(c_{02} - c_{01})\right] \tag{7-8}$$

is employed.

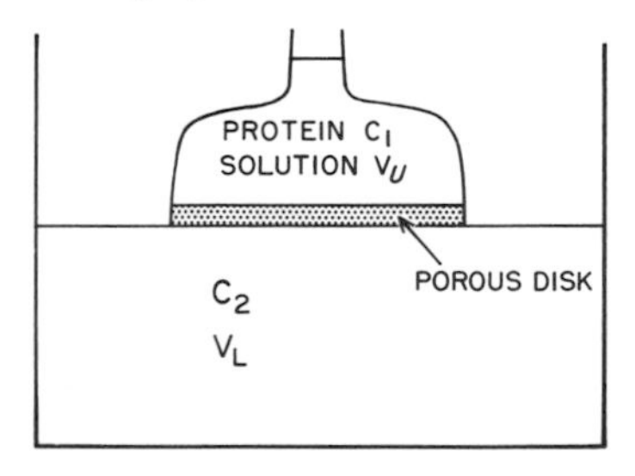

Fig. 7-1 Northrup-Anson cell for the measurement of diffusion coefficients.

As seen in Eq. 7-7, this determination of D requires only the ratio of the concentrations in the two compartments to the original concentration. This is an important advantage, since it means that the concentration may be expressed in terms of any measurable parameter that is proportional to concentration. This may be ultraviolet absorbancy, biuret or Kjeldahl analysis, or biological activity if an assay is available (e.g., activity of an enzyme, plaque forming units or infectivity titer of a virus). For this reason the diffusion coefficient can be obtained for a macromolecule at relatively early stages of purification even if the substance measured comprises only a minute fraction of the total protein present. If the activity being measured should be bound to other components in the mixture, however, an erroneous result will obviously be obtained. Error may also occur with this method if the solutions in the chambers are not stirred sufficiently to prevent the formation of concentration gradients or if the rate of diffusion through the capillaries is not the same as that in free solution. Abnormal diffusion will result if the molecules adsorb to the walls of the channels or if the molecules are large and asymmetric, thereby passing through the narrow channel with a specific rotational alignment instead of a random orientation. More complex diffusion apparatuses have been designed to eliminate some of these sources of error.[1,2]

1-3 Free Diffusion

Diffusion coefficients may also be determined without the use of artificial barriers. In a typical experimental arrangement the protein solution (ordinarily in a buffered solvent) is placed in a cell and buffer is carefully layered on top to give an initially-sharp boundary between the protein solution and the solvent. Since the protein solution is denser than the buffer alone, such a system is stable under gravity, provided there are no thermal or mechanical disturbances. As time progresses, solute molecules move into the solvent region, causing a decrease in concentration on the solute side of the initial boundary and an increase on the solvent side. The concentration and concentration gradient curves as a function of distance x from the initial boundary for a typical diffusion process are shown in Fig. 7-2.

Free diffusion experiments may be carried out in a Tiselius-type electrophoresis apparatus (cf. Chapter 10) using a Schlieren or Rayleigh interference optical system. Both systems depend on the higher refractive index of the protein solution compared to the solvent; the difference is directly proportional to protein concentration. The Rayleigh interference pattern (Fig. 7-2*d*) shows refractive index as a function of distance and is similar in appearance to the c vs. x curves of Fig. 7-2*b*. The Schlieren pattern (Fig. 7-2*c*) gives the change in refractive index as a function of x ($\Delta n/\Delta x$ or dn/dx), which is proportional to dc/dx.

Utilization of the Tiselius-type apparatus for measurement of diffusion coefficients has markedly declined in recent years, as has its use for electrophoretic experiments. The difficulty in performing the experiment, the high concentrations of protein required, and the development of ultracentrifugal methods, have all contributed to this decline. Measurement of diffusion coefficients from optical patterns obtained at low speed in the ultracentrifuge is often quite simple and straightforward to perform, and this procedure is currently being employed with increasing frequency. Experimental details and recommended methods for analyses of the optical patterns obtained are detailed by Markham.[3]

The solution of Fick's second law (Eq. 7-5) for the concentration gradient $(\partial c/\partial x)$ is

$$\frac{\partial c}{\partial x} = \frac{c_0}{\sqrt{4\pi\ Dt}} \exp(-x^2/4\ Dt) \tag{7-9}$$

where c_0 is the total solute concentration difference across the boundary. This is the well known equation of a Gaussian curve. The x coordinate is defined so that $x = 0$ at the maximum (H_m) of $\partial c/\partial x$, and since $d/dx(\partial c/\partial x) = 0$ at the maximum, one may solve for the value of H_m:

$$H_m = c_0/\sqrt{4\pi\ Dt} \tag{7-10}$$

The area under the gradient curve is given by the integral

$$A = \int_{-\infty}^{+\infty} \left(\frac{\partial c}{\partial x}\right) dx$$

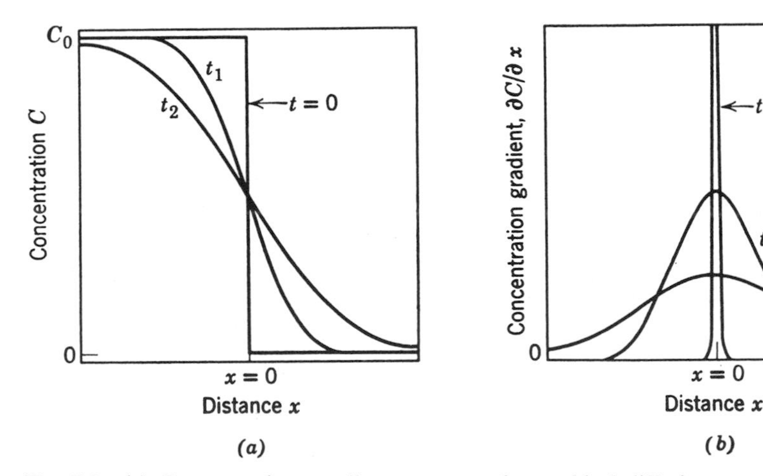

Fig. 7-2 (*a*) Concentration vs. distance curves for an ideal diffusion process at several time intervals. The initial (time = 0) curve is a sharp step function and succeeding curves become increasingly shallow with increasing time. (*b*) Concentration gradient vs. distance plot of the process depicted in (*a*). These curves are derivatives (dc/dx) of the c vs. x curves in (*a*). The maximum occurs at the inflection point of the c vs. x curve. [From C. Tanford[29]]

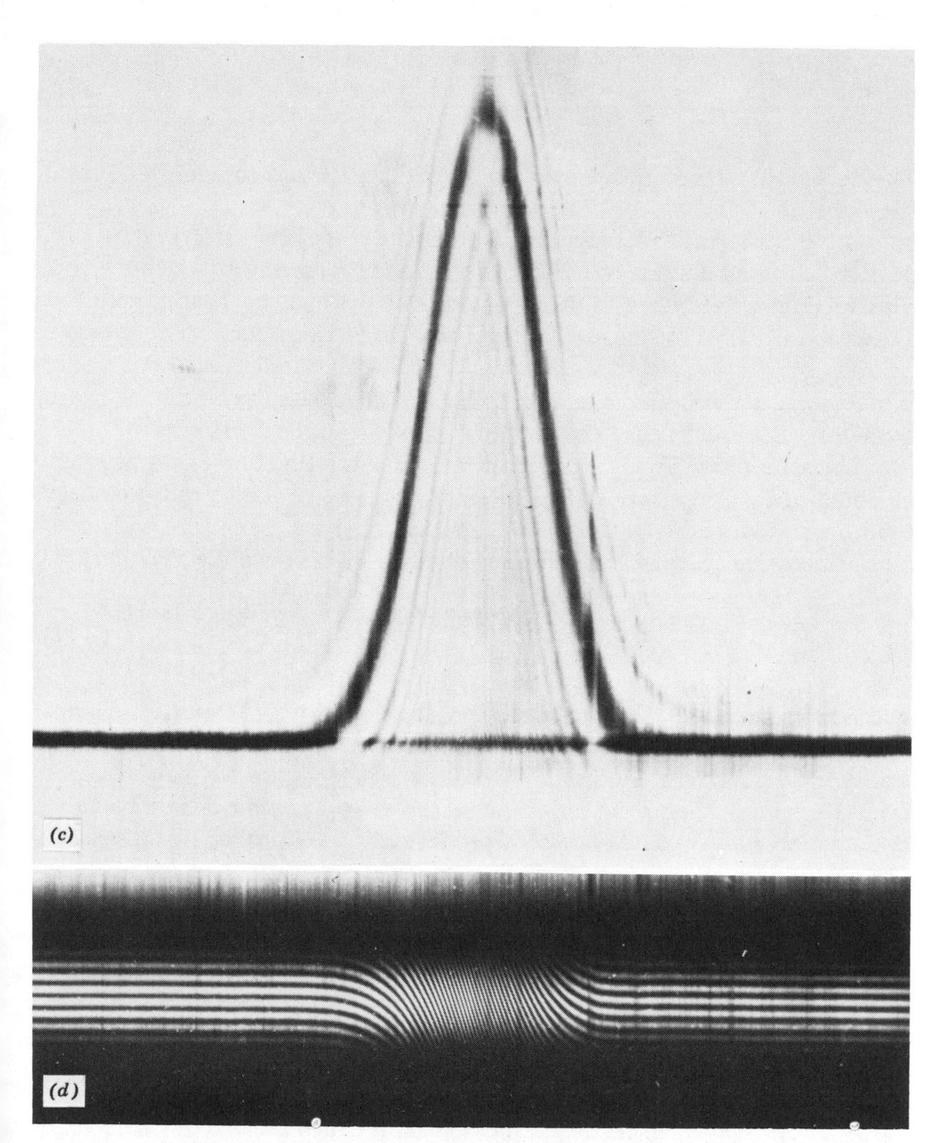

Fig.7-2 (*continued*) (*c*) An experimental diffusion curve obtained with the Schlieren optical system. The refractive index gradient of the solute is superimposed upon that of the reference solvent (an essentially horizontal baseline). The pattern may be transformed into a dc/dx vs. x plot if the refractive increment of the protein and optical constants of the instrument are known. (*d*) A Rayleigh interference pattern of the same solution shown in (*c*). This pattern can be transformed into a c vs. x (or kc vs. x) plot once the optical constants are calibrated. [From R. Markham[3]]

and is equal to c_0. Equation 7-10 may then be rearranged to permit determination of D by the "height-area method":

$$D = \frac{A^2}{H_m{}^2 4\pi t} \tag{7-11}$$

This equation can be applied by plotting $H_m{}^2$ vs. $1/t$ and computing D from the slope.

Other equations for computing D from concentration gradient data may be derived, but in practice the so-called method of moments is to be preferred. This method is described with an illustrative example by Markham.[3] The evaluation of the diffusion coefficient from the integral curve (i.e., where a quantity proportional to concentration has been measured as a function of x) is quite accurate and simple to apply. The required data are readily obtained from Rayleigh interference or ultraviolet scanning optical patterns.

The solution of Fick's second law to give c as a function of x and t may be obtained by integration of Eq. 7-9 with appropriate initial and boundary conditions. The result in the case of an initial sharp boundary with concentration c_2 below the boundary (positive x) and c_1 above the boundary is

$$2(c - \bar{c})/\Delta c = \frac{2}{\sqrt{\pi}} \int_0^{x/(4Dt)^{1/2}} \exp(-\alpha^2)\, d\alpha \tag{7-12}$$

where $\alpha = x/2\sqrt{Dt}$. In the equation, $\bar{c} = (c_2 + c_1)/2$ and $\Delta c = c_2 - c_1$. The right-hand side of this equation is the well-known error function, in this case written erf $[x/(4Dt)^{1/2}]$. For each value of $c = f(x, t)$ the left-hand side of Eq. 7-12 is known, and the argument of the error function, which contains D as the only unknown, may be obtained from tables of probability functions. The entire data set may be handled in a variety of ways. For example, an ideal single solute of constant D ought to give a horizontal straight line when D vs. c is plotted for each time series. Except for sometimes unavoidable distortions at the boundary extremes, a straight line in this plot is evidence for the Gaussian character of the diffusion curve. A smooth increase or decrease of D with c suggests that D may be concentration dependent. Heterogeneous systems are marked by a minimum in the D vs. c curve. Evaluation of the concentration dependence of D is best achieved by applying Eq. 7-12 (or Eq. 7-9) to multiple experiments in which Δc is kept small while $\bar{c}$ is varied.

A convenient and surprisingly accurate way to evaluate D from the integral curves is the probability paper method described in detail by Markham.[3] The c vs. x data obtained at a given time are expressed as the percent change in c across the boundary and plotted on arithmetical probability paper as shown in Fig. 7-3. The value of the standard deviation, σ, located

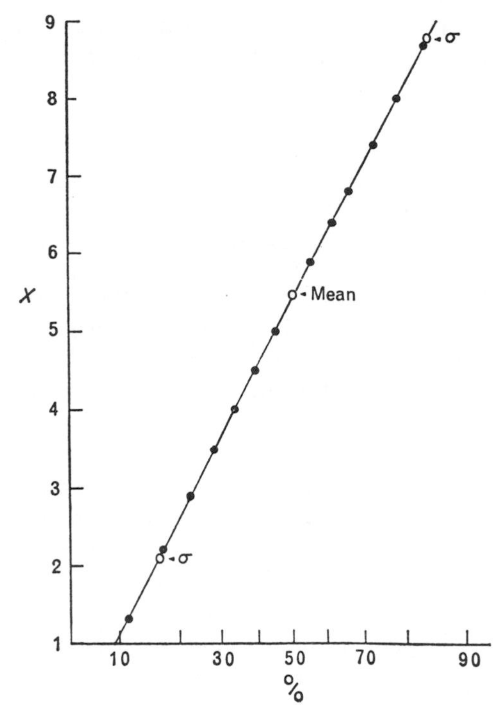

Fig. 7-3 Determination of the diffusion coefficient from a plot of percent c_0 (the initial plateau solute concentration) vs. x on arithmetical probability paper. The positions of the mean and the standard deviation (at 16 and 84%) are indicated. [From R. Markham[3]]

at the 16 and 84% position on the plot, is read from the curve. The standard deviation is related to D by the equation

$$D = \sigma^2/2t \tag{7-13}$$

The value of D is most reliably obtained as the slope of a plot of σ^2 against $2t$. Again, non-Gaussian behavior is detected as curvature in the probability plot. Deviation at the very extremes of the plot are often caused by convective disturbances due to insufficient density stabilization and may generally be disregarded.

It is customary to convert a measured diffusion coefficient D to standard conditions, $D_{20,w}$, the value expected if the experiment had been conducted in pure water at 20°C. The conversion is made with the equation:

$$D_{20,w} = D\,\frac{293}{T}\,\frac{\eta}{\eta_{20,w}} \tag{7-14}$$

where T is the temperature of the measurement, η is the viscosity of the solvent at that temperature, and $\eta_{20,w}$ is the viscosity of water at 20°C.

1-4 Relation of D to the Frictional Coefficient

The diffusion coefficient is related to molecular parameters through the Einstein-Sutherland equation:

$$D = kT/f \tag{7-15}$$

where f is the frictional coefficient of the particle. This relation is derived by comparison of Fick's first law of diffusion and the thermodynamic equation for diffusion expressed in terms of chemical potential. Equation 7-15 may be better understood by thinking of the diffusion of a molecule in terms of a balance between the driving force of thermal energy (kT) and the opposing energy of resistance (resistive force F_r × distance). F_r may be represented in the usual way as a frictional coefficient multiplied by the velocity of movement. Hence, at equilibrium, $kT = f \times$ *velocity* × *distance*. The relationship to the diffusion coefficient, which from Fick's law has the units of *velocity* × *distance*, is thus indicated. Although Eq. 7-15 suggests a simple dependence of D on temperature, it should be pointed out that the dependence is more complicated since f is related to the solvent viscosity which also depends on T. For nonideal solutes Eq. 7-15 should be replaced by

$$D = \frac{kt}{f}\left[1 + c\frac{d \ln \gamma}{dc}\right] \tag{7-16}$$

where γ is an activity coefficient which usually decreases with increasing concentration. The frictional coefficient may also be concentration dependent, but a model to predict the magnitude of the concentration dependence is still lacking. For nonideal systems the diffusion coefficient is best obtained by extrapolating the results to infinite dilution.[1]

1-5 Stokes' Law

In general, the frictional coefficient is a complicated function of the size, shape, and hydration of a macromolecule. However, for the simple case of a rigid (nondeformable) sphere (whether hydrated or not), it reduces to

$$f = 6\pi\eta r \qquad \text{(Stokes' law)} \tag{7-17}$$

where η is the viscosity of the solvent and r is the radius of the sphere. Measurement of D can be used to obtain the molecular weight of nonhydrated spheres, provided that the partial specific volume $\bar{v}$ is known, that is, the volume occupied by 1 g of the solute in the solution. The experimental measurement of $\bar{v}$ will be discussed later in this chapter. The relation of the volume of a sphere V to $\bar{v}$ is:

$$V \text{ (cc/molecule)} = \bar{v} \text{ (cc/g)} \times M \text{ (g/mole)}/N \text{ (molecules/mole)}$$

where M is the molecular weight and N is Avogadro's number. Since $V = 4\pi r^3/3$, the molecular weight may then be determined from

$$M = 4\pi N r^3/3\bar{v} = \frac{4\pi N}{3\bar{v}}\left(\frac{f}{6\pi\eta}\right)^3 \tag{7-18}$$

This relation is valid only when there is no bound solvent, which would cause an increase in hydrodynamic volume of the particles. For spheres hydrated with an amount w (in g/g protein) of solvent of density ρ, M can be calculated from Eq. 7-18 after replacing $\bar{v}$ by $(\bar{v} + w/\rho)$. If w is not known, an estimate of M for proteins (taken as spheres) can be obtained by using a typical value of 0.3 for w.

For many particles, Eq. 7-18 does not give a good approximation to M. If the molecule is anisometric (nonspherical) or is hydrated, f will be greater than that of a nonhydrated sphere with the same molecular weight, and correlation with molecular weight and dimensions is more difficult. For such molecules the degree of deviation from Stokes' law behavior is often expressed by a frictional ratio f/f_0. Here, f is the actual frictional coefficient of the molecule, obtained from the measured D, and f_0 represents the frictional coefficient of an anhydrous sphere of the same molecular weight, that is,

$$f_0 = 6\pi\eta\sqrt[3]{\frac{3\bar{v}M}{4\pi N}}$$

The frictional ratio is sometimes factored into terms for hydration and for asymmetry of shape:

$$f/f_0 = (f/f_0)_H (f/f_0)_A \tag{7-19}$$

The frictional ratio for hydration may be readily expressed in terms of the degree of hydration w, as follows:

$$(f/f_0)_H = \left(1 + \frac{w}{\bar{v}\rho}\right)^{1/3} \tag{7-20}$$

In general, the diffusion coefficient may be expressed as a function of molecular weight, hydration, and asymmetry. By employing simplifying assumptions regarding two of these unknowns (or by independent measurement), the equation may be solved for the third unknown. In more complete analyses, diffusion data are used in combination with data obtained by other methods, as discussed later in this chapter.

2 SEDIMENTATION VELOCITY[7–10]

We have discussed the fact that the diffusion of particles in solution depends upon size and shape parameters. These properties also determine the rate at which the particles will move through solution in an externally-applied field, such as a centrifugal field. Proteins and other macromolecules are sufficiently large that they can be caused to redistribute in solution under the influence of attainable centrifugal fields. In the analytical ultracentrifuge the movement of macromolecules in centrifugal fields up to about 300,000 × gravity is observed by means of optical systems that have been devised to measure optical characteristics of the solutions.

2-1 The Ultracentrifuge

Since the first ultracentrifuge was developed by Svedberg during the early 1930s, the field has grown enormously, and hundreds of ultracentrifuges are in use throughout the world. Although other types of instruments are being used, the machine in greatest use is the commercially available Model E Ultracentrifuge (Spinco Division of Beckman Instruments). The mechanics of the Model E are shown diagrammatically in Fig. 7-4. In practice the cell (Fig. 7-5), filled with the solution to be studied, is placed on one side of the rotor (Fig. 7-6) and a counterweight or cell containing solvent is placed on the other side. After coupling the rotor to the drive shaft within the main chamber, the chamber is closed and evacuated, and the rotor is accelerated to the desired speed (as high as 69,000 rpm). Photographs taken at various time intervals measure either the absorbancy (absorption optical system), refractive index (Rayleigh interference optical system), or refractive index gradient (Schlieren optical system) in the cell as a function of the distance r from the center of rotation.

2-2 Basic Equations of Sedimentation

The centrifugal force acting on a particle in solution is given by the effective mass of the particle times the centrifugal acceleration. From Archimedes' principle

$$\text{effective mass} = (m - v\rho)$$

where $v\rho$ represents the mass of solvent displaced by the particle of mass m; v is the volume change produced when the particle is added to a large amount of solvent; and ρ is the density of the solvent. If the centrifuge is rotating at a rate of ω radians/sec, the centrifugal acceleration is $\omega^2 r$ in units of cm/sec^2.

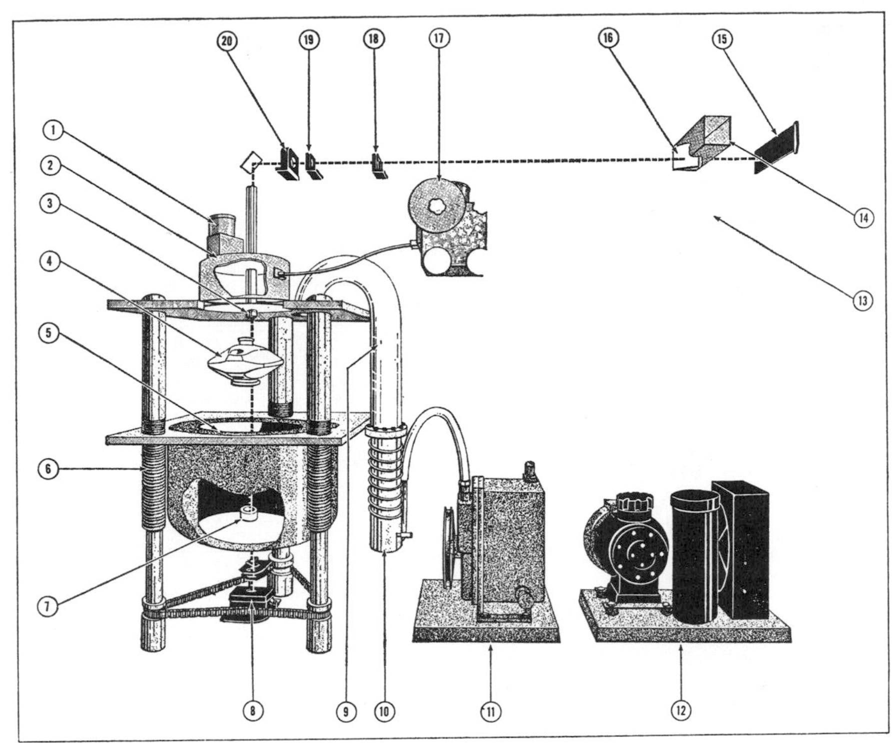

Fig. 7-4 Spinco Model E ultracentrifuge with Schlieren optical system. 1, motor housing; 2, drive gear housing; 3, condensing lens; 4, rotor; 5, rotor chamber; 6, lift rod; 7, collimating lens; 8, light source; 9, connection to pump; 10, oil diffusion pump; 11, vacuum pump; 12, refrigerator; 13, temperature control; 14, eyepiece; 15, photographic plate; 16, mirror; 17, speed control; 18, cylindrical lens; 19, camera lens; 20, phaseplate. (Courtesy of Spinco Division of Beckman Instruments.)

Hence, the centrifugal force F_c is

$$F_c = (m - v\rho)(\omega^2 r) \tag{7-21}$$

The movement of the particles is opposed by frictional resistance; the frictional force F_r is proportional to the velocity of the particle:

$$F_r = f \times velocity = f(dr/dt) \tag{7-22}$$

where the frictional coefficient f is a characteristic of the particle and of the solvent. This parameter is the same as that previously introduced in our discussion of diffusion (cf. Eq. 7-15). Under the conditions of a sedimentation velocity experiment, the molecules reach an almost instantaneous maximum velocity, that is, they are no longer accelerated, and the net force is zero (neglecting the small force that changes the velocity with increasing r). Thus, $F_c = F_r$. Multiplying through by $N/\omega^2 r$ and rearranging gives

$$Nm\left[1 - \frac{v\rho}{m}\right] = Nf \quad (\text{velocity}/\omega^2 r) \tag{7-23}$$

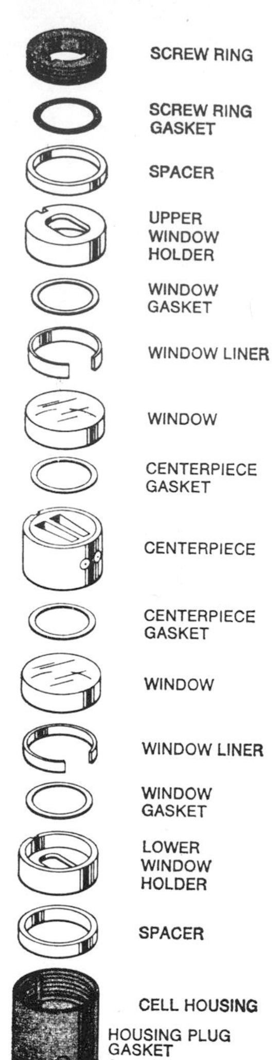

Fig. 7-5 The basic components of a double sector analytical ultracentrifuge cell. (Courtesy of Spinco Division of Beckman Instruments.)

The sedimentation coefficient s is defined as the velocity per unit field

$$s = (\text{velocity})/\omega^2 r = (dr/dt)/\omega^2 r = (d \ln r/dt)/\omega^2 \tag{7-24}$$

Substituting into Eq. 7-23 and using the fact that $Nm = M$ (molecular weight) and $v/m = \bar{v}$ (partial specific volume), one obtains

$$M = Nfs/(1 - \bar{v}\rho) \tag{7-25}$$

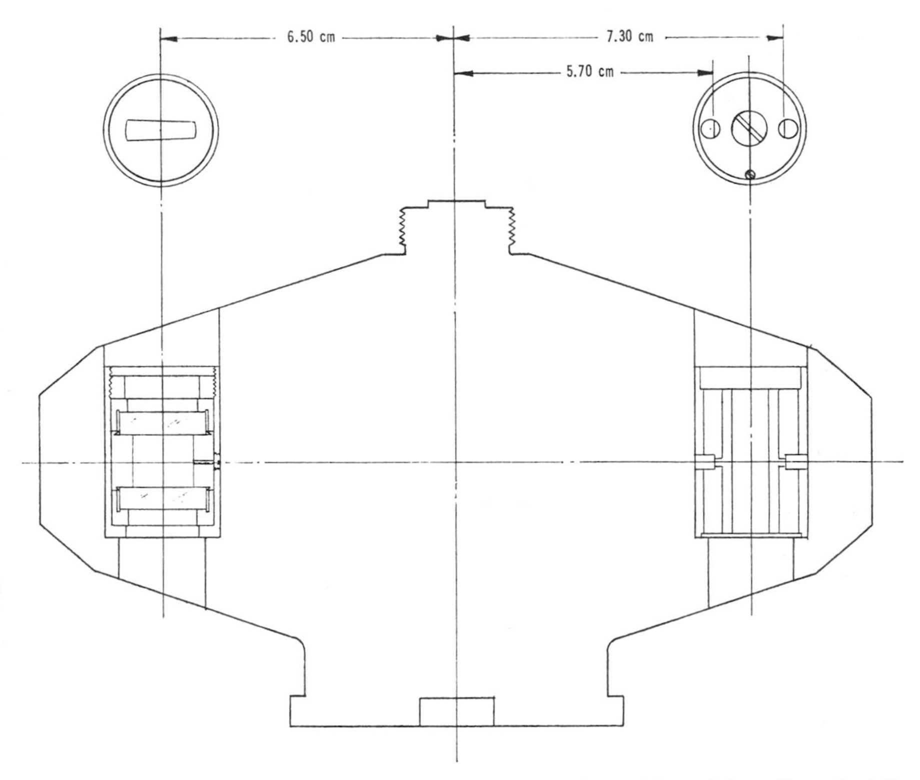

Fig. 7-6 Cross section of an analytical rotor showing the position of the cell on the left and a counterbalance on the right. Light passing the knife edges in the counterbalance holes produces a superimposed image on the photographic plate from which radial distances can be calculated. (Courtesy of Spinco Division of Beckman Instruments.)

This is the basic equation which relates the sedimentation coefficient to the molecular properties M, f, and $\bar{v}$. According to this equation, determination of s yields the ratio M/f. With the assumption that the particle is a rigid, nonhydrated sphere, a molecular weight can be calculated directly by using the Stokes' law expression for f (Eq. 7-17). In general this assumption is not valid, and a molecular weight determined in this way will represent only a minimum value.

The frictional coefficient appearing in Eq. 7-25 is identical with that in the diffusion equation ($D = kT/f$); therefore, the sedimentation and diffusion expressions may be combined to give

$$M = RTs/(1 - \bar{v}\rho)D \qquad \text{(Svedberg equation)} \qquad (7\text{-}26)$$

The use of sedimentation and diffusion data obtained from homogeneous solutions of a macromolecule will, therefore, give a true molecular weight independent of shape or degree of hydration. Once M is known, either s or D can be used to calculate the frictional coefficient, f. This quantity will give the dimensions of an equivalent particle, such as a prolate ellipsoid, provided the hydration is known. Sedimentation data can also be combined with viscosity measurements. Both methods are discussed later.

The sedimentation coefficient is usually expressed as $s_{20,w}$, referring to the standard condition of water as the solvent and 20°C. The conversion is made as follows:

$$s_{20,w} = s \frac{\eta}{\eta_{20,w}} \frac{(1 - \bar{v}\rho)_{20,w}}{(1 - \bar{v}\rho)} \tag{7-27}$$

Sedimentation coefficients are usually reported in Svedberg units ($1S = 10^{-13}$ sec). Most proteins have sedimentation coefficients in the range of 1 to 30S.

2-3 The Differential Equation of the Ultracentrifuge

The equations presented in the previous section cannot completely describe the concentration or concentration gradient curve as a function of r and t due to the assumptions implicit in the derivation. For example, the effects of diffusion and radial dilution in a sector cell were not considered.

The appropriate equation may be derived in a manner quite analogous to that used to obtain the diffusion equation, except that a term for the transport of mass by sedimentation must be included in the expression for dm/dt.

Sedimentation experiments in the ultracentrifuge are conducted in a sector-shaped cell so that concentration buildup at the cell walls and consequent convective disturbances may be avoided. The cell geometry is shown schematically in Fig. 7-7. The quantities r_m and r_b refer to the position of the solution meniscus and cell bottom, respectively. The amount of mass transferred per unit time due to sedimentation across the cell area ϕra at r is the product of the area, concentration, and velocity of molecules at r, namely

$$\frac{dm_s}{dt} = c(\phi ra) \frac{dr}{dt} \tag{7-28}$$

or

$$\frac{dm_s}{dt} = c(\phi ra) s\omega^2 r \tag{7-29}$$

since $dr/dt = s\omega^2 r$ by definition. That transported in the backward direction by diffusion is given by Fick's first law (Eq. 7-1) with the area (ϕra) replacing A. The net transport per unit time across r is the sum of Eqs. 7-29 and 7-1:

$$\frac{dm}{dt} = \phi ra \left[cs\omega^2 r - D \frac{\partial c}{\partial r} \right] \tag{7-30}$$

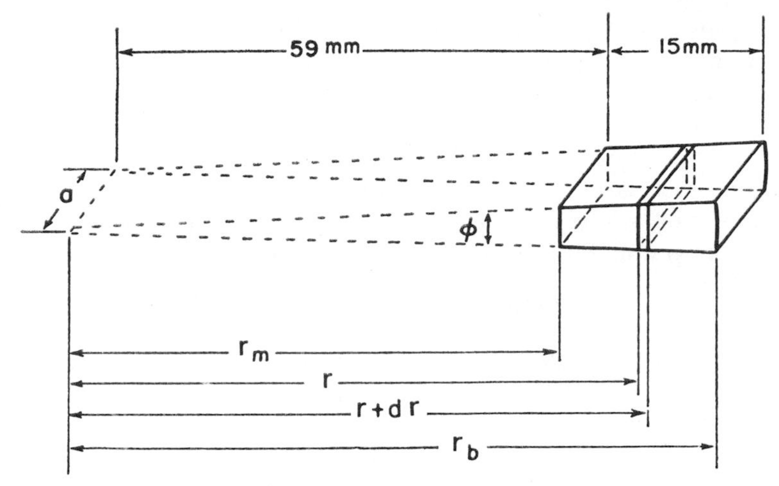

Fig. 7-7 Geometry of a sector-shaped ultracentrifuge cell. In the diagram, ϕ is the sector angle, a the thickness of the liquid column, r_m the position of the meniscus, and r_b the position of the base of the liquid column.[8]

As in the case of diffusion, the rate of mass transfer at $(r + dr)$ is given by $(dm/dt) + (\partial^2 m/\partial t\, \partial r)\, dr$, and the accumulation within the volume element between r and $r + dr$ is obtained by subtracting Eq. 7-30. The result is

$$\frac{\partial^2 m}{\partial t\, \partial r} = -\frac{\partial}{\partial r}\left[r\left(cs\omega^2 r - D\frac{\partial c}{\partial r}\right)\right]\phi a\, dr \tag{7-31}$$

The rate of concentration increase, $\partial c/\partial t$, is obtained by dividing by the volume, $(\phi r a)\, dr$, and the basic equation of the ultracentrifuge is obtained:

$$\frac{\partial c}{\partial t} = \frac{1}{r}\frac{\partial}{\partial r}\left[\left(D\frac{\partial c}{\partial r} - cs\omega^2 r\right)r\right] \tag{7-32}$$

Clearly, the functional form of D and s (e.g., as a function of c) must be obtained before a solution of Eq. 7-32 can be attempted. If s and D are not concentration dependent, they will also be independent of r and Eq. 7-32 may be written in the form

$$\frac{\partial c}{\partial t} = D\left[\frac{\partial^2 c}{\partial r^2} + \frac{1}{r}\frac{\partial c}{\partial r}\right] - s\omega^2\left[r\frac{\partial c}{\partial r} + 2c\right] \tag{7-33}$$

Solutions to Eqs. 7-32 and 7-33 have been obtained for a variety of initial and boundary conditions. In general, the solutions will require some approximations and often take the form of a series solution that is most conveniently evaluated numerically with the aid of a computer. The situation is further complicated for multicomponent systems that involve interactions between components (e.g., by chemical equilibria or dependence of s and D for one component on the concentration of other components). Although discussion of these solutions is beyond the scope of this text, many significant results have been obtained from them, some of which will be stated here without proof. The monographs by Fujita[9] and by Cann[11] may be recommended for those who wish to examine some of these results in more detail.

A limiting solution is obtained at $t = \infty$, representing an equilibrium state by integrating Eq. 7-32 after setting $\partial c/\partial t$ equal to 0. The result is

$$[D(\partial c/\partial r) - \omega^2 rsc]r = \text{constant} \tag{7-34}$$

where the right-hand side of the equation represents the constant of integration. Since the left-hand side of the equation is proportional to the number of molecules crossing any plane at r in a centrifugal direction, and since transport across the meniscus or bottom of the solution is 0, the integration constant must be 0. The equation

$$D(\partial c/\partial r) = \omega^2 rsc \tag{7-35}$$

therefore pertains to the equilibrium state. When this equation is compared with the expression for $(\partial c/\partial r)$ in the equilibrium state derived by a thermodynamic approach (cf. Chapter 8) one can show that

$$s/D = M(1 - \bar{v}\rho)/RT \tag{7-36}$$

This is again the Svedberg equation as obtained previously (Eq. 7-26) but, in this case, without the assumption of a functional form for D. The identity of the result by the two types of derivation provides justification for use of the Einstein-Sutherland equation ($D = kT/f$).

Another useful result obtained from Eq. 7-32 is the description of the concentration dependence of molecules in the plateau region that exists ahead of the sedimenting boundary in a sedimentation velocity experiment. In this region $\partial c/\partial r = 0$ and the plateau concentration c_p can be obtained from

$$\begin{aligned}\frac{dc_p}{dt} &= -\frac{1}{r}\frac{\partial}{\partial r}(\omega^2 sc_p r^2)\\ &= -2\omega^2 sc_p\end{aligned} \tag{7-37}$$

When Eq. 7-37 is integrated from $t = 0$ to time t, one obtains

$$\ln\frac{c_p}{c_0} = -2\omega^2 st$$

or (7-38)

$$c_p = c_0 \exp(-2\omega^2 st)$$

Alternatively, Eq. 7-24 can be integrated over t from zero to t and over r from r_m to $\bar{r}$, where $\bar{r}$ represents the average distance moved by molecules in the plateau region. This integration yields

$$\ln \frac{\bar{r}}{r_m} = \omega^2 st$$

Combination of this result with Eq. 7-38 gives the relation

$$c_p = c_0 \frac{{r_m}^2}{\bar{r}^2} \tag{7-39}$$

which is known as the radial dilution law. Since the concentration in the plateau region continuously decreases during sedimentation, it follows that if s is concentration dependent, the rate of sedimentation in the plateau region will vary with time. This may introduce a small error in the determination of s, but usually the effect is small enough so that it may be ignored.

2-4 Experimental Determination of S

In a normal sedimentation velocity experiment the centrifuge cell is filled with an initially homogeneous solution. Thus the concentration has a constant value (c_0) over the length of the cell. Upon application of the centrifugal field, macromolecules sediment away from the center of rotation and begin to pile up at the bottom of the cell while being depleted at the top of the cell (meniscus). Figure 7-8*a* illustrates the progress of such a high-velocity sedimentation run. A boundary is formed between the upper depleted region and the solution in the plateau region ahead of the boundary, and from its rate of movement down the cell, one may determine the sedimentation coefficient of the molecules. Hypothetical curves of c vs. r for sedimentation are shown in Fig. 7-8*b*. The sedimenting particles also undergo diffusion, and this produces a broadening of the sedimentation profile, which increases as the run progresses. Figure 7-8*c* shows the corresponding gradient (dc/dr) curves for this velocity run. Sedimentation is followed with one or more of the three optical systems supplied with the commercial ultracentrifuge. Figure 7-9 shows the type of photograph that would be obtained in a typical sedimentation velocity experiment using the different optical systems.

The position of the boundary is usually taken as the distance r of the maximum of the Schlieren curve from the center of rotation; the 50% point is used as the position of the boundary for the interference or absorption curves. A small error is introduced in the calculation of s when this definition of boundary position is used; however, it can be eliminated by making a mass transport calculation, as discussed below. The determination of r from a

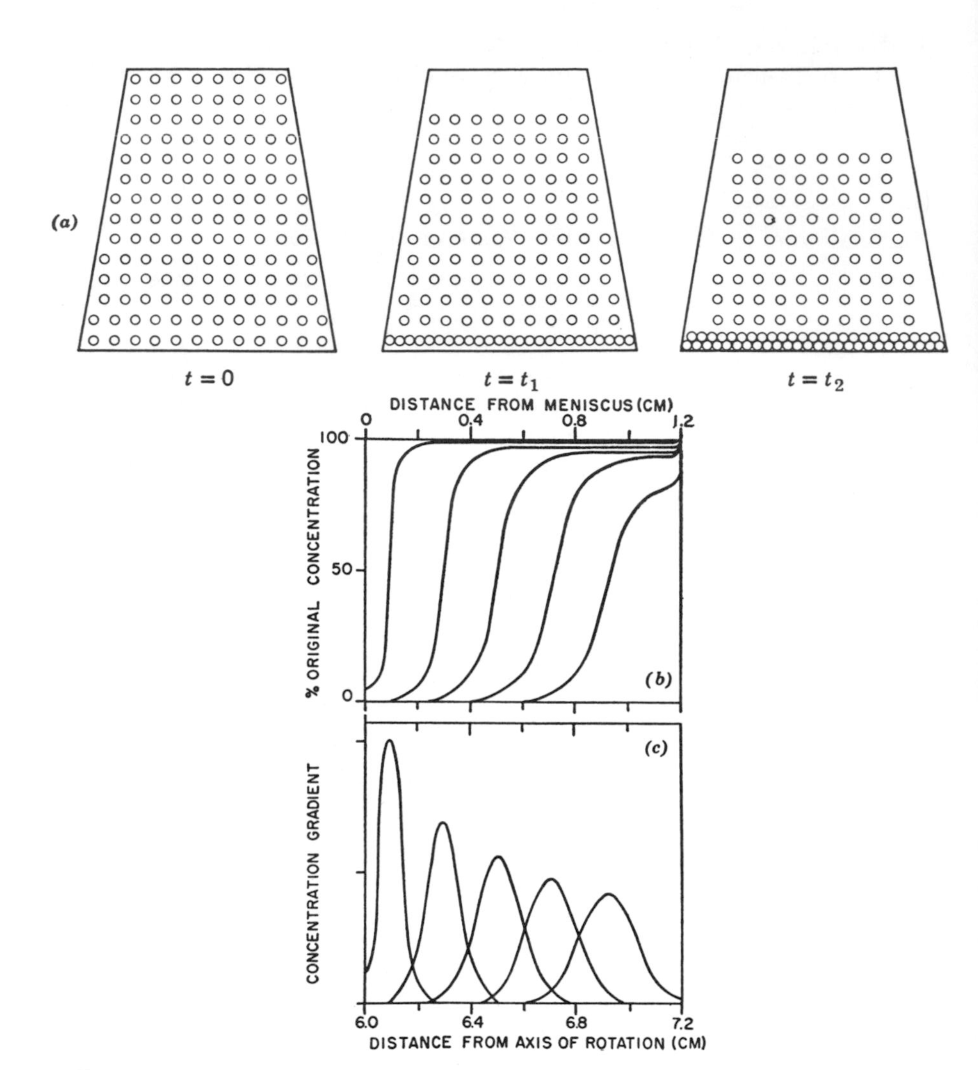

Fig. 7-8 (*a*) Idealized diagram showing the progress of a sedimentation velocity experiment.[29] Diffusion is assumed to be negligible. (*b*) Concentration vs. distance curves for a sedimentation velocity experiment.[8] With increasing time, the curves broaden due to diffusion and the concentration in the plateau region diminishes slightly because of radial dilution. (*c*) Concentration gradient curves corresponding to (*b*) above.[8]

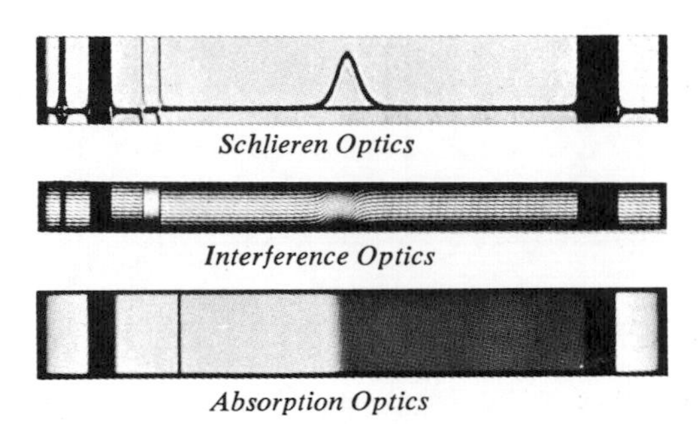

Fig. 7-9 Sedimentation velocity patterns obtained with the Model E ultracentrifuge. The top photograph is taken with the Schlieren optical system, the middle pattern is obtained with the Rayleigh system, and the lower pattern is from the ultraviolet photographic system. [Courtesy Beckman Instruments.]

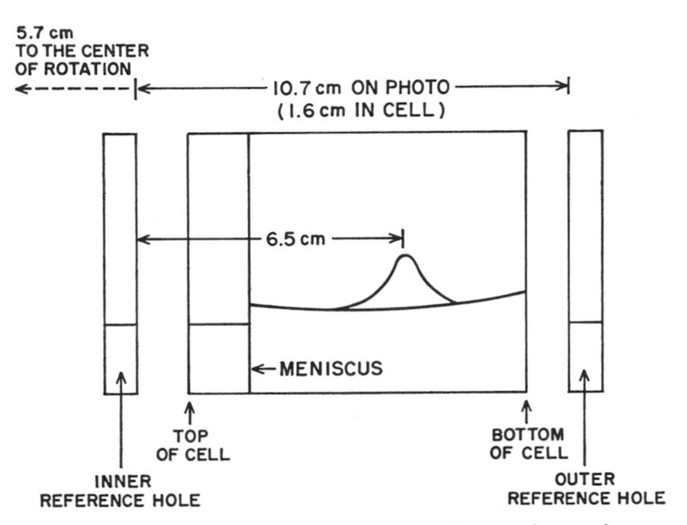

Fig. 7-10 Calculation of the radial position corresponding to the maximum ordinate of the Schlieren peak. Since 10.7 cm on the photograph corresponds to 1.6 cm in the cell, the distance of the maximum ordinate (shown at 6.5 cm) from the inner reference edge is $6.5 \times 1.6/10.7 = 0.95$ cm. The inner reference edge is 5.70 cm from the center of rotation, thus the maximum ordinate is $5.70 + 0.95 = 6.65$ cm from the center of rotation.

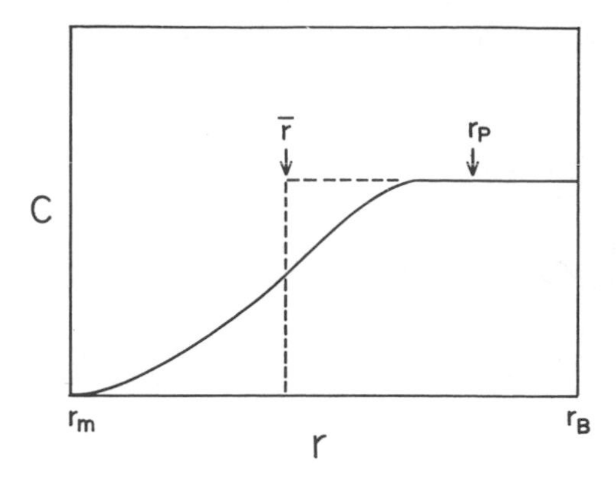

Fig. 7-11 Calculation of $\bar{r}$ from a concentration distribution of a solute during a sedimentation velocity experiment. The dotted line at $\bar{r}$ is chosen such that the amount of mass between $\bar{r}$ and r_p is identical to that between r_m and r_p along the solid curve.

Schlieren photograph is shown in Fig. 7-10. A plot of $\ln r$ vs. t gives a straight line with slope $\omega^2 s$, in accordance with Eq. 7-24, and thus s is obtained.

The above procedure affords a reliable measurement of s for components which exhibit reasonably sharp boundaries. However, in the case of slowly sedimenting species, for boundaries of complex shape or for interacting systems, it may be preferable to use an alternate method in which the average sedimentation coefficient of molecules in the plateau region is determined. This method does not depend on the shape of the boundary. Instead, the principle of conservation of mass is employed to obtain the necessary relationships.

The sedimentation of molecules in the plateau region may be described by the movement of a hypothetical infinitely sharp boundary located at $\bar{r}$ (Fig. 7-11). The value of $\bar{r}$ may be found by setting up an equation for total mass between r_m and r_p for the hypothetical case and comparing this with the experimental curve. For the infinitely sharp boundary, the concentration is 0 between r_m and $\bar{r}$ and has a value c_p between $\bar{r}$ and r_p. The total mass between the meniscus and r_p is the concentration times the volume elements summed over all volume elements, namely,

$$\text{total mass between } r_m \text{ and } r_p = c_p \int_{\bar{r}}^{r_p} \varphi r a \, dr$$

$$= \varphi a \frac{c_p}{2} (r_p{}^2 - \bar{r}^2) \qquad (7\text{-}40)$$

For the experimental curve

$$\text{mass} = \varphi a \int_{r_m}^{r_p} cr\, dr \tag{7-41}$$

where the integral may be evaluated numerically. Equating the two expressions and solving for $\bar{r}^2$ gives

$$\bar{r}^2 = r_p^{\,2} - \frac{2}{c_p} \int_{r_m}^{r_p} cr\, dr \tag{7-42}$$

A variety of alternative expressions for evaluating $\bar{r}$ have been tabulated by Trautman.[15] The most commonly employed alternative to Eq. 7-42 is

$$\bar{r}^2 = \frac{\int_{r_m}^{r_p} r^2(\partial c/\partial r)\, dr}{\int_{r_m}^{r_p} (\partial c/\partial r)\, dr} \tag{7-43}$$

which may be used to obtain $\bar{r}$ from Schlieren patterns provided that c_m is 0. The integrals are evaluated by numerical integration. The resulting values of $\bar{r}$ as a function of time are then substituted into Eq. 7-24 to obtain the sedimentation coefficient. In the case of polydisperse systems, an average s value ($\bar{s}$) will be obtained as defined by

$$\bar{s} = \frac{\sum_i c_i s_i}{\sum_i c_i} \tag{7-44}$$

where s_i is the sedimentation coefficient of the ith species and c_i is the concentration of the ith species in the units recorded by the optical system (e.g., based on absorbancy for the ultraviolet optical system or refractive index increment with the interference or Schlieren optical system).

2-5 Sedimentation of Polyelectrolytes

The sedimentation of charged molecules such as proteins may be complicated by two factors. The first, known as the primary charge effect, results from the large difference in sedimentation coefficient between the charged macromolecule and its counterions (which have negligibly small s values). Because of the requirement for electrical neutrality within each volume element, the counterions will tend to retard sedimentation of the macromolecule. In extreme cases the sedimentation coefficient may drop to about half its expected value. The problem is handled effectively by adding a supporting electrolyte to the solution. For most proteins the primary charge effect becomes negligible in buffers of ionic strength 0.1 to 1.0. The possible existence of a primary charge effect can be checked by running the sample as a function of ionic strength, provided that conformational changes in the protein due to the added salts do not occur.

A secondary charge effect results when the positive and negative ions of the supporting electrolyte have different (though small) s values. In this case, local electric fields are established due to charge separation. This acts as an additional force on the charged macromolecule, and s may decrease or increase, depending on the sign of the charge on the protein. This secondary charge effect can be largely circumvented in practice by carrying out the centrifugation at a pH near the isoelectric point of the protein and/or by using a supporting salt, such as NaCl, for which s of the anion and cation are similar. These charge effects are described in more detail in the reviews by Svedberg and Pedersen[7] and by Schachman.[8]

2-6 Interpretation of Sedimentation Velocity Patterns in Simple 2-Component Systems

Sedimentation velocity experiments are often performed with the intent of demonstrating that a highly purified protein is homogeneous. Such a demonstration is not as simple as is often supposed, and we present here some minimum criteria which, if fulfilled, lend credence to the claim of homogeneity. Clearly, a single solute component will migrate as a single boundary. The shape of the boundary, however, is dependent on several factors. If both s and D are independent of concentration, the shape of the boundary is given by the solution of Eq. 7-33. If sedimentation is sufficiently rapid so that the boundary is not disturbed greatly by a long period of restricted diffusion near the meniscus, the shape of the boundary is essentially identical to that expected for free diffusion. Consequently, the gradient curve $\partial c/\partial r$ should have a Gaussian shape and the concentration curve will have the form of the probability error function. Lamm (see, for example, Schachman's monograph[8]) has shown that the diffusion coefficient (D_{sed}) obtained from the sedimenting boundary by standard methods, as for free diffusion, will be related to the true D by

$$D = D_{\text{sed}}(1 - \omega^2 st) \tag{7-45}$$

for ideal solutions with a negligible period of restricted diffusion. It follows that D calculated in this manner should be identical to that obtained by diffusion alone for a homogeneous solute when s is independent of c. The boundary, however, may be skewed due to the effect of restricted diffusion at the meniscus, the imbalance resulting from the fact that molecules cannot diffuse beyond the meniscus. In this case the analysis is much more complex. It is preferable to carry out the experiment by forming an initially sharp boundary in a synthetic boundary cell and then apply the above criteria.

In cases where sedimentation is markedly concentration dependent, boundary shape is more difficult to interpret. In general, s values are found to decrease with increasing concentration. This lowered velocity at higher solute concentration is due in part to backward flow of solvent as particles move down the cell, as well as to density effects. However, the principal cause is apparently the effectively higher viscosity experienced by the particle because of its neighbors. The total effect often exceeds that expected on the basis of solution and solvent viscosity measurements. Because of this phenomenon, molecules in the trailing portion of a boundary tend to move at a greater velocity per unit field than molecules on the leading edge or in the plateau region. The result is that a "sharper" boundary is obtained than when the expected diffusion-broadening is taken into account. This phenomenon is referred to as "hypersharpening." If a series of experiments are conducted to ascertain the functional relationship between s and c (e.g., see Schachman[8]), the shape of the boundary can be correctly determined by solution of Eq. 7-32. Another approach is to obtain data at various total sample concentrations, extrapolate to zero concentration, and analyze the shape of the extrapolated c vs. r curve. For most, but not all, proteins, the concentration dependence of s is sufficiently small that it may be neglected when experiments are run at low concentrations (e.g., 0.05–0.5 mg/ml). These concentrations are generally too low for sensitive detection by the Schlieren optical system of the ultracentrifuge but are adequate when the ultraviolet system is used.

Boundary spreading may be conveniently analyzed by the probability paper method discussed earlier in this chapter and compared with the results obtained in a free diffusion experiment. It must be stressed that since heterogeneity tends to broaden the boundary, and s vs. c dependence tends to sharpen it, a mixture of the two effects may lead to fortuitous cancellation. Whenever D calculated in a sedimentation experiment is larger than that for free diffusion, heterogeneity in the sample is demonstrated. The finding that diffusion spreading is as narrow or narrower in a sedimentation velocity than in a diffusion experiment, on the other hand, is not in itself demonstration of homogeneity (with respect to sedimentation coefficients), due to possible hypersharpening. Several further experiments can then be performed; for example, the data for D_{sed} may be extrapolated to zero concentration. The latter analysis is seldom done today, probably because most investigators find alternate criteria (e.g., sedimentation equilibrium or electrophoretic methods) more convenient and reliable.

When dealing with real systems, it often happens that significant quantities of material sediment in the "presumed" plateau region or trail behind the boundary in a somewhat continuous manner. This may be difficult to detect directly, especially when the Schlieren optical system is employed. It is wise, therefore, to confirm that the concentration in the plateau region obeys the radial dilution law. If this procedure were employed more routinely, numerous interpretive errors could be avoided.

2-7 Multicomponent Mixtures

The presence of two or more solute components that are not in chemical equilibrium and that exhibit no dependence of their sedimentation on concentration gives rise to sedimentation velocity patterns that are simply the sum of those of the individual components analyzed separately. The interpretation of such patterns is then quite straightforward and, if the several boundaries are well separated during part of the experiment, the individual values of s as well as relative concentration (in the optical units employed) are easily obtained. It is also possible, though seldom done, to analyze overlapping boundary curves by numerical methods.

If one or more of the solute components exhibits s vs. c dependence, interpretation of the sedimentation pattern is more complicated. In a two-solute system, for example, s of each component will depend both on its own concentration and on that of the second component. It follows that the sedimentation coefficient of the slower sedimenting species is less in the plateau containing both types of species than it is in the plateau region between the boundaries. This results in a higher concentration of the slower moving component in the plateau region between the boundaries. Consequently, the apparent concentration of the faster species, based on the concentration difference between the two plateaus, is too low, while the apparent concentration of the slower moving species is high. This phenomenon, often termed the Johnston-Ogston effect, is described in detail in the review by Schachman.[8] Reliable estimates of sedimentation coefficients and relative concentration of components in such systems are obtained by extrapolation of results to zero concentration.

2-8 Sedimentation Velocity of Chemically Reacting Systems

Many highly purified proteins tend to undergo self-association, that is, some degree of reversible aggregation or dissociation, depending on the solvent system employed. Macromolecular equilibria of the type $n\mathrm{A} + m\mathrm{B} \rightleftharpoons A_n\mathrm{B}_m$ also occur. In some cases, one may wish to study these systems by sedimentation velocity methods in order to obtain thermodynamic parameters. More commonly, however, self-association is an inadvertent occurrence in the experimental system, even for highly purified proteins. In this case it is of great importance to recognize the problem, for otherwise the sedimentation pattern may be erroneously interpreted in terms of impurities. Clearly, the word homogeneity cannot be applied in the classical sense to such systems. Description of the shape of sedimentation boundaries in chemically interacting systems is difficult. However, such analyses are now being done on data from sedimentation velocity as well as other moving boundary methods. A brief discussion of these approaches is presented here; for more details the interested student is referred to the monographs by Fujita[9] and by Cann.[11]

In an equilibrium system the relative concentrations of species can be expressed as functions of total concentration. Since concentration varies across a boundary, the equilibria will be perturbed and the final shape of the boundary will depend in a complex way on the balance between the forces acting to separate various species and the tendency to reestablish equilibrium. Three general kinetic cases may be distinguished: those where the rate for establishing equilibrium is slow, similar in magnitude, or rapid compared to the time of the sedimentation velocity experiment. In the former case the observations of the equilibrium mixture at a particular concentration are similar to those of the noninteracting mixture discussed before. However, in general, the relative concentration of the slower sedimenting species will be expected to increase with decreasing concentration. This effect is opposite to that produced by the Johnston-Ogston effect.

In cases where equilibrium in an associating system is rapid compared to sedimentation, the characteristics of the sedimentation pattern cannot be easily predicted. The number of boundaries observed, for example, will depend on the mode of association (e.g., monomer–dimer or monomer–trimer), the total concentration, and the possible involvement of small nonsedimenting species (e.g., H^+ or salts) in the equilibrium. Fortunately, detection of self-association and partial characterization of these systems by sedimentation velocity is possible without taking boundary characteristics into account. The average sedimentation coefficient of molecules in the plateau region may be computed by determination of $\bar{r}$ as described earlier. Although radial dilution occurs in the plateau region, it is normally not sufficient to markedly disturb the equilibrium and thus may be neglected (or an adjustment can be made by use of an average concentration for the plateau). The average sedimentation coefficient determined in this manner for the associating system is related to the parameters of the system by the equation

$$\bar{s} = \frac{\sum_i c_1{}^i K_i s_i}{\sum_i c_1{}^i K_i} \tag{7-46}$$

where c_1 is the concentration of monomer, K_i is the equilibrium constant for the ith association, and s_i is the sedimentation coefficient of the ith species. The equilibrium condition, where the total concentration c is given by

$c = \sum c_1{}^i K_i$, must be fulfilled simultaneously. Comparison of these equations indicates that s will increase as a smooth function of c for an associating system, in contrast to the decrease expected for mixtures with s vs. c dependence. In simple cases such as monomer–dimer equilibria, Eq. 7-46 can be used directly to obtain the value of the equilibrium constant.

Determination of the s vs. c dependence is highly recommended as a routine test for association, regardless of the appearance of the boundary. The sedimentation velocity runs should be carried out at relatively large differences in initial concentrations to accentuate possible variation in $\bar{s}$.

2-9 Separation Cell Techniques

It is not always possible to determine the concentration of a substance of interest by optical methods. However, with the use of the separation cell, sedimentation coefficients can be determined from any measurable property (e.g., enzyme activity or radioactivity). In this type of cell the solution column is separated into two compartments by a partition which allows sedimenting molecules to pass but prevents mixing from occurring at the end of the experiment. Runs are made for different time periods, and the sedimentation coefficient is obtained from the ratio of the concentration (or other measured property depending on concentration) in the two compartments to that in the original solution, with the equation:

$$s = -\frac{1}{2\omega^2 t} \ln \left[\frac{r_0{}^2}{r_p{}^2} + \frac{c_t}{c_0}\left(1 - \frac{r_m{}^2}{r_p{}^2}\right)\right] \tag{7-47}$$

where r_p and r_m refer to the position of the partition and meniscus, respectively, c_0 is the original concentration, and c_t is the concentration in the upper compartment at time t. This type of relation is derived from the mass transport equations for the ultracentrifuge.[8]

2-10 Binding of Small Molecules to Macromolecules

The binding of small molecules (inorganic ions, coenzymes, substrates, inhibitors, dyes, detergents) to macromolecules is of great interest and can be measured in several ways in the ultracentrifuge. If the small molecule absorbs strongly in the ultraviolet, the ultraviolet optical system can be used. In general, the small molecule will have an absorption spectrum different from that of the protein, and therefore, with a monochromator to measure absorbancy at different wavelengths, the c vs. r curve of both components in the cell can be obtained and the binding determined (e.g., the binding of methyl orange to bovine serum albumin[12]). The binding of nonabsorbing molecules can be determined by interference optical methods if the extent of binding is sufficient to produce a measurable change in refractive index.[13] In other cases the separation cell is used and the small molecule is assayed by direct analysis. Radioactive labeling or microbiological assays (for vitamins or cofactors) can be used.

2-11 Conformational Studies

Protein denaturation produced by such agents as acid, base, detergents, urea, or other chemical treatments usually produces a change in molecular size through aggregation or dissociation, or a change in frictional coefficient due to shape alterations. These effects often can be conveniently followed by sedimentation velocity measurements. A sensitive procedure to measure small differences in sedimentation coefficient directly is available.[14]

3 ZONE CENTRIFUGATION

Sedimentation velocity centrifugation in a preformed density gradient has proved to be a powerful technique for the separation of macromolecules. In ordinary preparative centrifugation (carried out in an angle head rotor with separation made between the "pellet" and supernatant fractions) or in sedimentation velocity with a separation cell, only a fraction of the trailing component may be isolated in pure form. Separation of species differing widely in sedimentation coefficient can be achieved in some cases, but resolution is far from the boundary resolution in sedimentation velocity. It is the purpose of zone centrifugation to afford complete separation of species at a resolution comparable to the separation of boundaries in sedimentation velocity centrifugation.

3-1 Zone Centrifugation in Preformed Density Gradients

Figure 7-12 illustrates the preparation of a linear density gradient (5–20% sucrose) and its use in zone centrifugation. A rotor of the swinging bucket type is used in which the containers holding the gradient tubes swing out as the rotor is brought up to speed. Thus most molecules sediment down the tube without hitting the wall instead of tumbling down the walls as in the angle head rotors used in ordinary preparative centrifugation. It is essential that the gradient be steep enough so that convection does not occur as the macromolecules move down the gradient. This means that at any level the

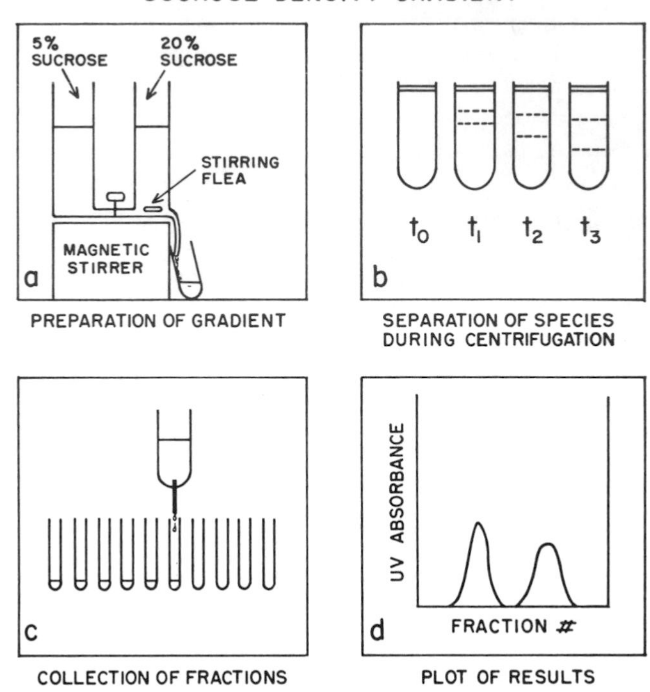

Fig. 7-12 (*a*) A nearly linear gradient in density is established by allowing 5% sucrose to flow by gravity into 20% sucrose (with mixing) as the mixture slowly flows into a centrifuge tube (for swinging bucket rotor). (*b*) Position of two macromolecular species with different sedimentation coefficients as a function of time of centrifugation. Time $\mathbf{t}_0$ represents the sample at the start of centrifugation. The sample solution, which must be less dense than 5% sucrose, has been carefully layered on top of the gradient. (*c*) Collection of samples as drops from a hole punctured in the bottom of the tube. Fractions may be analyzed for any suitable property (absorbance, enzymatic activity, radioactivity). (*d*) Plot of absorbance vs. fraction number. Fractions are plotted from left to right in the order of collection, thus the plot represents a direction of sedimentation from right to left.

density of the gradient solution plus the sedimenting molecules must be less than the density of solution just below. This prevents density inversion and consequent macroscopic convective disturbances. Diffusion, of course, acts to destroy the gradient; however, because the gradient is continuous, the concentration gradient of sucrose at any point is relatively low, and thus the rate of transport of sucrose by diffusion is extremely slow.

After the desired time of centrifugation, which depends upon the sedimentation rates of the species being separated, the distribution of materials in the gradient is determined. The gradient is quite stable and may be handled with minimum precautions; however, analyses must be made before the protein "zones" have undergone too much diffusion. Fractions can be collected (for example) by puncturing the bottom of the tube with a syringe needle and are analyzed for such properties as absorbancy, radioactivity, and enzymatic activity. The ability to obtain concentration profiles in any measurable unit characteristic of the protein is a major advantage of the technique. Figure 7-13 demonstrates the use of zone centrifugation for characterization of enzymes at different stages of purification.

Preparative ultracentrifuges are available with swinging bucket rotors capable of speeds up to 60,000 rpm; centrifugal fields are achieved that are comparable to those obtained in the optical analytical ultracentrifuge. Since solvents used for zone centrifugation are of higher density and viscosity than those normally chosen for analytical work, the rate of protein migration is decreased but diffusion is also decreased by the same degree. The layered zone will diffuse at both its leading and trailing edge, consequently resolution will be somewhat less than that obtained for boundaries in normal sedimentation velocity. With narrow starting zones, however, resolution in zone centrifugation can be quite impressive, especially for large proteins.

The calculation of sedimentation coefficient in zonal gradient centrifugation is possible from theoretical considerations[16] or may be estimated by comparing the position of an unknown protein in the gradient with the position of calibration proteins of known sedimentation coefficient run under identical conditions.

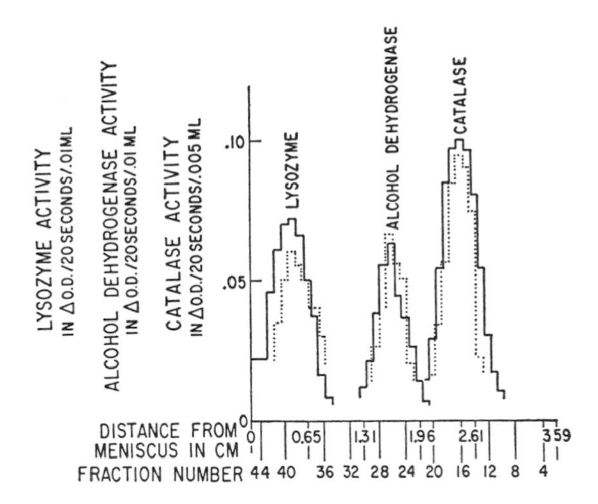

Fig. 7-13 Separation of three enzymes by sucrose gradient sedimentation of the purified enzymes (solid lines) and in a crude bacterial extract (dotted lines). Experimental details are given by Martin and Ames.[16]

3-2 Band Centrifugation

The term band centrifugation is used to refer to a technique first described by Vinograd and his associates[17] for zone centrifugation in an optical ultracentrifuge without employing preformed gradients. The sample is placed in a small chamber connected to the main cell chamber by a small capillary. At low speed the sample solution flows through the small channel and is layered on top of a denser solvent. As the sample migrates into the solvent, it is stabilized against convection by a diffusion gradient near the starting zone and later by redistribution of the salts in the centrifugal field. This procedure requires only a fraction of the material necessary for a normal sedimentation experiment and has been principally applied to the study of nucleic acids and viruses. Few applications to proteins have thus far been reported.

4 THE VOLUME OF PROTEINS IN SOLUTION[18,19]

Interpretation of diffusion and sedimentation data requires accurate knowledge of solute partial specific volumes. Proper evaluation of the $(1 - \bar{v}\rho)$ term is also required for the determination of molecular weights by the sedimentation equilibrium methods to be discussed in the next chapter. It can be shown that for a simple two-component system containing one solute dissolved in one solvent, $\bar{v}$ is the thermodynamically defined partial specific volume of the solute, that is, the volume increase obtained by adding 1g of dry isoionic protein to a large excess of solvent. Although the two-component approximation may be experimentally valid for some protein solutions, it is obvious that it cannot be theoretically precise since all proteins are charged and experiments are usually conducted in the presence of a moderate concentration of supporting electrolyte and buffer to minimize charge effects. We will first discuss the two-component approximation and then show how precise molecular parameters may be obtained for all systems by replacing the $(1 - \bar{v}\rho)$ term with a density increment.

The partial specific volume is obtained by measurements of the apparent specific volume ϕ, which is related to density measurements by

$$\phi = \frac{1}{W_s}\left(\frac{1}{\rho} - \frac{1 - W_s}{\rho_0}\right) \tag{7-48}$$

where W_s is the weight fraction of solute and ρ and ρ_0 are the density of the solution and solvent, respectively. For solutes with specific volumes that are concentration independent, ϕ is equal to $\bar{v}$. Otherwise the apparent specific volume must be extrapolated to infinite dilution to obtain the thermodynamic specific volume. Similarly, if the molecular weight of a solute is known, one may obtain the apparent molal volume (i.e., the volume occupied in solution by 1 mole of solute) Φ, from the relation:

$$\Phi = \frac{1000}{c}\left(\frac{\rho - \rho_0}{\rho}\right) + \frac{M}{\rho} \tag{7-49}$$

where c is the concentration in moles per liter of the solute with molecular weight M. A number of important observations related to the volume occupied by proteins in solution are discussed in detail by Cohn and Edsall[18] and will be summarized here.

1. A comparison of the partial molal volume of many organic molecules provides justification for assigning molal volume increments to certain groups and to atoms. For example, two compounds differing only by a CH_2 group will differ in volume by about 16.1 cc/mole.
2. The apparent volume of a molecule is always about 13 to 14 cc/mole greater than the sum of the volumes of the constituent atoms, regardless of the size of the molecules. This additional volume is termed the "covolume."
3. If the molecule is charged, the volume is less than that calculated as the sum of atoms plus the covolume due to closer packing of water molecules in the hydration shell of the ion than in bulk solvent. The magnitude of this effect, called electrostriction, depends on the ion in question and on its separation from nearby charged groups. The electrostriction produced by amino acids averages about 13.3 cc (11.5–14.7 cc).

On the basis of the empirical observations discussed above, it is possible to obtain estimates of partial molal volumes of amino acid residues in proteins. This may be done by taking the sum of the group volumes or by adjustment of measured values for free amino acids to account for elimination of water in the peptide bond and for electrostriction and covolume effects. Volume measurements on series of homopolymers may also be used. The results for amino acid residue volumes are given in Table 7-1. If it is assumed that the volume of a protein is the sum of the amino acid residues (the covolume is a negligible fraction of the total volume), then the partial specific volume may be calculated from amino acid composition data by

$$\bar{v} = \frac{\sum_i \bar{v}_i W_i}{\sum_i W_i} \tag{7-50}$$

where W_i and $\bar{v}_i$ are the weight fraction and partial specific volume of the ith residue. When this calculation is compared to measured values of $\bar{v}$ for most proteins, the agreement is quite good, from which it may be inferred that few "holes" exist in the protein structure and nearly all atoms are in van der

Table 7-1 Partial specific volumes of amino acid residues.[18]

Amino acid	$\bar{v}_{residue}$
Glycine	0.64
Alanine	0.74
Serine	0.63
Threonine	0.70
Valine	0.86
Leucine, Isoleucine	0.90
Proline	0.76
Methionine	0.75
Phenylalanine	0.77
Cystine	0.61
Tryptophan	0.74
Tyrosine	0.71
Histidine	0.67
Arginine	0.70
Lysine	0.82
Aspartic acid	0.60
Glutamic acid	0.66
Glutamine	0.67

Waals contact with each other or with solvent molecules. This agreement has also prompted many investigators to calculate $\bar{v}$ rather than measure it; the result is subject to considerable error and should be used only when direct determination is not possible.

In two-component systems the molecular weight of the solute determined by using $\bar{v}$ in conjunction with s and D (for example) refers to the anhydrous molecule, even though the molecule is hydrated under the conditions of measurement. This can be understood if we assume that $\bar{v}$ for the hydrated protein ($\bar{v}_H$) of hydrated molecular weight M_H can be expressed as the weighted average of $\bar{v}$ and the partial volume of the solvent. It can be shown that

$$M_H(1 - \bar{v}_H\rho) = M(1 - \bar{v}\rho) + xM_s(1 - \bar{v}_s\rho) \tag{7-51}$$

where x is the number of moles of solvent bound per mole of protein, M_s is the molecular weight of the solvent, and $\bar{v}_s$ is the partial volume of the solvent. At infinite dilution $\bar{v}_s$ is equal to $1/\rho$, and the second term on the right-hand side of Eq. 7-51 vanishes.[8]

In systems of three or more components, solvent bound to the protein may have a density different from that of the bulk solvent. For example, the preferential binding of solvent components such as ions or the preferred binding of one component of mixed solvents (e.g., of urea or of water in 8 *M* urea solutions) will lead to substantial errors in molecular weights determined on the basis of an ordinary value of $\bar{v}$ (calculated or measured). Such multicomponent systems have been treated rigorously by thermodynamic methods and an unambiguous approach based on experimental measurements has been reviewed by Casassa and Eisenberg.[19] For simplicity we limit our discussion here to the case of a single macromolecular component, usually termed component 2, in a solution containing any number of substances of low molecular weight (e.g. salts, urea) in addition to the solvent.

Component 2 has a molecular weight that is the sum of those of the individual amino acid residues. It is the species that exists (though it may be hydrated) for the isoionic protein in pure water, with only H^+ and OH^- available as counterions. One may also define a component 2* that, when added to a dialysis bag containing solvent in dialysis equilibrium with more solvent outside the osmotic membrane, causes no net transport of diffusible species. Component 2* of molecular weight M^* may be viewed in a simplified physical way as a particle consisting of component 2 plus the quantity of diffusible substances that represents the imbalance between diffusible components "bound" to component 2 and those present in bulk solvent (i.e., it contains only the amount preferentially bound, not the total solvent components bound in the hydrodynamic particle). It can be shown that for these multicomponent systems, the term $(1 - \bar{v}\rho)$ applicable to two-component systems should be replaced with $1 - \phi'\rho$ in which ϕ' is defined by

$$\phi' = \frac{1}{\rho_0}\left(1 - \frac{\Delta^*\rho}{c_2}\right) \tag{7-52}$$

where ρ_0 is the solvent density, $\Delta^*\rho$ is the difference in the density of solution and solvent at dialysis equilibrium, and c_2 is the concentration of component 2 in weight per volume units. Substitution of the right-hand side of the above equation for ϕ' gives

$$1 - \phi'\rho = \frac{\Delta^*\rho}{c_2} \tag{7-53}$$

More precisely, the density increment should be expressed as $d\rho/dc$ at constant chemical potential, but we will assume here that ϕ' is independent of concentration so that $\Delta^*\rho/c_2$ is equal to the thermodynamic density increment.

By measuring the dry weight of known volumes of solution and dialyzate, the concentration $c_2{}^*$ of component 2* is obtained, and the analog to ϕ' for the specific volume of this component ϕ^* may be defined as

$$\phi^* = \frac{1}{\rho_0}\left(1 - \frac{\Delta^*\rho}{c_2^*}\right) \tag{7-54}$$

It can be shown that

$$M_2(1 - \phi'\rho) = M_2\frac{\Delta^*\rho}{c_2} = M_2^*(1 - \phi^*\rho) = M_2^*\frac{\Delta^*\rho}{c_2^*}$$

In a multicomponent system the molecular weight obtained (from s/D or from sedimentation equilibrium) therefore depends on the manner in which the buoyant density term is determined. Once $\Delta^*\rho$ is known, the experimental data give M/c unambiguously; the resulting value of molecular weight then depends on the concentration measurement. Clearly other more arbitrary values can be obtained in addition to M_2 and M_2^*. For example, if c is expressed in terms of an extinction coefficient that has not been related to dry weight, one can still obtain molecular weight data that permit the calculation of the number of moles of component 2 in the same solvent, provided the extinction coefficient is measured identically. In this manner the moles of coenzyme or inhibitor bound per mole of enzyme can be calculated without determining the molecular weight of component 2 *per se.*

The measurement of density increments is not easy. $\Delta^*\rho$ may be obtained pycnometrically if substantial amounts of protein are available.[20] The use of a magnetic balance[21] or of a frequency densitometer[22] requires special apparatus, though less protein is needed. The Cartesian diver method developed by Hunter[23,24] should prove particularly useful for density measurements on moderate amounts of protein. The determination of density by the location of droplets in a density column of organic liquids[25] is often hazardous because of minor redistribution of some components between phases.

Experimental measurement of the concentration of component 2 is itself difficult to achieve in some cases. If the amino acid composition is known (including amide nitrogen) and if the solvent is free of ammonia and other nitrogen containing compounds, c_2 can be obtained from a nitrogen determination of a known volume of the solution used for the density experiments. The absorbancy due to protein may usually be obtained with considerable accuracy and may be used to measure c_2 if the extinction coefficient has been related to c_2 in the solvent of interest. This in turn will require a dry weight measurement of the isoionic protein. These dry weights may be difficult to obtain without special precautions[24,26] because of the extreme tendency for many dried proteins to absorb water or because the protein tends to precipitate at the isoionic point.

5 VISCOSITY[27–29]

Viscosity is a measure of the work that must be done to produce a given rate of flow in a liquid, that is, it represents the resistance to liquid flow. The viscosity of a liquid is, in general, increased by the presence of a solute, the magnitude of the effect depending upon the concentration and structural properties of the solute particles. Viscosity measurements constitute another hydrodynamic method for the study of macromolecules.

5-1 Measurement of Viscosity

A convenient apparatus for the measurement of viscosity is the Ostwald viscometer shown in Fig. 7-14. The solution introduced in the lower bulb is drawn up by suction past the scratch mark (*a*) then allowed to flow back. The time at which the meniscus passes the two scratch marks *a* and *b* is carefully noted. The time interval, called the outflow time, is dependent on the geometry of the apparatus and is related to the viscosity as follows:

$$\eta = \frac{Pgr^4t\pi}{8lV} \qquad \text{(Poiseuille's equation)} \qquad (7\text{-}55)$$

where P is the pressure head, g is the acceleration of gravity, l and r are the length and radius of the capillary, respectively, and V is the volume between a and b. It is interesting to note that the outflow time is inversely proportional

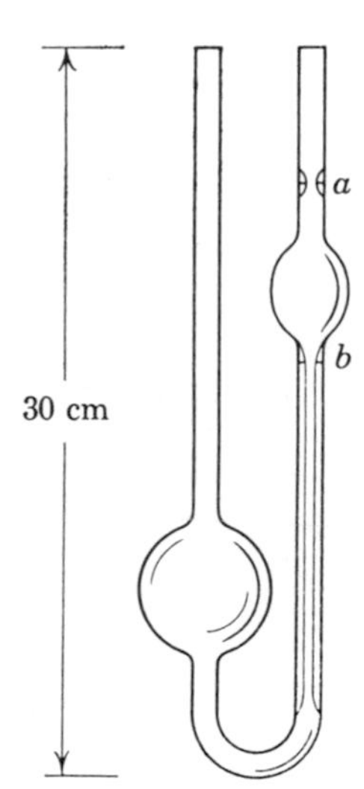

Fig. 7-14 Diagram of a typical Ostwald viscometer. (From H. D. Crockford and S. B. Knight, *Fundamentals of Physical Chemistry*, Wiley, New York, 1959.)

to r^4, and therefore the rate of flow of liquid becomes negligible for small capillaries. It is on this basis that convective flow can be eliminated in chromatography, zone electrophoresis, or in the Northrup diffusion cell (cf. diffusion).

Although it is difficult to measure absolute viscosities by this method, the ratio of η for an unknown solution to η_0, the viscosity of a reference standard (such as water), is readily obtained from Eq. 7-55, using the fact that P is proportional to the density ρ:

$$\eta/\eta_0 = (\rho/\rho_0)(t/t_0) \tag{7-56}$$

where t and t_0 are the outflow times for the unknown and the standard solutions, respectively, both measured at the same temperature in the same viscometer. If η_0 is known at that temperature, η is readily calculated. The solution densities are determined by weighing in a pycnometer.

5-2 Effect of Macromolecules on Liquid Flow

In normal Newtonian liquid flow there are no discontinuities in the velocity profile (i.e., no slippage), and therefore the velocity of the liquid directly adjacent to a solid is identical with that of the solid. The velocity profile of the flow in a capillary is given by a parabolic curve going from zero velocity at the walls to a maximum at the center of the capillary. When macromolecules are present in the solution, the flow pattern is disturbed where the macromolecules intersect the flow lines.

5-3 Ways of Expressing Viscosity Data

In general, the viscosity of a solution relative to that of the solvent may be expressed in powers of concentration:

$$\eta/\eta_0 = 1 + Ac + Bc^2 + Cc^3 + \cdots \tag{7-57}$$

where the coefficients A, B, C, . . . , which depend on the nature of the solute, are empirical parameters. Experimental data are usually expressed in terms of one of the following:

$$\text{Relative viscosity} \quad \eta_{\text{rel}} = \eta/\eta_0 \tag{7-58}$$

$$\text{Specific viscosity} \quad \eta_{\text{sp}} = \eta_{\text{rel}} - 1 \tag{7-59}$$

$$\text{Reduced viscosity} \quad \eta_{\text{red}} = \eta_{\text{sp}}/c \tag{7-60}$$

$$\text{Intrinsic viscosity} \quad [\eta] = \lim_{c \to 0} (\eta_{\text{sp}}/c) \tag{7-61}$$

5-4 Spherical Particles

A theoretical treatment of viscosity by Einstein showed that for a dilute solution of large, rigid spheres, the relative viscosity is

$$\eta/\eta_0 = 1 + 2.5\Phi \tag{7-62}$$

where Φ is the volume fraction of particles. According to this equation, the viscosity increment produced by spherical particles does not depend on their size but only on the total (hydrodynamic) volume they occupy in solution. This simple volume dependence may be understood by realizing that the important factor increasing the viscosity is the number of flow lines intersected by the solute molecules, a function of their total volume in solution. For particles free of bound solvent the value of Φ may be calculated from the concentration and partial specific volume of the solute, that is, $\Phi = c\bar{v}$.

5-5 Hydration

Proteins in aqueous solution are normally hydrated and thus their hydrodynamic volume (time average of molecular volume plus volume of immobilized solvent) is greater than that indicated by the partial specific volume $\bar{v}$. The effect of particle hydration on viscosity can be directly calculated in terms of its effect on the solute volume fraction. For example, if w grams of solvent of density ρ are bound per gram of anhydrous solute, this bound solvent will contribute an additional factor of $(w/\rho)c$ in units of cc/cc solution to Φ. Hence, for spheres

$$\eta_{\text{sp}} = 2.5[c\bar{v} + (w/\rho)c] \tag{7-63}$$

or with extrapolation to zero concentration,

$$[\eta] = 2.5\bar{v}\left[1 + \frac{w}{\bar{v}\rho}\right] \tag{7-64}$$

If the solute molecules are known to be spherical by other criteria, the viscosity determination will give an approximate value for hydration.

5-6 Other Particles

The Einstein relation does not hold if the particles are anisometric (nonspherical) or deformable. (Another source of deviation is the electroviscous effect that occurs if the particles possess charged groups, but this can be minimized by addition of salt.) Calculation of the intrinsic viscosity has been possible for a number of structural models (e.g., long thin rods, random coils). One of the earliest treatments is due to Simha who considered the variation in $[\eta]$ with axial ratio for ellipsoids of revolution and suggested that a variable coefficient ν, termed the shape factor, be used instead of the value 2.5 in the Einstein relationship. Thus intrinsic viscosity would be expressed as

$$[\eta] = \nu\bar{v}$$

The coefficient ν approaches 2.5 as the axial ratio approaches unity. Since $[\eta]$ also depends on hydration, a new coefficient ν', which is dependent both on shape and hydration, may be defined (the terminology used here is that given by Edsall[28]). Consequently for any real protein (with concentration measured in grams per 100 ml):

$$[\eta] = \lim_{c \to 0} [(\eta/\eta_0 - 1)/c] = \nu'\bar{v} \quad (7\text{-}65)$$

The numerical value of ν' and ν differ by the hydration term $(1 + w/\bar{v}\rho)$. If the particle is assumed to be spherical ($\nu = 2.5$), an upper limit to the hydration term may be obtained from the viscosity measurements. Evaluation of the shape factor will be discussed later.

Although viscosity measurements can yield information on hydrodynamic parameters of molecules, they are often more valuable for comparative studies. The method is simple and convenient for monitoring changes in shape, such as a native protein → random coil transition, because the intrinsic viscosity is more sensitive to expansion and contraction of molecular domains than are s and D, for example. As will be discussed later, the intrinsic viscosity of some models (i.e., flexible polymer) may be related semi-empirically to the molecular weight of the polymer.

It is possible that asymmetrical particles will tend to assume a preferential orientation in the velocity gradient during a viscosity experiment. Also, high velocity gradients tend to cause deformation in flexible polymers. Use of Simha's equations requires that no preferred orientation or deformation occurs under the conditions of the experiment. To eliminate this source of error, viscosity may be measured as a function of the shear gradient and extrapolated to zero shear. On the other hand the study of orientation in a velocity gradient or of particle deformation may be used to obtain additional information about proteins. For example, non-Newtonian viscosity caused by orientation of anisometric particles can be used to measure the principal component of the rotational diffusion coefficient.

6 ROTATIONAL DIFFUSION[28]

In addition to the translational movement of molecules in solution caused by the Brownian motion of the solvent, each solute molecule is also caused to rotate relative to its center of mass. This motion is described in terms of a rotational (or rotary) diffusion coefficient Θ, expressing the average angle of rotation per unit time. In general Θ, which is related to the dimensions of the particle, is a tensor quantity with principal components in three directions; in anisometric molecules only one component is usually measurable. Like the translational diffusion coefficient, Θ may be expressed in terms of a frictional coefficient

$$\Theta = kT/f_{\mathrm{rot}}$$

The determination of Θ is particularly valuable in the case of molecules of high axial ratio (length/width) such as long thin rods, since it can give a precise measure of the length of the particle in solution.

The rotational diffusion coefficient can be obtained from a variety of measurements. Those with the widest application have been flow birefringence, electric birefringence, and polarization of fluorescence.

The first two methods named depend upon the differences with direction (anisotropy) of refractive index produced when certain kinds of molecules are oriented in solution. The phenomenon is produced by molecules that are intrinsically optically anisotropic or by anisometric particles that are themselves isotropic, provided that the refractive index of the particles differs from that of the solvent. Proteins in general fall into the latter category; thus proteins of nonspherical shape can be studied by the birefringence methods. Although electric birefringence yields rotational diffusion coefficients, its more important application is the study of dipole moments, and discussion of the method is deferred to Chapter 10. Fluorescence methods for the measurement of rotational diffusion are not limited to nonspherical proteins and are considered in Chapter 9.

6-1 Flow Birefringence Measurements

The solution to be studied is placed between two concentric cylinders, one of which (usually the outer cylinder) is rotated while the other remains stationary. This sets up a velocity gradient (v varying linearly from zero at the inner cylinder to a maximum at the outer cylinder) that produces a torque on the molecules tending to align their long axes along the flow lines, perpendicular to the radial direction. The higher the axial ratio of the particle (length/width), the greater the degree of alignment will be.

The alignment of the molecules is determined by passing parallel light from a polarizing prism (the polarizer) through the solution, and then through another polarizing prism (the analyzer) placed at right angles to the first. When the particles are randomly oriented, the solution is optically isotropic; the incident light then passes through without loss of linear polarization and can be blocked by the analyzer (Fig. 7-15*a*). However, as the outer cylinder

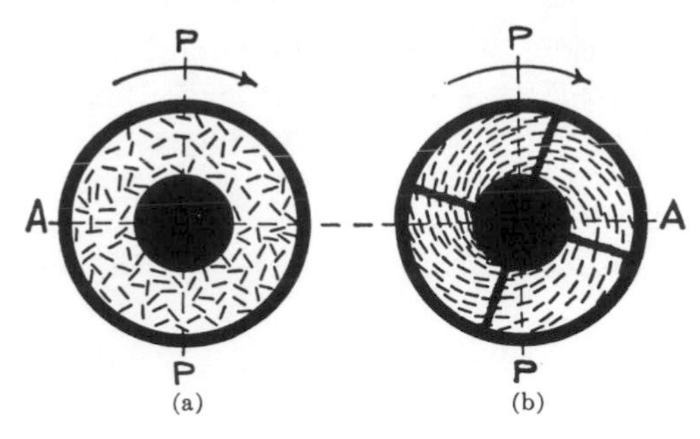

Fig. 7-15 Flow birefringence: determination of the extinction angle χ from the position of the cross of isocline with respect to the polarizer (P) and analyzer (A) planes.[18]

is rotated and the particles align along the flow lines, the solution becomes birefringent, that is, the refractive index of the solution is no longer the same in all directions. The transmitted light then becomes elliptically polarized, and some light passes the analyzer. For a very slight degree of orientation, weak light comes through everywhere except at the four positions at about 45° to the polarizer axis where the average orientation of the molecules is at 45° to that axis. These four dark areas, which correspond to no birefringence, form a cross called the *cross of isocline*. As the orientation is increased the dark cross, which can be seen more clearly because more light is passing through around it, moves toward the cross formed by the polarizer and analyzer (Fig. 7-15*b*).

The position of the cross is measured by the extinction angle χ, which ranges from 45° (no orientation) to zero when the dark cross nearly coincides with the cross formed by the polarizer and analyzer (complete orientation). The cross becomes sharper with increasing orientation until it becomes vanishingly thin as complete orientation is approached. Determination of Θ does not require complete orientation of the particles, but is made from a plot of χ vs. the shear gradient G, which depends on the speed of the outer cylinder and the distance between the cylinders.

6-2 Relation of Θ to Length

In general Θ is a complex function of the molecular dimensions; however, if the particle can be reasonably approximated by a simple model, it may be possible to obtain the principal dimension of the molecule. For a prolate ellipsoid of revolution with large axial ratio ($a/b > 5$), for example, the length ($2a$) can be obtained from Θ_b representing the component for rotation about the b axis, with Perrin's equation:

$$\Theta_b = \frac{3kT}{16\pi\eta a^3}\,[2\ln(2a/b) - 1] \tag{7-66a}$$

This equation may be contrasted with that for a sphere of radius r which has only one component of Θ:

$$\Theta = kT/8\pi\eta r^3 \tag{7-66b}$$

At the present time rotational diffusion of spherical macromolecules can be studied only by fluorescence methods (cf. Chapter 9).

7 SUMMARY AND INTERPRETATION OF HYDRODYNAMIC METHODS

The techniques presented in this chapter may be individually applied to the study of protein structure in a variety of ways. In addition to the determination of the molecular weight (also obtainable by methods to be discussed in the following chapter) one may obtain the frictional ratio f/f_0, the viscosity factor ν, and the rotational frictional coefficient f_{rot}. These factors depend on the shape and hydration of the real hydrodynamic particle in a complex way. In order to interpret the results it is necessary to relate these parameters to a theoretical model whose properties closely approximate those of the protein in solution.

We discuss briefly below the two structural models, other than the sphere, that are most frequently encountered in the study of protein systems, the prolate ellipsoid of revolution and the random coil. Analysis is normally carried out to determine the best fit between observed data and the parameters of the standard structural models. If the model chosen to represent the protein is grossly incorrect, the same model parameters will usually not fit (for example) the experimentally determined value of f/f_0 and ν simultaneously. Some uncertainty may remain, however, because of inaccuracy in the data or because two or more quite dissimilar models may fortuitously give a similar fit of the data. Additional proof for the validity of a chosen hydrodynamic model may be obtained by light scattering or low-angle X-ray scattering. These methods are discussed in the next chapter.

7-1 The Prolate Ellipsoid Model

The prolate ellipsoid is shown in Fig. 7-16. It is the solid volume swept out when an ellipse of major axis a and minor axis b is rotated about the a axis. The model thus appears cigar-shaped in contrast to the oblate ellipsoid formed by rotation about the b axis, which looks like a flying saucer. At one extreme ($a \gg b$) the prolate ellipsoid becomes an approximation for a long thin rod structure, and at the other ($a \rightarrow b$) it represents a nearly spherical shape. Most proteins in their native form in solution fall into the general category of the prolate ellipsoidal shape.

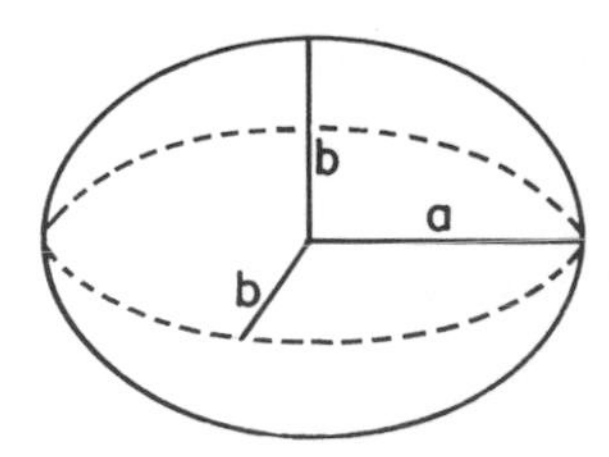

Fig. 7-16 The prolate ellipsoid model approximation to the shape of proteins. The model is obtained by rotation of an ellipse of major axis **a** and minor axis **b** about the **a** axis.

The dependence of the frictional coefficient of prolate ellipsoidal particles on axial ratio has been derived by Perrin. The result, expressed as the asymmetry part of the frictional ratio (cf. Eq. 7-19), is

$$(f/f_0)_A = \frac{[1-(a/b)^2]^{1/2}}{\left(\dfrac{a}{b}\right)^{2/3} \ln\left[1+\left(1-\left[\dfrac{a}{b}\right]^2\right)^{1/2}\right]\Big/(a/b)} \tag{7-67}$$

The effect of asymmetry in prolate and oblate ellipsoids on the viscosity increment ν has also been determined. For example, for prolate ellipsoids of very large axial ratio:

$$\nu = \frac{(a/b)^2}{15(\ln 2a/b - \frac{3}{2})} + \frac{(a/b)^2}{5(\ln 2a/b - \frac{1}{2})} + \frac{14}{15} \tag{7-68}$$

Tables of ν based on Simha's equations (see Reference 28) may be used with the experimental determinations of D and $[\eta]$ to determine (a/b) on the basis of a prolate ellipsoid model, provided an estimate for hydration can be made. If it is assumed that there is no hydration [i.e., $(f/f_0)_H = 1$ and $\nu = \nu'$], an upper limit for (a/b) can be established, since both factors act in the same direction.

Conversely, an upper limit value for hydration can be ascertained from the viscosity data by assuming the spherical model (cf. Eq. 7-64). A third alternative, which is sometimes employed, is to take a reasonable value for hydration of a protein (e.g., 20–40%) and then calculate the axial ratio using Eqs. 7-67 and 7-68. The merits and caution required in this approach are discussed in detail by Tanford.[29] The numerical values of axial ratio and hydration for various values of f/f_0 and ν are conveniently represented by the contour maps shown in Fig. 7-17. Table 7-2 illustrates the values obtained for f/f_0 and ν for some typical proteins.

Determination of another hydrodynamic property (sedimentation coefficient) or determination of molecular weight by another method (e.g., sedimentation equilibrium, Chapter 8) will eliminate the ambiguity over shape and hydration in the preceding analysis. Scheraga and Mandelkern[30] have defined a parameter β, which depends only on axial ratio of the hydrodynamically equivalent ellipsoid, and is related to the experimental results by

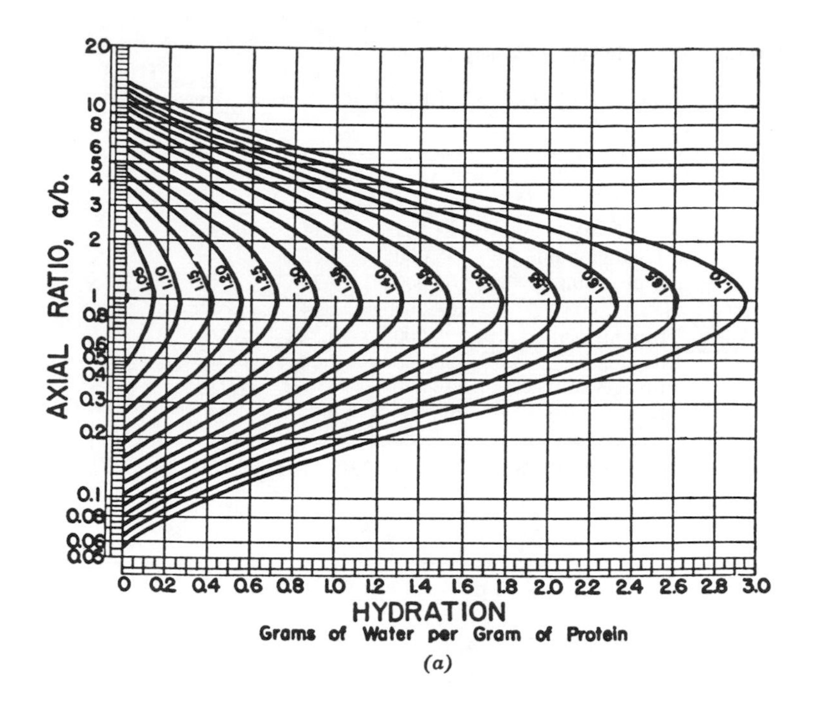

(*a*)

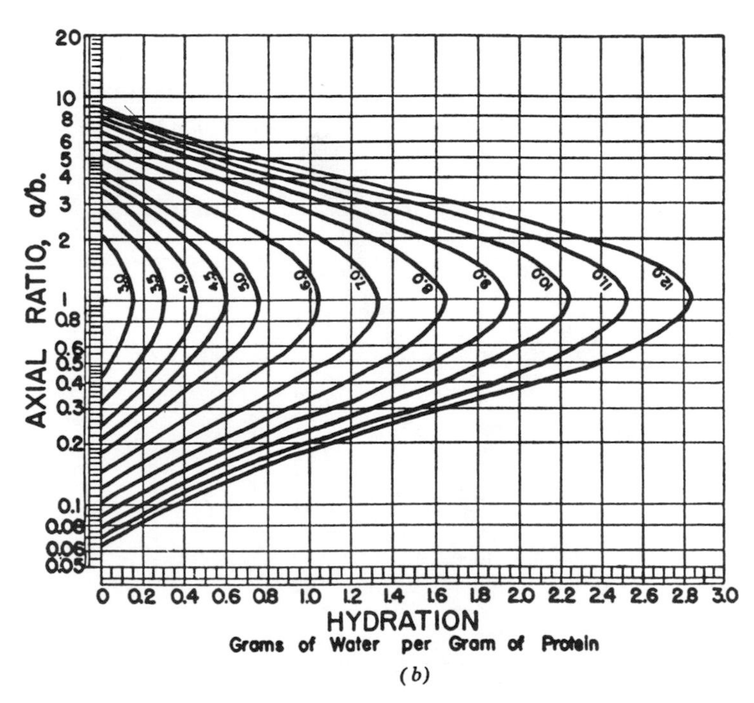

(*b*)

Fig. 7-17 (*a*) Values of axial ratio and hydration for various values of the frictional ratio $\mathbf{f}/\mathbf{f}_0$. (*b*) Values of axial ratio and hydration for various values of the viscosity factor. [J. L. Oncley, *Ann. N. Y. Acad. Sci.*, **41,** 121 (1941).]

Table 7-2 Values of f/f_0 and ν for some typical proteins.[a]

	M	f/f_0	ν
Ribonuclease	13,683	1.14	4.5
β-Lactoglobulin	35,000	1.25	4.5
Serum albumin	65,000	1.35	5.0
Hemoglobin	68,000	1.14	4.8
Tropomyosin	93,000	3.22	70
Fibrinogen	330,000	2.34	38
Collagen	345,000	6.8	1660
Myosin	493,000	3.53	298

[a] Taken from Tables 21-1 and 23-2 of Reference 29.

$$\beta = \frac{Ns[\eta]^{1/3}}{(1 - \bar{v}\rho)M^{2/3}} \tag{7-69}$$

Tables of β as a function of (a/b) are available. Table 7-3 presents the results for a prolate ellipsoid of revolution. Unfortunately the parameter β is rather insensitive to axial ratio until a becomes much larger than b. Consequently, extremely accurate data (often beyond the precision realistically obtainable) are required to compute a/b for small axial ratios. This insensitivity to a/b may, on the other hand, be used to advantage for the calculation of an approximate molecular weight from $[\eta]$ and s or D by assuming a typical value for β.

Table 7-3 Values of β as a function of axial ratio for prolate ellipsoids of revolution.

Axial ratio	$\beta(\times 10^6)$
1	2.12
2	2.13
3	2.16
4	2.20
5	2.23
10	2.41
15	2.54
20	2.64
30	2.78
50	2.97
100	3.22
200	3.48
300	3.60

If the rotational diffusion coefficient is known, an equation analogous to Eq. 7-69, which is more sensitive to small axial ratios,[30] may be employed.

After determining the nature of the ellipsoid that best fits the experimental data, one may calculate the hydrodynamic volume V_E of the particles and compare this with $\bar{v}$ to assess the magnitude of hydration. Hydration is often thought of in terms of a partially immobilized surface layer; however, no detailed physical description of hydration for a real protein is yet available. Solvent may be present in cracks or crevices or may represent a surface phenomenon of graded solute–solvent interactions. Consequently, comparison of V_E and $\bar{v}$ should not be interpreted too strictly.

7-2 Random Coils

A polypeptide chain in the random coil form (lacking any secondary or tertiary structure) occupies a much larger volume than it does as a compact sphere, and consequently exhibits lower rates of sedimentation and diffusion and increased intrinsic viscosity. Statistical approaches to the molecular dimensions of random coil chains have received intensive study: the results are interestingly reviewed and referenced by Tanford.[31] Several factors complicate the direct theoretical calculation of hydrodynamic parameters of protein random coils. One problem is that solutions of denatured polypeptide chains generally show nonideal behavior, that is, solute–solute interactions influence solution properties. This is an area of solution physical chemistry that is not well understood. Another consideration is the fact that amino acid composition of the polypeptide influences the time average dimensions of the random coil. For example, the greater rotational freedom about glycyl residues reduces the volume of chains that contain large amounts of glycine, while the opposite is true for proteins with high concentrations of proline. However, since many proteins contain "typical" amino acid compositions and reflect similar nonideal solution effects in the random coil form, empirical equations can be used to relate hydrodynamic properties to molecular weight of the random coil. Within these limitations, it is possible to use hydrodynamic data to test whether a protein of known molecular weight is in a random coil conformation or, if the random coil conformation seems certain, to estimate the polypeptide molecular weight. Care must be taken, since fortuitous agreement with a conformation of entirely different shape is possible.

Viscosity studies that provide a measure of molecular volume are the most sensitive single hydrodynamic test for the random coil conformation.

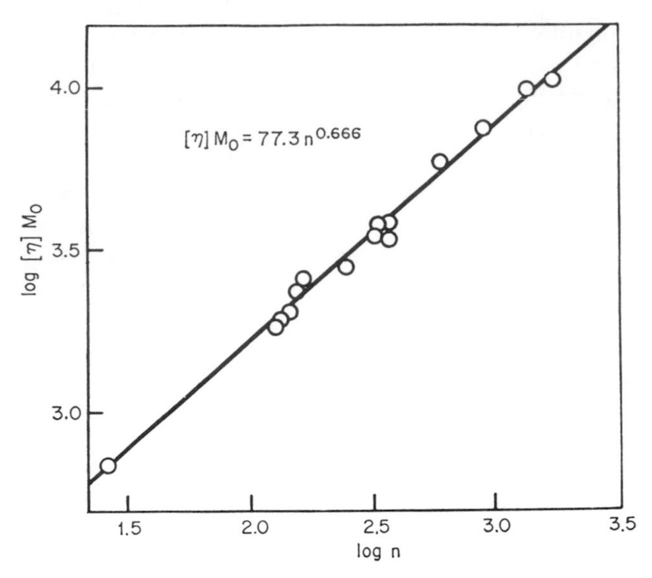

Fig. 7-18 Dependence of the intrinsic viscosity on chain length for polypeptide chains in 5 to 7.5 *M* guanidine-HCl at 25°C.[31]

Figure 7-18 shows a plot of log $[\eta]M_0$ vs. log n for some protein polypeptide chains in guanidine-HCl, where M_0 and n represent the mean residue weight and the number of residues in the polypeptide chain, respectively. The least squares line fits the equation given in the figure. Small peptides and very long peptide chains would be expected to show deviations from this empirical equation. Some numerical data on $[\eta]$ for denatured proteins (random coil chains) are given in Table 7-4 and compared to those for native proteins. Intrinsic viscosity data are also given for denatured proteins covalently restrained from full expansion by disulfide bridges. The extent to which such crosslinkages restrict molecular expansion to the full volume occupied by the random coil depends on the number of crosslinks and their locations.

Data for the sedimentation rate of several proteins in the random coil conformation in guanidine-HCl have been obtained by Tanford and his co-workers.[32] Accurate determination of s values for these substances requires extrapolation to zero concentration because of a large s vs. c dependence and other experimental difficulties. The data shown in Table 7-5 are plotted in Fig. 7-19 as a function of chain length. The least squares line follows the equation

$$s^0/(1 - \phi'\rho) = 0.286n^{0.473}$$

where s^0 is the value of s extrapolated to zero concentration, ϕ' is the partial volume factor (Eq. 7-52), ρ is the solution density, and n is the number of

Table 7-4 Intrinsic viscosities of protein polypeptide chains in concentrated guanidine hydrochloride solutions[b] [31]

Protein	Native state		Denatured state, no crosslinks[a]				Denatured state, SS bonds intact
	Mol wt	$[\eta]$ (cc/g)	Mol wt	M_o	n	$[\eta]$ (cc/g)	$[\eta]$ (cc/g)
Insulin	5,700	—	2,970	113	26	6.1	
Ribonuclease	13,700	3.3	13,700	110	124	16.6	9.4
Lysozyme	14,300	2.7	14,300	111	129	17.1	6.5
Hemoglobin	64,500	3.6	15,500	108	144	18.9	
Myoglobin	17,800	3.1	17,200	112	153	20.9	
β-Lactoglobulin	36,800	3.4	18,400	113	162	22.8	19.1
Chymotrypsinogen	25,700	2.5	25,700	105	245	26.8	11.0
Phosphoribosyl transferase	210,000	—	35,000	109	320	31.9	
Glyceraldehyde-3-phosphate dehyd.	145,000	—	36,300	110	331	34.5	
Tropomyosin	76,000	45	38,000	114	333	33	
Pepsinogen	40,000	—	40,000	107	376	31.5	27.2
Aldolase	160,000	4.0	40,000	108	370	35.5	
Serum albumin	69,000	3.7	69,000	114	605	52.2	22.9
Paramyosin	220,000	103	100,000	114	880	65.6	
Thyroglobulin	660,000	4.7	165,000	121	1364	82	
Myosin	600,000	217	200,000	115	1739	92.6	

[a] Disulfide bonds were prevented from forming by the presence of excess β-mercaptoethanol, or by protection by alkylation, following reduction of all existing bonds.

[b] GuHCl concentration varied from 5 to 7.5 M. Temperature was 25°C.

Table 7-5 Sedimentation coefficients (s^0) in 6 M guanidine hydrochloride containing a sulfhydryl reagent at 0.1 M, 25° [a]

Protein	Mol wt	Residues per chain	Apparent specific vol (ϕ'), (cc/g)	s^0_{25}, S
Ribonuclease	13,680	124	0.69	0.59
Hemoglobin	15,500	144	0.74	0.45
Myoglobin	17,200	153	0.74	0.46
β-Lactoglobulin	18,400	162	0.74	0.61
Immunoglobulin (fragment I)	25,000	230	0.73	0.57
Chymotrypsinogen	25,700	245	0.71	0.65
Glyceraldehyde 3-phosphate dehydrogenase	36,300	331	0.73	0.76
Pepsinogen	40,000	365	0.74	0.76
Aldolase	40,000	365	0.74	0.725
Immunoglobulin	40,000	372	0.73	0.74
Serum albumin	69,000	627	0.725	1.03
Myosin	197,000	1790	0.71	1.88

[a] From Reference 32.

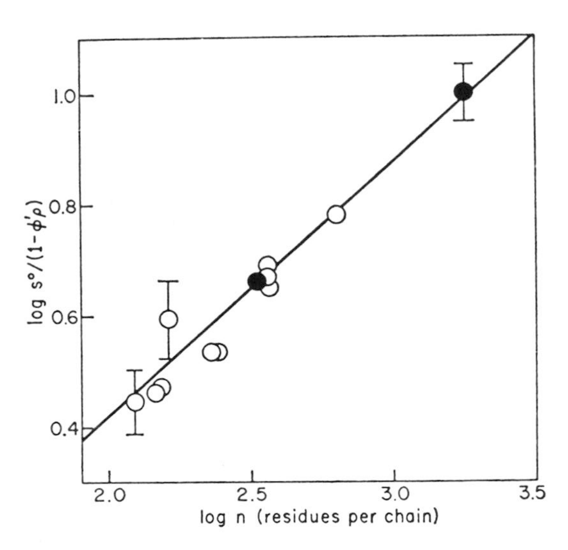

Fig. 7-19 Dependence of the sedimentation coefficient on chain length for polypeptide chains in guanidine-HCl.[32]

residues per chain. As is the case with viscosity measurements, the presence of disulfide crosslinks gives rise to less expansion and hence a larger sedimentation coefficient.

REFERENCES

Diffusion

1. L. J. Gosting, "Measurement and interpretation of diffusion coefficients of proteins," *Adv. Protein Chem.*, **11,** 429 (1956).
2. H. Svensson and T. E. Thompson, "Translational diffusion methods in protein chemistry," in *Analytical Methods of Protein Chemistry*, P. Alexander and R. J. Block, Eds., Pergamon Press, New York, 1961.
3. R. Markham, "Diffusion," in *Methods in Virology*, Vol. 2, K. Maramorosch and H. Koprowski, Eds., Academic Press, New York, 1967, p. 275.
4. L. G. Longsworth, "Diffusion in liquids," in *Physical Techniques in Biological Research*, Vol. 2, Part A, D. H. Moore, Ed., Academic Press, New York, 1968.
5. A. L. Geddes and R. W. Pontius, "Determination of diffusivity," in *Technique of Organic Chemistry*, Vol. 1, *Physical Methods*, Part 2, A. Weissberger, Ed., Wiley-Interscience, New York, 1959.
6. R. B. Setlow and E. C. Pollard, *Molecular Biophysics*, Addison-Wesley, Reading, Mass., and London, 1962.

Sedimentation Velocity

7. T. Svedberg and K. O. Pedersen, *The Ultracentrifuge*, Oxford University Press, Oxford, 1940.
8. H. K. Schachman, *Ultracentrifugation in Biochemistry*, Academic Press, New York, 1959.
9. H. Fujita, *Mathematical Theory of Sedimentation Analysis*, Academic Press, New York, 1962.
10. J. H. Coates, "Ultracentrifugal analysis," in *Physical Principles and Techniques of Protein Chemistry*, Part B, S. J. Leach, Ed., Academic Press, New York, 1970.
11. J. R. Cann, *Interacting Macromolecules; The Theory and Practice of Their Electrophoresis, Ultracentrifugation, and Chromatography*, Academic Press, New York, 1970.
12. H. K. Schachman, L. Gropper, S. Hanlon, and F. Putney, "Ultracentrifuge studies with absorption optics. II. Incorporation of a monochrometer and its application to the study of proteins and interacting systems, "*Arch. Biochem. Biophys.*, **99,** 175 (1962).
13. R. H. Haschemeyer and R. E. Nadeau, "Molecular substructure of fibrinogen," *Biochem. Biophys. Res. Commun.*, **11,** 217 (1963).
14. E. G. Richards and H. K. Schachman, "Ultracentrifuge studies with Rayleigh interference optics. I. General applications," *J. Chem. Phys.*, **63,** 1578 (1959).
15. R. Trautman, "Computational methods of ultracentrifugation," in *Ultracentrifugal Analysis in Theory and Practice*, J. W. Williams, Ed., Academic Press, New York, 1963, p. 203.

Sucrose Gradient Centrifugation

16. R. G. Martin and B. N. Ames, "A method for determining the sedimentation behavior of enzymes: Application to protein mixtures," *J. Biol. Chem.*, **236,** 1372 (1961); C. R. McEwen, "Tables for estimating sedimentation through linear concentration gradients of sucrose solution," *Anal. Biochem.* **20,** 114 (1967).

Band Centrifugation

17. J. Vinograd, R. Bruner, R. Kent, and J. Weigle, "Band-centrifugation of macromolecules and viruses in self-generating density gradients," *Proc. Nat. Acad. Sci., U.S.*, **49,** 902 (1963).

Protein Volumes

18. E. J. Cohn and J. T. Edsall, *Proteins, Amino Acids, and Peptides*, Reinhold, New York, 1941.
19. E. F. Casassa and H. Eisenberg, "Thermodynamic analysis of multicomponent solutions," *Adv. Protein Chem.*, **19,** 287 (1964).
20. N. Bauer and S. Z. Lewin, "Determination of density," in *Physical Methods of Organic Chemistry*, 3rd ed., Vol. I, Part I, A. Weissberger, Ed., Wiley-Interscience, New York, 1959.
21. D. V. Ulrich, D. W. Kupke, and J. W. Beams, "An improved magnetic densitometer: The partial specific volume of ribonuclease," *Proc. Nat. Acad. Sci., U.S.*, **52,** 349 (1964).
22. Data sheet, Digital Densimeter, Anton Paar K.G., Austria.
23. M. J. Hunter, "A method for the determination of protein partial specific volumes," *J. Phys. Chem.*, **70,** 3285 (1966).
24. M. J. Hunter, "The partial specific volume of bovine plasma albumin in the presence of potassium chloride," *J. Phys. Chem.*, **71,** 3717 (1967).
25. K. Linderstrøm-Lang and H. Lanz, *C.R. Trav. Lab. Carlsberg, Ser. Chim.*, **21,** 315 (1938).
26. R. Goodrich and F. J. Reithel, "Macromolecule dry weight determination with a vacuum balance," *Anal. Biochem.*, **34,** 538 (1970).

Viscosity

27. J. T. Yang, "The viscosity of macromolecules in relation to molecular conformation," *Adv. Protein Chem.*, **16,** 323 (1961).
28. J. T. Edsall, "The size, shape, and hydration of protein molecules," in *The Proteins*, Vol. 1, H. Neurath and K. Bailey, Eds., Academic Press, New York, 1953, p. 549.
29. C. Tanford, *Physical Chemistry of Macromolecules*, Wiley, New York, 1961.
30. H. A. Scheraga, *Protein Structure*, Academic Press, New York, 1961.
31. C. Tanford, "Protein denaturation," *Adv. Protein Chem.*, **23,** 121 (1968).
32. C. Tanford, K. Kawahara, and S. Lapanje, "Proteins as random coils. I. Intrinsic viscosities and sedimentation coefficients in concentrated guanidine hydrochloride," *J. Amer. Chem. Soc.*, **89,** 729 (1967).

VIII Equilibrium Methods

1 SEDIMENTATION EQUILIBRIUM[1–4]

The method of sedimentation equilibrium has continuously gained in popularity, particularly since the late 1950s, and now appears to be the method of choice for characterizing protein systems in terms of molecular weights. Determinations have been made for molecules as small as sucrose (molecular weight = 320) and as large as tobacco mosaic virus (molecular weight = 40×10^6). This method is probably the most accurate way to obtain molecular weights of macromolecules; however, the long time usually required to reach equilibrium precludes its use for unstable materials. The use of short liquid column heights coupled with more accurate optical measurements and methods of analysis have decreased the time required for sedimentation equilibrium to as little as a day or two for larger molecules (molecular weight = ~500,000) or just a few hours for substances with molecular weights of the order of 20,000 to 70,000, a range that includes many proteins.

A major advantage of this technique is that it is firmly based on thermodynamic principles. Recent experimental innovations now permit precise data to be obtained from small volumes of dilute protein solutions. Theoretical developments have been made that permit characterization of polydisperse protein solutions in terms of a molecular weight distribution (MWD). The MWD of such preparations can be obtained, provided that the distribution is determined by a limited number of unknown parameters that are sufficiently unique to avoid nearly-degenerate solutions of the basic equations. Sedimentation equilibrium is also a powerful technique for the study of macromolecular interactions, the binding of small molecules to macromolecules, and the interaction of small molecules with each other (these problems may in many respects also be considered as special cases of molecular weight distribution determination). Kinetic studies of systems in which the molecular weight distribution is altered considerably during the time required to reach equilibrium are, however, excluded from study.

1-1 Derivation of Basic Equations

In the sedimentation velocity experiments described earlier, the centrifugal force is sufficiently great so that the region near the meniscus is essentially depleted of the sedimenting species and normally a sedimenting boundary is observed. In contrast, in sedimentation equilibrium a lesser redistribution of molecules is obtained by using a centrifugal field low enough so that the sedimenting force is closer in magnitude to the force of diffusion.

The experimental observations are as follows: Starting from an initial constant concentration throughout the cell at $t = 0$, sedimentation causes a slow decrease of concentration near the meniscus and an increase at the bottom of the cell, as illustrated in Fig. 8-1. This sets up a concentration gradient

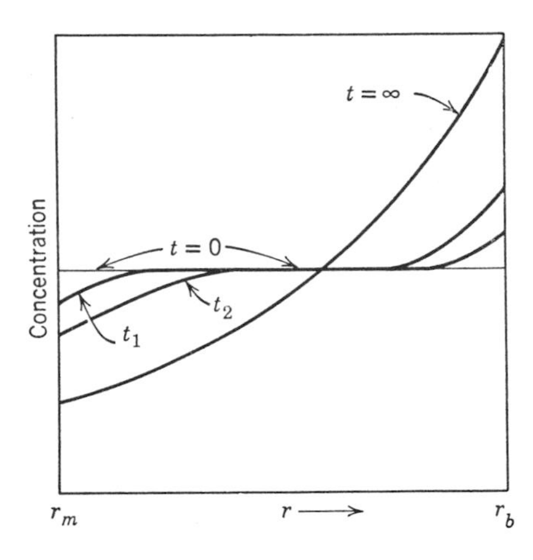

Fig. 8-1 Diagrammatic representation of concentration as a function of distance at several times during a low-speed sedimentation equilibrium experiment.[22]

leading to back diffusion toward the meniscus. Eventually the state of sedimentation equilibrium is reached, as illustrated by the curve labeled $t = \infty$, where mass transport toward the bottom of the cell by sedimentation is exactly equal to that toward the meniscus by diffusion. In the experiment illustrated in Fig. 8-1, the concentration at the meniscus is not zero, but in other cases sedimentation equilibrium experiments may be run at higher speeds so that the meniscus concentration becomes zero within experimental error. Under these circumstances the measurable region is limited to the lower portion of the cell.

Full appreciation of the value of the sedimentation equilibrium method can be gained by rigorous derivation of the pertinent equations. Such a treatment is beyond the scope of this text, but can be found elsewhere, as in the extensive review by Fujita.[3] Derivations presented here will attempt to present the general principles in terms of simple systems. In some cases these will be extended by analogy. The equations which apply to a homogeneous ideal system will be derived first.

The basic principle of any equilibrium system is that the free energy be constant throughout. In sedimentation equilibrium the free energy is a function of the distance from the center of rotation r, the pressure P, and the concentration c, that is, $G = f(r, P, c)$. Consequently,

$$dG = \left(\frac{\partial G}{\partial r}\right)_{P,c} dr + \left(\frac{\partial G}{\partial P}\right)_{r,c} dP + \left(\frac{\partial G}{\partial c}\right)_{r,P} dc \qquad (8\text{-}1)$$

The partial derivatives are evaluated from standard thermodynamic relations (e.g., see Lewis and Randall[5]). They are related to physical parameters as follows:

$$\left(\frac{\partial G}{\partial r}\right)_{P,c} = -M\omega^2 r; \qquad \left(\frac{\partial G}{\partial P}\right)_{r,c} = V_m; \qquad \left(\frac{\partial G}{\partial c}\right)_{r,P} = \frac{RT}{c}$$

where V_m is the volume of 1 mole of redistributing solute. For a two-component system V_m is equal to $M\bar{v}$ where $\bar{v}$ is partial specific volume. The use of concentration rather than activity in the expressions implies ideality. For noncompressible solvents the density, ρ, is independent of r, and $dP = \rho\omega^2 dr$. Substitution of these quantities into Eq. 8-1 and recognition that at equilibrium $dG = 0$ yields the basic equilibrium expression

$$M = \frac{RT}{(1 - \bar{v}\rho)\omega^2} \frac{1}{rc} \frac{dc}{dr} \qquad (8\text{-}2)$$

It is convenient to rewrite Eq. 8-2 with the quantities measured from an equilibrium pattern on the left:

$$\frac{1}{rc} \frac{dc}{dr} = \frac{M(1 - \bar{v}\rho)\omega^2}{RT} \equiv 2H \equiv 2AM \qquad (8\text{-}3)$$

This equation serves to define the convenient quantities H and A.

1-2 Application to Monodisperse Systems

Equation 8-3 may be used to calculate molecular weights from equilibrium patterns obtained with the Schlieren optical system (Fig. 8-2). Although dc/dr (with c in refractive index units) may be obtained directly from the pattern, the concentration itself is not directly determinable. The difference in concentration between any pair of r positions, however, is simply the area

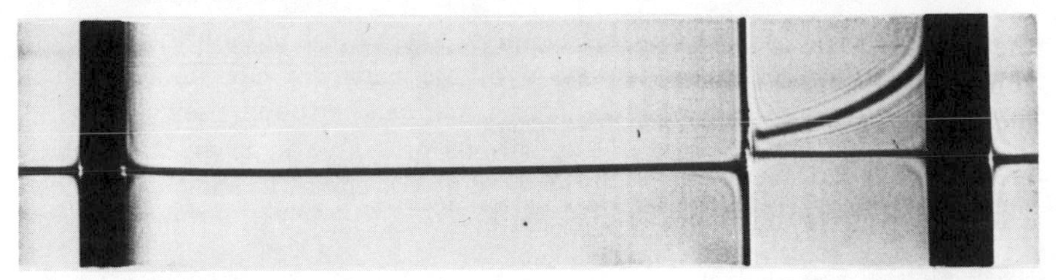

Fig. 8-2 Sedimentation equilibrium of neurophysin as viewed with the Schlieren optical system in a double sector cell (short column).

from equilibrium to baseline curves between those positions. Consequently, a plot of $(1/r)(dc/dr)$ against Δc, where Δc is the difference in concentration between the point r and some arbitrary reference position near the meniscus, will give a straight line with slope $2H$ (Fig. 8-3). Alternative methods of obtaining molecular weight from Schlieren data are given (for example) by Schachman[2] and Baldwin and Van Holde,[6] but the procedure outlined above is probably the most reliable. Upward or downward curvature of the plot indicates heterogeneity or nonideality, respectively.

Since $dc/c = d \ln c$ and $2r\, dr = dr^2$, Eq. 8-3 may be rewritten as

$$\frac{d \ln c}{dr^2} = H = AM \tag{8-4}$$

This equation indicates that a plot of $\ln c$ against r^2 will give a straight line with slope H for a homogeneous ideal system. Use of Eq. 8-4 implies that data proportional to concentration (i.e., $d \ln kc = d \ln c$ where k is an arbitrary constant) are available as a function of r. In most cases this is the most desirable form in which to obtain data for analysis, and it is therefore important to suggest here, primarily by reference, how these data may be obtained.

The most straightforward way to obtain the required data is with the ultraviolet scanning system, which gives absorbance vs. r directly. Since macromolecular samples are often contaminated with nonsedimenting material absorbing at the experimental wavelength (e.g., 280 nm), it is wise to check

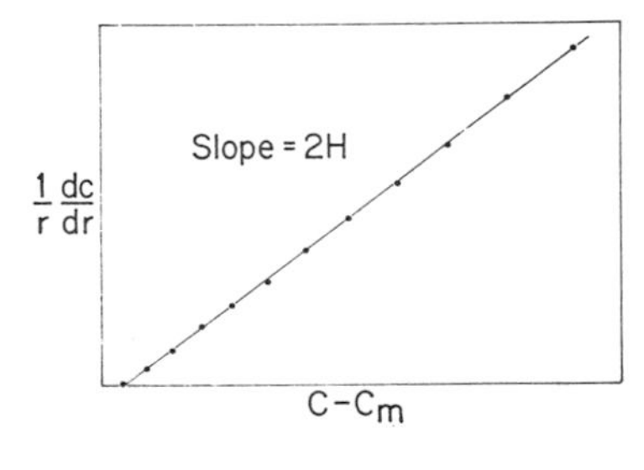

Fig. 8-3 A plot of the data obtained from Fig. 8-2 according to "method 2" of Van Holde and Baldwin.[6]

the baseline level by pelleting the protein at high speed after the equilibrium run is completed. The range of concentration suitable for absorbance measurements can be extended by employing cells of different pathlength and in some cases by varying the wavelength setting of the monochromator.

Despite the convenience of the absorption optical system in sedimentation equilibrium studies, the Rayleigh interference system is often preferred for these analyses, primarily because of its high accuracy. Unfortunately, the final equilibrium fringe pattern obtained measures concentration differences (in refractive index units) between arbitrary r positions, not absolute concentrations. Richards and Schachman[7] circumvented this problem by separately determining the initial concentration (in fringes). The number of fringes across a boundary formed between solvent layered over solution in a synthetic boundary cell (cf. Section 3) was counted, and then the shift in fringes with time was observed at some convenient radial position during the equilibrium experiment. Absolute concentration at the reference position can thus be obtained, and the concentration at all other positions is subsequently obtained by addition or subtraction of the appropriate Δc value.

An alternative procedure was introduced by LaBar.[8] He first depleted the meniscus by high-speed centrifugation, then slowed down the centrifuge to the equilibrium speed and followed the fringe shift at the meniscus as back diffusion occurred until equilibrium was established. The value for c_m can be added to the Δc data obtained from the final equilibrium pattern to give the desired c vs. r data. This principle has been extended to obtain data of greater reliability.[9,10]

A convenient and popular sedimentation equilibrium molecular weight method employing the Rayleigh interference system is the meniscus depletion method of Yphantis.[11] Yphantis used lower concentrations and higher speeds so that the concentration near the top of the column was experimentally indistinguishable from zero. In this case Δc data with reference to the meniscus concentration are identical with concentration data. This technique is somewhat limited because of the smaller concentration differences involved over the cell, and it may be subject to convective disturbances for large proteins because of the absence of a concentration gradient near the meniscus. Recent computerized analyses of data in which c_m is finite but small may provide the best method yet for obtaining maximal information from sedimentation equilibrium experiments.[12,13]

1-3 The Concentration Distribution of a Single Ideal Solute at Sedimentation Equilibrium

Equation 8-4 may be integrated between the value of c at the meniscus, c_m, and an arbitrary position, $c(r)$, to give the concentration distribution at equilibrium. The result is an exponential curve:

$$c(r) = c_m \exp [H(r^2 - r_m^2)] \tag{8-5}$$

The value of c_m may be expressed in terms of the original concentration (c_0), H, and cell loading conditions in the following manner. The total mass of solute in a sector-shaped cell of angle ϕ and thickness a is given by

$$\text{total mass} = \phi a \int_{r_m^2}^{r_b^2} c(r)\, dr^2 \tag{8-6}$$

In the equation $\phi a\, dr^2$ represents the infinitesimal volume element at r. At time = 0, before redistribution occurs, $c(r)$ is everywhere constant (c_0). At equilibrium, $c(r)$ is given by Eq. 8-5. When these expressions are substituted into Eq. 8-6 and the results set equal (for conservation of mass, assuming no material is lost on the bottom of the cell), then one obtains

$$\frac{c_b - c_m}{c_0(r_b^2 - r_m^2)} = H \tag{8-7}$$

In principle, if c_0 is determined independently, Eq. 8-7 could be used to compute M directly, although this is usually unwise because of the likelihood of error in c_b. This error results both from the difficulty in extrapolating data to the cell bottom and from the tendency of many proteins to aggregate and/or form a slight precipitate on the bottom of the cell, thus resulting in the inability to observe the total mass corresponding to c_0, and reducing the observed value of M.

Equation 8-7 can be combined with Eq. 8-5 to describe $c(r)$ in terms of the single unknown, H. The derivation is as follows. c_b is determined from Eq. 8-5 after substracting c_m from both sides:

$$c_b - c_m = c_m\{\exp [H(r_b^2 - r_m^2)] - 1\} \tag{8-8}$$

Elimination of ($c_b - c_m$) and substitution by combination of Eqs. 8-7 and 8-8 yield an expression for c_m. Substitution of this expression into Eq. 8-5 gives the desired result:

$$c(r) = \frac{Hc_0(r_b^2 - r_m^2)}{\{\exp [H(r_b^2 - r_m^2)] - 1\}} \exp [H(r^2 - r_m^2)] \tag{8-9}$$

This equation is useful to simulate the anticipated equilibrium concentration distribution if an approximate value of M is known. It is employed to estimate the appropriate speed and c_0 for a given column height that will produce optimal experimental results (e.g., fringe density or absorbance at the column base and concentration at the meniscus).

1-4 Selected Molecular Weight Distributions Encountered in the Study of Protein Structure

As stated in the introduction, sedimentation equilibrium is a uniquely powerful tool for obtaining information on systems containing more than one solute component. Before discussing further the theoretical and experimental aspects of sedimentation equilibrium, it is instructive to consider the types of systems that are of particular interest in protein structure analysis and for which sedimentation equilibrium may provide important information. Sedimentation equilibrium is most often applied to determine molecular weights of highly purified macromolecules that the investigator believes to be homogeneous. For a single solute the molecular weight distribution, MWD (or $f(M)$ as it is sometimes denoted), is characterized by a single value, the molecular weight of the solute. In practice this ideal situation is realized far less frequently than is generally supposed. When assumptions of homogeneity are made, significant errors often result. In these days of sophisticated purification procedures the problem for proteins usually is not the presence of contaminating macromolecules. Rather, the assumption that a purified protein exists in solution as a single component is most often invalidated as a result of intrinsic properties of the purified protein.

A commonly observed phenomenon with proteins is their tendency towards aggregation. This association may be simply a dimerization of a monomeric form; it may involve a stepwise polymerization (i.e., 1-mer, 2-mer, 3-mer, . . . , n-mer at rapidly decreasing concentrations) at the other extreme or anything in between. These aggregating systems may be classed in two general types (or a mixture of the two):

1. Monomer and all possible aggregates are in reversible thermodynamic equilibrium. Such systems are said to be "self-associating," and MWD will be concentration dependent. If it is known at any concentration, MWD may be computed for any other at constant environmental conditions from the equilibrium constants.

2. Multiple solute species are not in equilibrium and the mixture results from partial perturbation to (or from) a "frozen-in state" established during past history or resulting from a slow time-dependent kinetic process. Nonequilibrium aggregates can often be separated from the monomer or incompetent monomer separated from associated species (e.g., by molecular exclusion chromatography).

In addition to the types of aggregation described above, oligomeric proteins may dissociate into one or more subunits. Again, both equilibrium and nonequilibrium situations may be encountered. In cases where the protein is composed of nonidentical polypeptide chains, an additional complication is encountered because the molecular weights of different solute species are then not necessarily integral multiples of a basic (monomer) value.

As an illustration, consider a preparation of protein A containing 10% by weight of a dimer, A_2. An experimental determination of the monomer molecular weight which ignores the presence of the dimer will be in considerable error (e.g., 10% if the technique yields a value of M_w). If the form of MWD is known to be of the monomer–dimer type, MWD is characterized by two unknowns for a given concentration, namely, the molecular weight of the monomer and the weight fraction of monomer (or the equilibrium constant for reversible dimerization). If the only desired result is the molecular weight of the monomer, sedimentation equilibrium techniques may be applied in a fairly straightforward way to obtain this value without specifying the second unknown. One procedure that can be used[11] is to employ relatively high centrifugal fields so that the concentration near the meniscus is nearly zero. Extrapolation of the point-average molecular weight data to zero concentration will then approach the value for the lowest molecular weight species present. The procedure works well even if several higher aggregates than the monomer are present, provided that the concentration of monomer remains substantial. For the analogous case of 10% dissociation, the molecular weight of the principal species cannot be obtained by extrapolation and more sophisticated procedures must be used for analysis. If only two species are present (e.g., 90% dimer, 10% monomer) the 2-species plot suggested by Roark and Yphantis[13] or other procedures to be discussed below must be used for reliable results.

Oligomeric proteins that undergo association generally possess multiple weak bonding sites, possibly leading to an indefinite association. Often the concentration of monomer will be dominant and the monomer molecular weight obtainable with ease. The more general case can be solved only by assuming that successive steps in the polymerization are related by a single intrinsic equilibrium constant (e.g., see Reference 14).

Continuous or near continuous distributions sometimes may be encountered in protein studies, as might be the case when using synthetic polypeptides for model studies. If the form of the distribution is known, MWD may often be obtainable in terms of only a few unknown parameters. For example, if a distribution is known to be Gaussian in form, it is characterized by the molecular weight at the maximum in the curve and by the $\frac{1}{2}$ width of the curve at some total known concentration.

The above examples are sufficient to demonstrate the variety of problems that may be approached by sedimentation equilibrium studies. It should be emphasized that although the evaluation of arbitrary broad distributions in a more general sense (i.e., limited *a priori* knowledge concerning the *form* of MWD) now appears possible, a generalized approach to deduce MWD is not available. The student must therefore remember that the best approach to the molecular weight problem will require careful definition of the desired answer and some data on the possible form of MWD.

1-5 Discrete Molecular Weight Distributions—Exponential Analysis

In a solution containing i kinds of solute molecules, Eq. 8-5 can be written for each of the i species. If concentration is expressed in units of the optical system employed,

$$c(r) = \sum_i c_i(r)$$

where the subscript i denotes the ith species. Equation 8-5 then becomes

$$c(r) = \sum_i c_{im} \exp\left[H_i(r^2 - r_m^{\ 2})\right] \tag{8-10a}$$

If the ith species has a value of H which is i times that of the first solute, the equation becomes

$$c(r) = \sum_i c_{im} \exp\left[iH(r^2 - r_m^{\ 2})\right] \tag{8-10b}$$

This condition would hold for self-associating systems if it is assumed that $\bar{v}$ is unchanged by polymerization. For ideal self-associating systems the relation between total concentration and that of each component may be expressed in terms of the monomer concentration c_1 and the equilibrium constants for the associations:

$$c = \sum_i K_i c_1^{\ i} \tag{8-11}$$

where K_i is the equilibrium constant for the ith association. By definition, K_1 is unity.

The sedimentation equilibrium distribution for the purely self-associating system in chemical equilibrium is given by

$$c(r) = \sum_i K_i c_{1m}^{\ i} \exp\left[iH(r^2 - r_m^{\ 2})\right] \tag{8-12}$$

More complex schema involving multiple monomeric forms with different association constants, association between nonidentical units, or inclusion of terms for incompetent species (i.e., solute components of essentially static proportions and unable to enter into the equilibrium due to partial denaturation, etc.) are developed in similar fashion.

The requirement for conservation of mass means that

$$\int_{r_m}^{r_b} c(r)\, dr^2$$

must be unchanged as the result of centrifugal distribution. Since the integral has the value $c_0(r_b^{\ 2} - r_m^{\ 2})$ before redistribution has occurred, then it follows that

$$c_0 = \frac{1}{r_b^2 - r_m^2} \int_{r_m}^{r_b} c(r)\, dr^2 \tag{8-13}$$

The right-hand side is evaluated by substituting the appropriate equation for $c(r)$. As Adams[15] has emphasized, it is necessary to differentiate between the non- and self-associating expression in developing the conservation of mass equations. For nonassociating systems, the analogy of Eq. 8-13 is written separately for each of the i components:

$$\begin{aligned} c_{i0} &= \frac{1}{r_b^2 - r_m^2} \int_{r_m}^{r_b} c_i(r)\, dr^2 \\ &= c_{im}\left\{\frac{\exp\,[H_i(r_b^2 - r_m^2)] - 1}{(r_b^2 - r_m^2)H_i}\right\} \end{aligned} \tag{8-13a}$$

while for associating systems we have instead

$$c_0 = \sum_i c_{1m}{}^i K_i \left[\frac{\exp\,[iH(r_b^2 - r_m^2)] - 1}{(r_b^2 - r_m^2)iH}\right] \tag{8-13b}$$

where the relation between the ith-mer and C_{i0} is found by solution of the nth order equation. A numerical solution for c_{im} in Eq. 8-13b can be obtained if the K_i are known, but no closed form solution is available (in terms of unknown K_i) for the general case with n greater than 4.

Sedimentation equilibrium data provide continuous values of $c(r)$ *vs.* r and hence the "in principle" solution to Eqs. 8-10 and 8-12 exists in the form of a least-squares analysis of a set of overdetermined equations. If any of the exponents (H_i) are unknown, nonlinear solution is required. Experimental error limits the number of unknown parameters that can be extracted from such "exponential analysis." The effect of such error has been considered by Haschemeyer and Bowers[16] who concluded that about five unknown parameters could be determined from a single typical experiment under favorable conditions. Suggestions of suitable methods to extend this type of analysis to include nonideality have been made.

Since aggregating (or dissociating) systems are of special interest in protein studies, it is appropriate to examine in more detail the similarity and distinction of the physical situation implied by Eqs. 8-10b and 8-11. For example, it is often necessary to determine the extent to which equilibrium exists in systems that are partially incompetent (e.g., some denaturation has occurred). According to these equations, any observed concentration distribution of the ith species can be interpreted by either a properly chosen equilibrium system or a noninteracting system (or intermediate case). The weight fraction of the ith component in the original solution (at c_0), however, would be different in the two cases, as calculated from Eqs. 8-13a and 8-13b.

The two cases may be readily distinguished for ideal solutes by performing two experiments at different initial loading concentrations (or at different speeds, etc.). For nonassociating mixtures, MWD as calculated from Eqs. 8-10b and 8-13a will give the same weight fraction of the *i*th component in both experiments, whereas for the equilibrium self-association the K_i's calculated from Eqs. 8-12 and 8-13b are required to be constant. If neither of these conditions is fulfilled, partial incompetence must be suspected. In all cases the results obtained for the monomer molecular weight and the general nature of the heterogeneity (e.g., presence of dimers) are correct and identical. If this result is the goal of an experiment, any method of analysis for MWD, whether specifically conceived for the equilibrium or nonequilibrium situation, is valid.

1-6 Discrete Molecular Weight Distributions—Local Average Methods

Exponential analysis is a "whole system" approach that treats all of the data from an equilibrium experiment at one time by a least-squares approach. In contrast, the "local averages" method may be applied to calculate various point-average molecular weights from the data, which in turn are manipulated to interpret the data in terms of a complete or partial MWD. The "local averages" method has been used much more extensively for analyzing real protein systems, and some ways in which this can be done will now be discussed after first defining a few of the molecular weight averages which are obtainable from the experimental data.

Molecular Weight Averages. Many different molecular weight averages may be defined, the most common of which are the number-average molecular weight M_n, the weight-average molecular weight M_w, the z-average molecular weight M_z, and the $z+1$-average molecular weight M_{z+1}. These are defined by the equations

$$M_n = \sum_i n_i M_i \Big/ \sum_i n_i = \sum_i c_i \Big/ \sum_i (c_i/M_i) \tag{8-14a}$$

where n_i is the number of particles of molecular weight M_i,

$$M_w = \sum_i c_i M_i \Big/ \sum_i c_i \tag{8-14b}$$

$$M_z = \sum_i c_i M_i^2 \Big/ \sum_i c_i M_i \tag{8-14c}$$

and

$$M_{z+1} = \sum_i c_i M_i^3 \Big/ \sum_i c_i M_i^2 \tag{8-14d}$$

The significance of these molecular weight moments lies in the fact that it can be shown mathematically that the application to multicomponent solutes systems of certain experimental methods of molecular weight determination results in a weighted average corresponding to one of the above definitions. For example, the average determined by osmotic pressure or by particle counting in the electron microscope is M_n, while light-scattering measurements give the weight-average molecular weight. For any heterogeneous ideal system, $M_n < M_w < M_z < M_{z+1}$. For a single solute, all are equal.

In general, sedimentation equilibrium results are obtained as the concentration distribution, and a plot of $\ln c$ vs. r^2 is constructed from the primary data. As an example of the above discussion of molecular weight averages, we show that the slope at any point on this curve is proportional to the weight-average molecular weight at that point. Equation 8-4 may be written for the ith species, multiplied on both sides by c_i, and summed over all i species to give

$$\frac{dc}{dr^2} = \sum_i \frac{dc_i}{dr^2} = A \sum_i c_i(r)M_i = Ac(r)M_w(r) \tag{8-15}$$

or

$$\frac{d \ln c}{dr^2} = AM_w(r)$$

$M_z(r)$ and $M_{z+1}(r)$ as well as other more complex molecular weight moments may be computed from the data as described (for example) in References 3, 12, and 13. M_n is most conveniently obtained from high-speed experiments where c_m is small.[11–13] The above references also describe procedures to obtain the values of these molecular weight averages for the original solution of concentration c_0, provided that association is absent. If sufficient precision is available to determine several of the molecular weight averages with accuracy, the definitions of the moments and a statement of conservation of mass may be solved as simultaneous equations to obtain the best fitting constants for an assumed MWD. For associating systems the moment and conservation of mass equations may be solved simultaneously for the K_i and M. Roark and Yphantis[13] have used moment data to compute M and n for a monomer association to an aggregate of n monomer units (n-mer). They also showed that the value of n and m for a monomer–n-mer–m-mer equilibrium could be obtained if the value of the monomer molecular weight was known. The treatment was also extended to nonideal systems.

1-7 The Steiner Relation Applied to Self-Association in Sedimentation Equilibrium

Self-associating systems have most often been studied by some modification of the approach originally applied to light scattering by Steiner in 1952 (see

Reference 14). The Steiner relation is obtained by differentiating

$$c = \sum_i c_1{}^i K^i$$

with respect to c_1 and multiplying the result by $c_1 M_1/cM_1$. The result,

$$\frac{d \ln c}{d \ln c_1} = \frac{1}{M_1} \frac{\sum_i ic_i M_1}{c} = \frac{M_w}{M_1} \tag{8-16}$$

is conveniently described in terms of the weight fraction of monomer c_1/c as

$$d \ln \frac{c_1}{c} = \left[\frac{M_1}{M_w} - 1\right] d \ln c \tag{8-17}$$

which can be integrated between any two values of c. If data are available approaching $c = 0$, then c_1/c must approach unity and the result is

$$c_1/c = \exp \int_{c=0}^{c} [M_1/M_w - 1] d \ln c \tag{8-18}$$

Consequently, if M_1 is obtained independently, the integral may be evaluated by numerical integration of a plot of $[(M_1/M_w) - 1]/c$ vs. c. Teller and coworkers[12] have plotted $(c_1/c) - 1$ vs. c_1 to obtain the dimerization constant of α-chymotrypsin as shown in Fig. 8-4. This type of plot has the significance

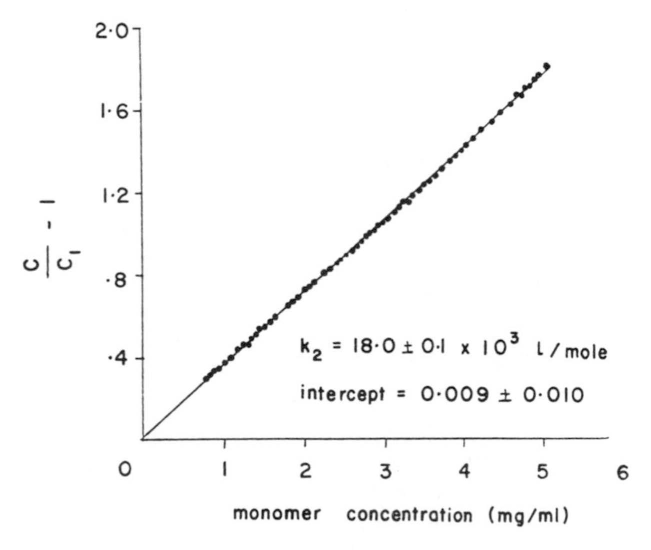

Fig. 8-4 Analysis of sedimentation equilibrium data for self-association of α-chymotrypsin according to Eq. 8-19. The resulting straight line with zero intercept is consistent with an ideal monomer–dimer equilibrium.[12]

nferred from

$$c = \sum_i c_1{}^i K_i$$

namely

$$c/c_1 - 1 = K_2 c_1 + K_3 c_1{}^2 + K_4 c_1{}^3 + \cdots \tag{8-19}$$

which, for association to the dimer only, has a slope of K_2 and zero intercept. Upward curvature signifies higher molecular weight species; downward curvature indicates nonideality. If data from two experiments run under different conditions fall on the same line, self-association is established. Partial incompetence or nonassociation will give different slopes for the two experiments, both of which, however, will be straight lines if monomer and dimer only are present. The dimerization constant may be determined in a more complex association as the limiting slope when $c \to 0$, and higher order constants may (in principle) be obtained from higher derivatives.

Relations similar to Eq. 8-17 may be written in terms of other molecular weight averages and used to give algebraic solution to more complex associations or as added assurance against errors. A general least-squares solution employing local molecular weight averages has been given by Van Holde et al.[14] for both ideal and nonideal systems where the monomer molecular weight is known. Equations for systems of the important type $A + B \rightleftharpoons AB$ have also been derived.

In summary, a variety of approaches are available to analyze MWD's involving discrete molecular species. The "best" approach is not yet clear, but depends on the complexity of the system and the information desired. A thorough analysis would include plotting the data (which must be very precise) in terms of several molecular weight averages as a function of concentration, employing multiple experiments encompassing a wide concentration range with overlapping values of c between experiments. In this way, pure association can be clearly established by the superposition of the data, and nonideality often will be detected.

Implicit in the foregoing discussion were several points applicable to frequently encountered problems.

1. The detection of heterogeneity in a highly purified protein sample is optimized by the highest centrifugal redistribution (highest speed) consistent with reliable data.

2. Limits of species detectability by the various procedures employed must be carefully determined.

3. Accurate high-speed equilibrium data in which c_m is nearly zero provide a set of molecular weight moments as a function of concentration, each of which extrapolates to the smallest molecular weight species present, provided that this species constitutes a major component and that its molecular weight is substantially lower than the next highest species present.

1-8 Continuous Distribution

The protein chemist will encounter continuous MWD's much less frequently than discrete distributions. The formulation of Scholte[17] appears promising and is directly applicable to discrete distributions. Scholte solved an appropriate set of linear equations by the technique of linear programming which permits constraint conditions to be applied to the unknowns, in this case that negative concentrations shall not be permitted. Solution by linear programming of the (ideal) "sum of exponentials problem" or the equations of Van Holde and associates,[14] when M_1 is known, may result in meaningful solutions for systems of considerably greater complexity than are currently being obtained. Whether or not this method can be extended to include non-ideality remains to be seen.

2 DENSITY GRADIENT SEDIMENTATION EQUILIBRIUM[18,19]

This technique, also termed isopycnic centrifugation, has most frequently been applied to the separation and characterization of nucleic acids and nucleoproteins. It is applicable to a number of problems confronting the protein chemist, particularly in the development of purification procedures. The method has been reviewed in detail,[18,19] and will only be briefly outlined here.

Macromolecules subjected to a centrifugal field in a solvent containing a density gradient will move away from or toward the center of rotation, depending on whether the effective density of the macromolecule in solution is greater than or less than the density of the surrounding solvent. If the density of the medium somewhere in the column is equal to the density of the macromolecule in that solvent, the macromolecules will sediment (or float) to that position and form a band there. The band will be nearly Gaussian and will be centered at a point where the net force on the molecule is zero. The width of the band is determined by the balance between diffusion spreading and the steepness of the gradient that serves to return molecules to the band center. From the shape of the band, the form of the density gradient, and the appropriate $\bar{v}$ term, molecular weight information for favorable cases may be obtained.

The density gradient can be prepared by establishing sedimentation equilibrium of a concentrated salt solution. Since the salt (e.g., KBr for proteins, or CsCl for nucleic acids) is of small molecular weight, only a modest redistribution will be obtained even at relatively high centrifugal fields. The concentration distribution of the salt produces the density gradient. If the

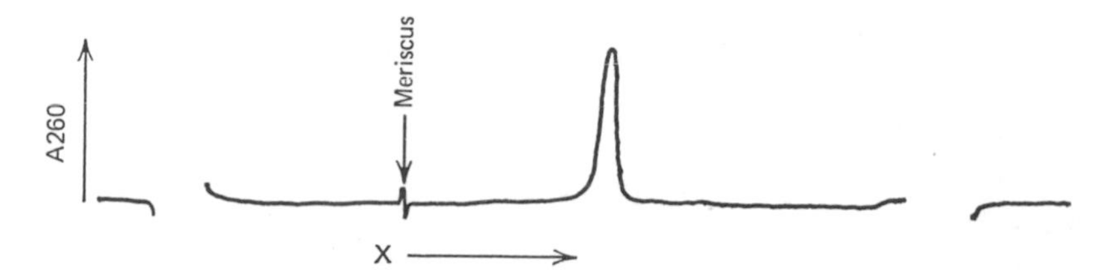

Fig. 8-5 Isopycnic banding of rat liver 40S ribosomal subunits at $\rho = 1.416$ in cesium sulfate in the Model E ultracentrifuge. The ultraviolet scanning trace represents absorbance at 260 nm (vertical axis) as a function of distance in the cell. (Courtesy of Mary G. Hamilton.)

salt concentration is chosen in such a way that the initial density closely approximates the banding density of the macromolecule of interest, the equilibrium gradient will be more dense at the bottom relative to the solute and less dense at the meniscus, and macromolecular banding will occur in the central portion of the gradient. If the macromolecular solute is uniformly mixed with the salt solution at time zero, it will move up from the bottom and down from the top as the salt gradient forms until equilibrium is obtained. This process is demonstrated for the banding of 40S rat liver ribosomal particles in the analytical ultracentrifuge in Fig. 8-5. The gradients are also sufficiently stable to permit banding of macromolecules in the preparative centrifuge (usually employing a swinging bucket rotor) and may then be collected and analyzed for biological activity, radioactivity, absorbance, or other properties, as described for sucrose gradient analysis (cf. Chapter 7).

2-1 Applications

Materials band in density gradient sedimentation equilibrium according to their densities, and thus the method can be used for identification of classes of macromolecules of different banding densities. For example, in cesium chloride DNA ($\rho \cong 1.7$), RNA ($\rho > 2.0$), nucleoprotein ($\rho = 1.5$ for ribosomes) and protein ($\rho = 1.3$) are readily distinguished. The method, however, is not usually effective for mixtures of proteins because their densities are so similar. In addition, the small size of most proteins (compared to many nucleic acids and viruses) leads to rather broad bands in such experiments (due to a high rate of diffusion) even at the highest possible centrifugal fields. The technique, however, may be used in the isolation of protein components of conjugated proteins, such as nucleoproteins or lipoproteins, that are first purified by density gradient methods. Banding is also carried out on preformed density gradients that are steeper than those made

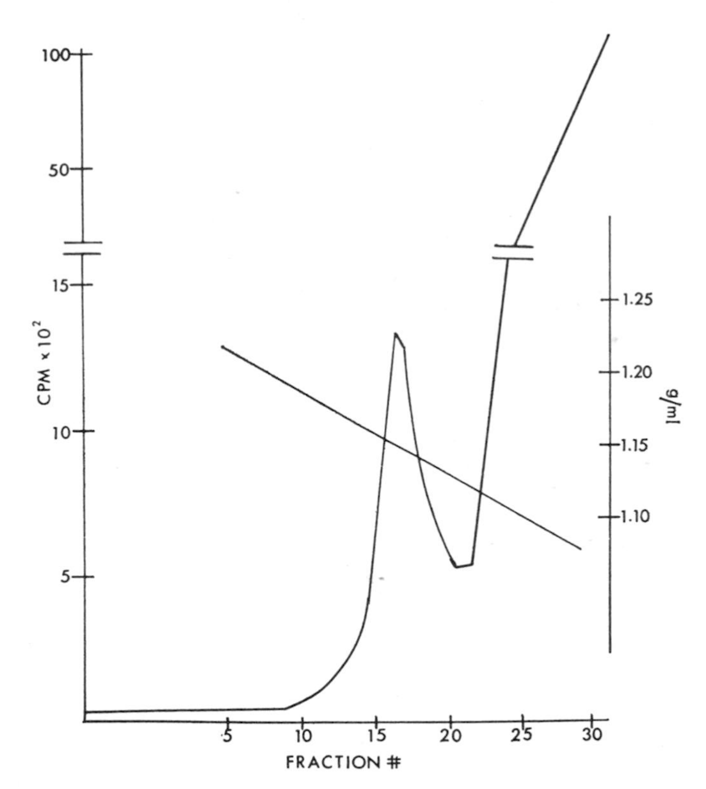

Fig. 8-6 Banding of ^3H-labelled hamster sarcoma virus in a preformed sucrose density gradient. The results are plotted as radioactivity vs. fraction number. (Courtesy of A. P. Albino, E. deHarven and F. K. Sanders.)

by sedimentation equilibrium alone. These gradients are designed to be relatively stable over the periods required to band the macromolecules. The isopycnic banding of hamster sarcoma virus in a preformed sucrose gradient (Fig. 8-6) is a good example of this method.

An elegant application of isopycnic equilibrium centrifugation to a purified protein is the determination by Zipser of the number of subunits in β-galactosidase. For these experiments, the banding density was first determined for the enzyme from bacteria grown in normal and in $D_2{}^{18}O$ medium. A mixture of radioactively labeled "light" enzyme and heavy, $D_2{}^{18}O$-labeled enzyme in excess was then dissociated and reassociated. The position of the isotopically labeled hybrid one-quarter of the way between the two original densities (Fig. 8-7) is consistent only with a four subunit model for the enzyme.

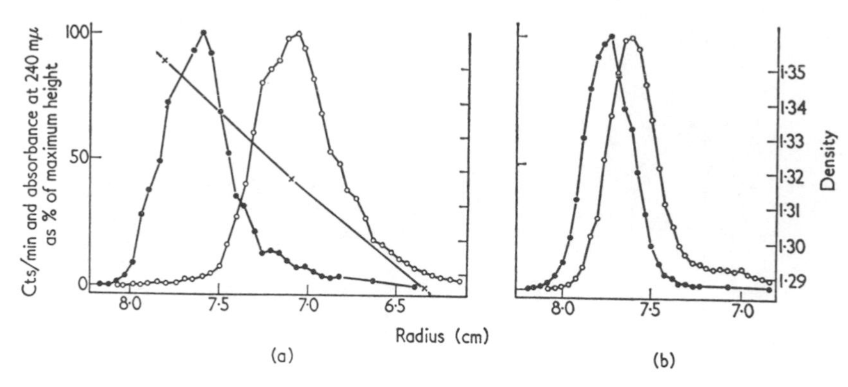

Fig. 8-7 (a) Isopycnic banding of β-galactosidase: Radioactive "light" enzyme (open circles); nonradioactive "heavy" enzyme (solid circles). The diagonal line (crosses) is the measured density with reference to the scale on the right. (b) A similar gradient of enzyme obtained by reconstitution of light enzyme subunits with an excess of heavy enzyme subunits. Most radioactive hybrid appears at a position $\frac{1}{4}$ of the way from heavy to light. This result indicates that the enzyme contains four subunits. [D. Zipser, *J. Mol. Biol.* **7,** 113 (1963).]

3 THE TRANSIENT STATE METHOD[2–4]

This method, also known as the approach to equilibrium or Archibald method, is not truly an equilibrium technique but so closely resembles sedimentation equilibrium that it is conveniently described in this section. The Archibald procedure is quite reliable when properly applied and has the major advantages that the experiments are carried out more rapidly than sedimentation equilibrium and that much larger molecules can be studied with ease.

The principle upon which the method is based is a consequence of the boundary conditions of the general differential equations of the ultracentrifuge. Archibald noted that the amount of mass transported across the column ends by sedimentation and diffusion is identically zero at all times. It can therefore be shown that since no mass is transported *across* the meniscus or bottom of the solution column, an equation identical to that for sedimentation equilibrium pertains *at* the column extremities at all times, namely,

$$\frac{1}{r_b c_b}\left(\frac{dc}{dr}\right)_b = 2H$$
$$\frac{1}{r_m c_m}\left(\frac{dc}{dr}\right)_m = 2H \qquad (8\text{-}20)$$

where the subscripts m and b refer to the position of the meniscus and bottom of the column, respectively. Accordingly, a value of molecular weight at these positions can be established whenever the concentration gradient and concentration at the meniscus and bottom can be accurately measured. For heterogeneous systems the result will be the weight average H (i.e., M_w is obtained if the $\bar{v}_i$ are equal) of species present at the time of measurement. Since heterogeneous solutes will be fractionated during centrifugation, the value of M_w depends on the time of measurement and, under conditions of constant speed, will decrease at the meniscus while increasing at the bottom as the experiment progresses. The weight-average molecular weight of the original solution may be obtained by extrapolation of the results to zero time.

The experimental procedures that have been employed are adequately referenced by Schachman,[2] Fujita,[3] and Coates.[4] Most typically, dc/dr is determined by extrapolation to the meniscus position, while the change in concentration Δc is obtained from the shift in Rayleigh fringes. The true concentration is then obtained by subtracting the fringe shift from the total fringe displacement across a boundary obtained in a separate synthetic boundary cell experiment. That experiment also serves to obtain the required relation between Schlieren units (i.e., area) and fringes. A typical pattern is shown in Fig. 8-8.

Experimentally, the synthetic boundary cell experiment is carried out in a double sector cell, as shown in Fig. 8-9. Initially, the cell is loaded with a small volume (e.g., 0.2 ml) of protein solution in the right sector, and with a larger volume (e.g., 0.4 ml) of dialysate in the left sector. As the ultracentrifuge is accelerated (to about 2000–8000 rpm), solvent flows across the lower

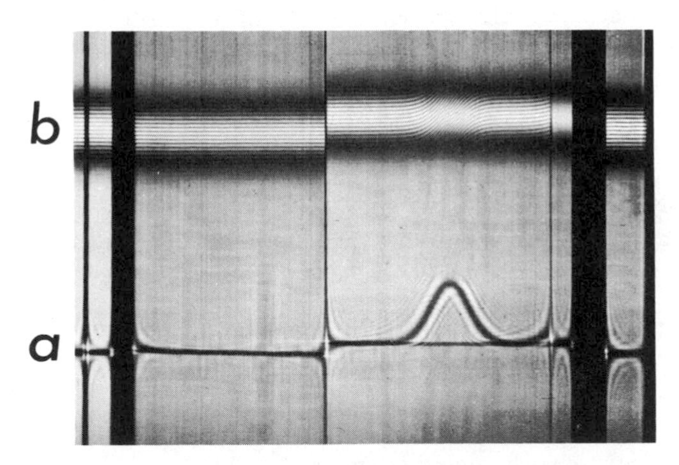

Fig. 8-8 Photographs obtained in the synthetic boundary cell determination of $\mathbf{c}_o$. (*a*) Schlieren photograph. (*b*) Rayleigh interference pattern. (From C. H. Chervenka, *A Manual of Methods for the Analytical Ultracentrifuge*, Spinco Division, Beckman Instruments, Inc., Palo Alto, Calif., 1968.)

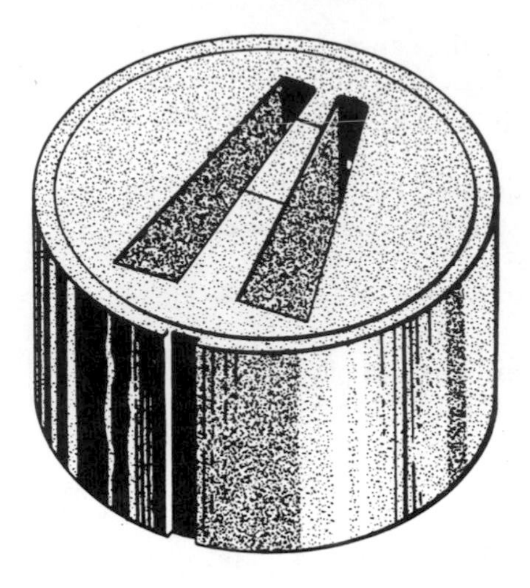

Fig. 8-9 A double sector synthetic boundary cell centerpiece. Note the two small connecting channels between compartments. Liquid flow across the lower channel occurs only in the presence of a centrifugal field. The upper channel permits the return flow of air. (Courtesy of Beckman Instruments Inc.)

capillary channel (scratch) connecting the two sectors due to the large hydrostatic pressure difference in the centrifugal field until the menisci are even. The upper scratch supports the return flow of air above the solution. The critical requirement for dialysis equilibrium in the determination of c_0 can hardly be overstressed, since even a small departure from equilibrium will contribute greatly to the apparent c_0 for buffers containing high concentrations of diffusible species.

In addition to the standard determination of the molecular weight of (nearly) homogeneous solute systems, the Archibald method may be applied to multicomponent systems. For example, in the study of rapid self-associating systems, the M_w vs. c data may be analyzed by the Steiner approach discussed with respect to sedimentation equilibrium. The test for incompetent species not entering into the equilibrium can be made by conducting experiments at different initial loading concentrations and speeds. A totally competent system is indicated by the requirement that the apparent value of M_w is a unique function of concentration only. Such an analysis was first applied to a study of the self-association of chymotrypsin by Rao and Kegeles.[20]

4 LIGHT SCATTERING[21–24]

Light scattering is a well-established method for the determination of the molecular weight of macromolecules in solution. The method can also be used to measure the second virial coefficient (a measure of excluded volume and Donnan effects in nonideal systems) and to investigate interactions of macromolecules with small ions and molecules. For large macromolecules the angular dependence of the scattered radiation provides an unambiguous determination of an important quantity known as the radius of gyration, which in turn may be used to calculate the principal dimension for some model shapes (e.g., the radius of a sphere, the length of a rod, the mean end-to-end distance of a random coil). Light-scattering data also help to determine shape, though not without ambiguity, unless the choice of models is limited by information obtained by other methods. The measurement of light scattering is relatively rapid compared to hydrodynamic measurements, thereby offering an advantage in kinetic studies involving changes in the molecular parameters discussed above.

A derivation of the light-scattering equations is beyond the intent of this book. A rigorous derivation may be found in some textbooks of electricity and magnetism and optics (e.g., see References 25 and 26). A reasonable plausibility derivation of the basic equation obtained by Lord Rayleigh in 1871 is given by Tanford.[22] Tanford also presents a valuable review of the treatment of light scattering by the fluctuation theory developed by Einstein and by Debye. This approach clearly shows the theoretical basis of the phenomenon in terms of equilibrium thermodynamics. Throughout the following discussion, unless otherwise stated, it will be assumed that the macromolecules are dilute enough to scatter independently; and that they are nonabsorbing, randomly oriented in space, and of constant refractive index throughout; and that secondary scattering and other fine details may be neglected.

4-1 The Equation for Identical Isotropic Particles Which Are Small Compared to the Wavelength of Light

The electric field E of light of frequency ν and maximum amplitude E_0 may be written in a manner expressing its periodicity in time t as

$$E = E_0 \cos(2\pi\nu t) \tag{8-21}$$

When the light strikes matter, it tends to displace the electrons from their equilibrium positions with respect to the nucleus to give rise to a dipole moment p, which is proportional to E,

$$p = \alpha E = \alpha E_0 \cos(2\pi\nu t) \tag{8-22}$$

The proportionality constant α is called the polarizability and its magnitude is a measure of the ease with which an electron can be displaced. The oscillating dipole (electron and nucleus) is itself a source of electromagnetic radiation of frequency ν equal to that of the incident light. The electric field of this scattered radiation E_s is proportional to d^2p/dt^2 and is given by

$$E_s = \frac{4\pi^2\nu^2\alpha^2 E_0 \cos(2\pi\nu t)}{c^2 r}$$

where c is the velocity of light and r is the distance of the detector from the dipole. The intensity of a light wave is given by E^2 so that the ratio of the intensity of scattered radiation i_s to the incident intensity I_0 is

$$\frac{i_s}{I_0} = \frac{E_s^2}{E^2} = \frac{16\pi^4\alpha^2}{\lambda^4 r^2}$$

where λ is the wavelength of light ($\lambda = c/\nu$). It is of interest to note the fourth power dependence of scattering on wavelength.

The polarizability itself cannot be measured but may be related to a number of experimentally measurable quantities (e.g., see Kauzmann[27]) including the refractive index n, in which case, to a good approximation,

$$n^2 = n_0^2 + 4\pi N\alpha$$

where n_0 is the refractive index of the solvent and N is the number of particles per cubic centimeter. The refractive index may be expressed as a Taylor series in concentration, which for dilute solutions is

$$n^2 = n_0^2 + 2n_0\left(\frac{dn}{dc}\right)c$$

Solving the above two equations for α, and recognizing that c/N, the mass of one particle, is also equal to the molecular weight divided by Avogadro's number ($c/N = M/\mathcal{N}$), we obtain

$$\alpha = \frac{M}{\mathcal{N}}\frac{n_0}{2\pi}\frac{dn}{dc} \tag{8-24}$$

Substitution into Eq. 8-24 gives the scattering for one particle as

$$\frac{i_s}{I_0} = \frac{4\pi^2 M^2 n_0^2}{\mathcal{N}^2\lambda^4 r^2}\left(\frac{dn}{dc}\right)^2 \tag{8-25}$$

or for N ($N = c\mathcal{N}/M$) particles per cubic centimeter, the scattering per unit volume becomes

$$\frac{i_s}{I_0} = \frac{4\pi^2 M n_0^2}{\lambda^4 r^2}\left(\frac{dn}{dc}\right)^2\frac{c}{\mathcal{N}} \tag{8-26}$$

The above result is for vertically polarized light (oscillating in the z direction, viewed in the xy plane). Horizontally polarized light will give rise to zero intensity at a 90° angle (no scattering in the direction of the oscillation), increasing to full intensity in the forward and backward direction. Unpolarized light may be considered as the sum of horizontally and vertically polarized light, and Eq. 8-26 must be corrected by the factor $(1 + \cos^2\theta)/2$ for unpolarized light. The result

$$\frac{i_s}{I_0} = \frac{2\pi^2 M n_0^2}{\lambda^4 r^2}\left(\frac{dn}{dc}\right)^2\frac{c}{\mathcal{N}}(1 + \cos^2\theta) \tag{8-27}$$

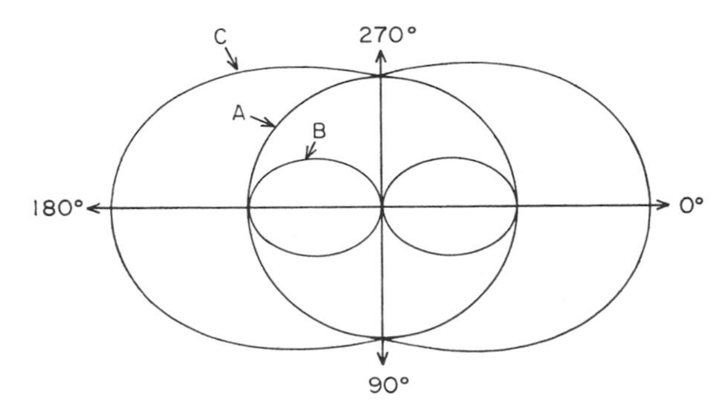

Fig. 8-10 The angular dependence of the intensity of light scattered by small particles. (*A*) Vertically polarized light. (*B*) Horizontally polarized light. (*C*) Unpolarized light (sum of *A* and *B*).

is the basic light-scattering equation for particles that are small compared to the wavelength of light (i.e., largest dimension ~1/20 of λ or less). The angular dependence of intensity is shown in Fig. 8-10. The above result is more appropriately derived from fluctuation theory, and it can be argued that if local fluctuations in polarizability did not exist (e.g., due to concentration, temperature, and pressure variations), volume elements could be paired off in a manner to give rise to complete destructive interference in all but the forward direction, as is the case for crystals.

Macromolecular solutions will also often exhibit nonideal behavior in typical light-scattering experiments, and it may be shown that for nonideal systems

$$\frac{i_s}{I_0} = \frac{2\pi^2 n_0^2 (dn/dc)^2 (1 + \cos^2\theta)c}{\mathcal{N}\lambda^4 r^2 (1/M + 2Bc + 3Cc^2 + \ldots)} \tag{8-28}$$

where B and C are the second and third virial coefficients, respectively.[22]

4-2 Scattering from Particles Comparable in Size to the Wavelength of Incident Light

In the previous section it was assumed that the scattering particles of interest were sufficiently small so that for all scattering elements within the particle the dipoles were oscillating in phase. The degree to which this assumption is valid, of course, decreases continuously with increasing size of the particle, but most authors agree that from a practical point of view the equations are not valid for particles of major dimension greater than about $\frac{1}{20} - \frac{1}{10}$ that of the wavelength of the incident light. For such particles the scattered radiation in the forward direction (0°) is unaffected, but that scattered in the other directions decreases with increasing angle due to destructive interference (if

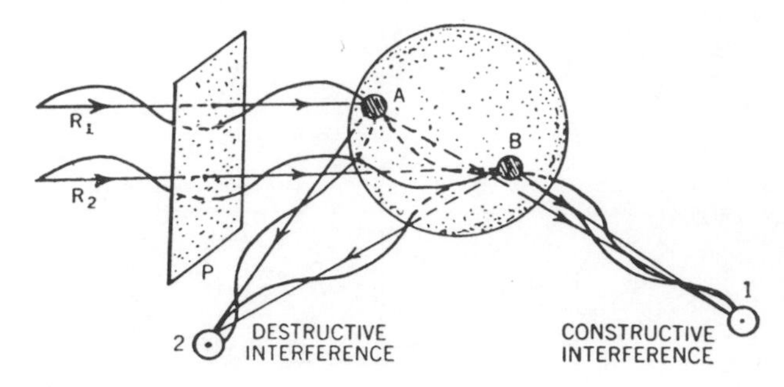

Fig. 8-11 Interference between light scattered from points A and B within the same particle. The interference is constructive in the forward direction and destructive in the backward direction. [From Oster[24]]

the particle has major dimension greater than $\lambda/2$, the degree of destructive interference will actually pass through a maximum between 0° and 180°). The physical basis for this is shown schematically in Fig. 8-11, and a typical scattering intensity envelope is depicted in Fig. 8-12. In order to describe the scattering from large particles, it is therefore necessary to correct Eq. 8-28 by multiplying it by a particle scattering factor, usually written $P(\theta)$, whose numerical value depends on the angle of observation, starting at 1 for zero angle and typically decreasing continuously in numerical value toward a minimum at 180° for particles of major dimension less than $\lambda/2$. As will be described later, the same factor is used in low-angle X-ray scattering, in which case multiple maxima and minima are expected. The complete light-scattering equation therefore is written

$$\frac{i_s}{I_0} = \frac{2\pi^2 n_0^2 (dn/dc)^2 (1 + \cos^2\theta)c}{\mathcal{N}\lambda^4 r^2 (1/M + 2Bc + 3Cc^2 + \cdots)} P(\theta) \tag{8-29}$$

In one sense $P(\theta)$ may be considered an empirical factor that corrects the scattering curve for large particles from what would have been observed had no destructive interference been present. However, $P(\theta)$ may also be directly related to a physical model for the particle and can be expressed in terms of the radius of gyration, the numerical value of which is obtainable

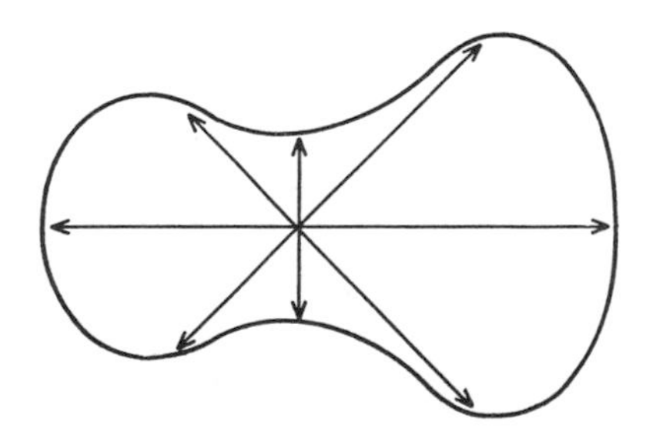

Fig. 8-12 The scattering envelope for large particles. The length of the vector is proportional to the intensity, which is decreased for backward scattering.

from the data. These concepts are elegantly developed and referenced by Geiduschek and Holtzer[21] and by Guinier and Fournet.[28] A few of the results will be presented here, following as closely as possible the terminology of Geiduschek and Holtzer.

The relationship of $P(\theta)$ to molecular parameters is conveniently expressed by the equation:

$$P(\theta) = \frac{1}{N_0^2} \sum_n^{N_0} \sum_m^{N_0} \frac{\sin hr_{nm}}{hr_{nm}} \qquad (8\text{-}30)$$

where N_0 represents the number of scattering elements (e.g., electrons, or groups of electrons close enough together to oscillate in phase). The distance between the nth and mth scattering element is denoted r_{nm}; the symbol h is used for the term $(4\pi/\lambda) \sin \theta/2$, where λ is the wavelength of light in the solvent medium. For a given model (e.g., sphere, rod, random coil), the value of $P(\theta)$ can be calculated from the equation by using atomic coordinates if they are known, or by assuming constant electron density and integrating appropriately over volume elements. For a sphere of radius r, for example, one finds

$$P(\theta) = 9\left[\frac{\sin hr - hr \cos hr}{(hr)^3}\right]^2 \qquad (8\text{-}31)$$

and for a true random coil of root-mean-square end-to-end distance $\overline{(r^2)}^{1/2}$

$$P(\theta) = \frac{2}{x^2}[x - 1 + e^{-x}], \qquad (8\text{-}32)$$

where

$$x = h^2\overline{(r^2)}/6$$

Other examples are given in the reviews cited.[21,28] It must be emphasized here that although $P(\theta)$ can always be calculated from any assumed model (at least by numerical approximation), the experimental data that give $P(\theta)$ numerically cannot in general be used to ascertain the correct model. We will later define more fully the limitations of light scattering in determining the general shape of macromolecules.

One of the most important attributes of $P(\theta)$ is that the function can also be expressed in terms of consecutive moments of the electronic mass distribution about the electronic center of mass.[21,28] For example, Eq. 8-30 may be expanded in a power series followed by a translation of the coordinate system to the electron center of mass. The result, derived in detail by Geiduschek and Holtzer[21] is

$$P(\theta) = 1 - \frac{h^2}{3}\left(\frac{\sum_n r_{on}^2}{N_0}\right) + \cdots \qquad (8\text{-}33)$$

where r_{on} is the distance of the nth scattering center from the origin. The term in parentheses is the square of the radius of gyration R_G of the particle. This important result in no way requires a knowledge of the particle shape. It is convenient here to note that the reciprocal of $P(\theta)$ is given by

$$P^{-1}(\theta) = 1 + \frac{h^2 R_G^{\ 2}}{3} - \cdots \tag{8-34}$$

so that R_G may be directly evaluated from the initial slope of a plot of $P^{-1}(\theta)$ vs. $\sin^2(\theta/2)$.

The relationship of the radius of gyration to particle dimensions has been calculated for several standard particle shapes (e.g., see References 21 and 28). For a sphere of radius r, the radius of gyration is given by

$$R_G^{\ 2} = \tfrac{3}{5} r^2 \tag{8-35}$$

For an ellipsoid of semiaxes a and b (Fig. 7-16), the relationship is

$$R_G^{\ 2} = (a^2 + 2b^2)/5 \tag{8-36}$$

Figure 8-13 illustrates how R_G changes with increasing axial ratio. For a long thin rod of length L,

$$R_G^{\ 2} = L^2/12 \tag{8-37}$$

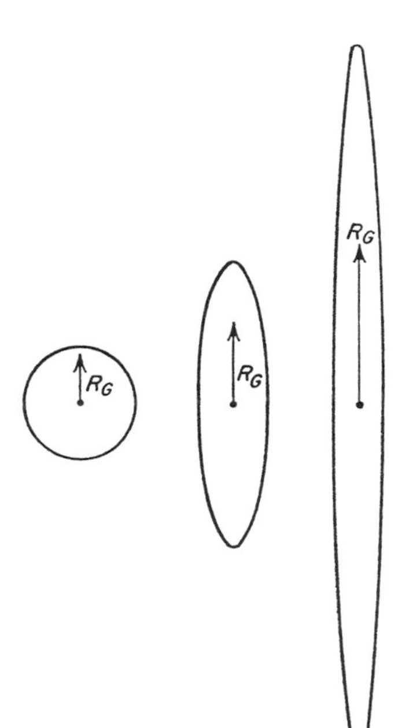

Fig. 8-13 Effect of increasing anisotropy on radius of gyration (drawn to scale for particles of equal volume). [From O. Kratky, *Progr. Biophys.* **13,** 105 (1963).]

and for a random coil of RMS (root-mean-square) end-to-end distance $\overline{(r^2)}^{1/2}$

$$\overline{(R_G{}^2)} = \overline{(r^2)}/6 \tag{8-38}$$

4-3 The Collection and Interpretation of Light-Scattering Data

The experimental procedures used for light scattering are discussed in several excellent reviews.[21,23,24] For the fullest interpretation, data must be obtained for each of four or more concentrations as a function of θ. Since the interpretation requires extrapolation both to zero concentration and zero angle, it is highly advantageous to acquire accurate data at small values of θ and c. Current developments in the adaptation of laser light sources to light scattering will hopefully permit measurement to well below 5°, a substantial advantage over the more usual limit of about 15° for the older commercially available apparatus. The independent measurement of the refractive index increment should be made in accord with the theoretical development of Eisenberg and Casassa.[29] The problems encountered in multicomponent solvent systems are quite similar to those discussed in the measurement of $\bar{v}$ for the interpretation of hydrodynamic and sedimentation equilibrium data (cf. Chapter 7).

It is convenient to express Eq. 8-29 in simpler form by defining the reduced Rayleigh ratio R_θ such that

$$R_\theta = \frac{i_s r^2}{I_0(1 + \cos^2\theta)} \tag{8-39}$$

Some authors prefer to have R_θ be the symbol for the Rayleigh ratio $(i_s r^2/I_0)$ itself, and the student should check in a particular article whether or not the $(1 + \cos^2\theta)$ term is included in the definition or not. The optical parameters are expressed in terms of a constant K defined by

$$K = \frac{2\pi^2 n_0{}^2 (dn/dc)^2}{\mathcal{N}\lambda^4} \tag{8-40}$$

Equation 8-29 may then be written as

$$R_\theta = KcP(\theta)\left(\frac{1}{M} + 2Bc + \cdots\right)^{-1} \tag{8-41}$$

or, in a more convenient form, as

$$\frac{Kc}{R_\theta} = P^{-1}(\theta)\left[\frac{1}{M} + 2Bc + \cdots\right] \tag{8-42}$$

For small values of θ and c, Kc/R_θ is linear in both $\sin^2(\theta/2)$ and c; hence, as Zimm showed, the experimental data, Kc/R_θ, may be plotted against

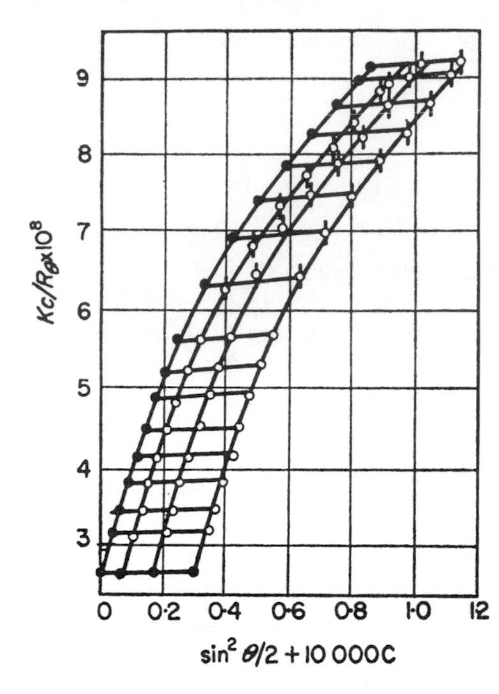

Fig. 8-14 Zimm plot of light scattering by tobacco mosaic virus. The curvature of the plot is more extreme than for a typical protein due to the high molecular weight and asymmetry of this virus. [H. Boedtker and N. S. Simmons, *J. Amer. Chem. Soc.*, **80,** 2550 (1958).]

$\sin^2(\theta/2) + kc$ for the determination of M and the radius of gyration. The constant k is arbitrary and is chosen in such a way that the points are conveniently spaced on the graphical scale used. Such a plot is shown in Fig. 8-14. Note that the angular dependent data at each concentration are extrapolated to zero angle and the concentration data at each angle are extrapolated to zero concentration. Both extrapolations intersect the ordinate at $Kc/R_\theta = 1/M$. $P^{-1}(\theta)$ may then be obtained from the slope of the $c = 0$ curve according to

$$P^{-1}(\theta) = KcM/R_\theta \tag{8-43}$$

The nonideality term $2B$ is given by the initial slope at small c of the $\theta = 0$ extrapolated curve. Finally, the square of the radius of gyration is given by

$$R_G{}^2 = \frac{3\lambda^2 M}{16\pi^2}\left[\frac{d(Kc/R_\theta)}{d\sin^2(\theta/2)}\right] \tag{8-44}$$

where the term in brackets is the initial slope of the $c = 0$ curve in the Zimm plot.

The final step in the general analysis is to obtain information concerning the shape of the molecule. A convenient way to do this is to plot $P^{-1}(\theta)$ against $h^2R_G{}^2$ and compare the results with calculated curves obtained for various models. Figure 8-15 shows the form of the curves expected for

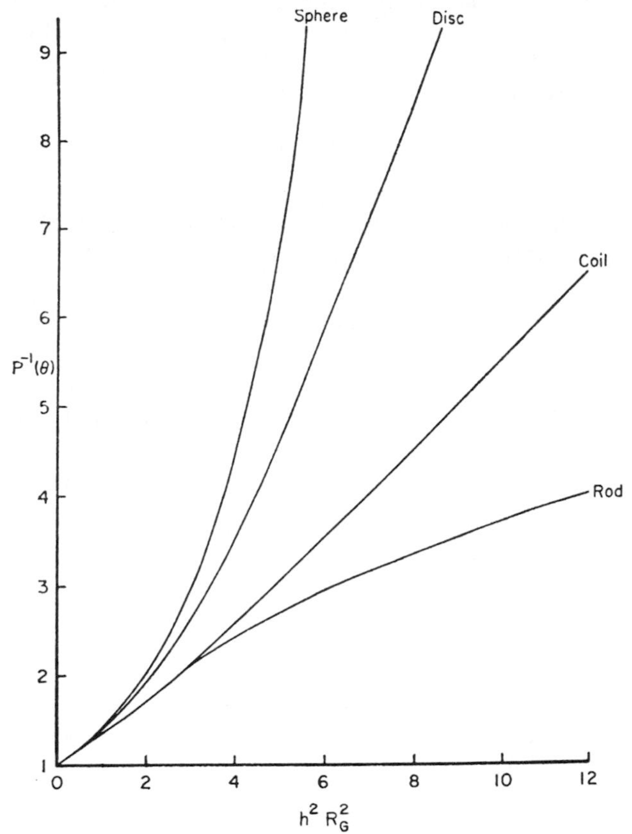

Fig. 8-15 Plot of $\mathbf{P}^{-1}(\theta)$ against $\mathbf{h}^2\mathbf{R}_G{}^2$ for various particle shapes.[21]

spheres, discs, coils, and rods. In accordance with Eq. 8-34 it is seen that all of the curves show an initial slope of $\frac{1}{3}$, a fact used to obtain R_G for any model. As discussed by Geiduschek and Holtzer,[21] shape comparisons are possible only for relatively large particles; for smaller particles the data fall near the origin of Fig. 8-15 and thus do not permit differentiation of shape. In most cases that have been studied the experimental data lie between model curves, and more complex analysis and/or additional details of shape from other experimental methods are required to resolve the ambiguity.

The complete analysis discussed above is not necessary for many problems. For example, in the study of smaller macromolecules for which $P(\theta)$ is essentially unity at all angles, measurements at a single angle (usually 90°) suffice to determine M and B, according to the equation

$$\frac{Kc}{R_{90}} = \frac{1}{M} + 2Bc + \cdots \tag{8-45}$$

For very large particles, the transmitted light I is sufficiently different from the incident light to permit measurement of the turbidity τ given by $-\ln(I/I_0)$ (analogous to absorbance in Beer's law). The turbidity can be related to molecular weight by summing the scattered light over all angles to give

$$\tau = 16\pi KcM/3 \tag{8-46}$$

for a unit volume (1 cm cube).[22]

For multicomponent solutes at low concentrations where nonideality may be neglected, one may sum the scattering contributions from each of the i species:

$$R_\theta = P(\theta)K \sum_i c_i M_i \tag{8-47}$$

This equation is valid provided that concentrations are in refractive index units or, if weight per volume units are used, that all components have the same refractive index increment (a reasonable assumption for protein mixtures). The light-scattering expression for low concentration then becomes

$$\frac{KcP(\theta)}{R_\theta} = \frac{\sum_i c_i}{\sum_i c_i M_i} = \frac{1}{M_w} \tag{8-48}$$

where M_w is the weight-average molecular weight. Light scattering may therefore be used to study macromolecular interactions (e.g., by analyzing self-association with the Steiner approach as discussed earlier). It can be shown that the radius of gyration that is experimentally determined for multicomponent systems represents a z-average value.

5 SMALL-ANGLE X-RAY SCATTERING[28,30,31]

We have seen in the previous section that for typical small proteins (maximum dimension 100 Å), light-scattering data can be used to obtain molecular weights but do not give information on size or shape, because $P(\theta)$ is near unity for these particles at the wavelength of ordinary light-scattering sources (e.g., 5460 Å). This problem could be circumvented by using shorter wavelength radiation, but it is not until one reaches the X-ray region that measurements can be made. This is due in part to a lack of sufficiently intense sources and detectors, but primarily to absorption of radiation in the intermediate wavelength region by matter. Longer wavelength X-rays are excluded for the same reasons. The most frequently used source is copper $K\alpha$ radiation ($\lambda = 1.54$ Å). The nature of the scattering (or diffraction) of X-irradiation can be predicted from Bragg's law, $\lambda = 2d \sin\theta$, where d is the distance between scattering points. The diffraction angle is calculated to be 0.45° for spacings of 100 Å and 0.045° for spacings of 1000 Å. This is the region in which measurements must be made. In addition, in the study of solutions information is obtained from the shape of the central maximum, and therefore measurements down to very small angles are required.

Small-angle X-ray scattering can give information on many types of systems, including analysis of packing arrangements in fibers and other macromolecular aggregates and determination of surface to volume ratios. Some of these applications are discussed in the published proceedings of a conference on small-angle X-ray scattering.[31] A related technique, low-angle neutron scattering also shows potential for these problems, for example in the study of quaternary structure of ribosomes.[32] In this section we confine our discussion to the study of solutions, the intent of which is to gain similar information to that obtained with light scattering on very large particles. The theory and practice of the technique are rather complex and only a brief overview will be presented here.

The interaction of radiation with matter in the absence of absorption may be expressed in terms of the same basic theories for all wavelengths. The resultant distribution of intensity is a function solely of the electronic geometry of the system and λ. The final equations, however (e.g., for X-ray diffraction and light scattering), do not necessarily resemble each other. One of the reasons for this is that in the more general treatment the approximation made to obtain the Rayleigh scattering formula for light assumes that the wavelength used is large compared to the dimensions of the particle, while in Thomson's classical formula for the scattering of *X*-rays the opposite is true.[26] Thomson's formula for the scattering from one electron may be written

$$I_e(h) = 7.9 \times 10^{-26} \frac{I_0}{r^2}\left(\frac{1 + \cos^2 2\theta}{2}\right) \tag{8-49}$$

where $I_e(h)$ is the intensity of scattered radiation at the angle h, where h is given by $(4\pi/\lambda) \sin \theta$ and the scattering angle is 2θ. [The angles of measurement are all so near 0° that $\frac{1}{2}(1 + \cos^2 2\theta)$ can be taken as unity.] The averaged intensity of scattered radiation from a randomly moving particle may be written in terms of the average value of the square of the structure factor $F(h)$, that is,[28,30]

$$\overline{I(h)} = I_e(h)\overline{F^2(h)} \tag{8-50}$$

Guinier has shown that the average squared structure factor may be expressed, to a good approximation, in terms of the radius of gyration R_G as follows:

$$\overline{F^2(h)} = n^2 \exp(-h^2 R_G{}^2/3) \tag{8-51}$$

where n is the total number of electrons. The result is exact at small angles.

At zero angle, $\overline{F^2}(0) = n^2$, so that by expanding the exponential in Eq. 8-51 we obtain

$$\frac{\overline{F^2}(h)}{\overline{F^2}(0)} = 1 - \frac{h^2 R_G{}^2}{3} + \cdots \qquad (8\text{-}52)$$

which is identical to the result obtained for $P(\theta)$ at small angles for light scattering. Equation 8-51 may be written in terms of the observed intensity to give a Gaussian curve

$$I(h) = I_0 \exp\left(-h^2 R_G{}^2/3\right) \qquad (8\text{-}53)$$

or in logarithmic form

$$\ln I(h) = \ln I_0 - \frac{h^2 R_G{}^2}{3} \qquad (8\text{-}54)$$

so that a plot of ln $I(h)$ vs. h^2 will be a straight line. The equation is approximate as noted before, and the radius of gyration should therefore be obtained from the limiting slope.

The experimental determination of the radius of gyration from the logarithmic plot is shown for several typical proteins in Fig. 8-16*a*. In cases where the line is not straight, extrapolation to the correct value at $\theta = 0$ is obtained by plotting the value of the slope against h^2 (usually a straight line at small h) and extrapolating to obtain the intercept at $\theta = 0$. The exact

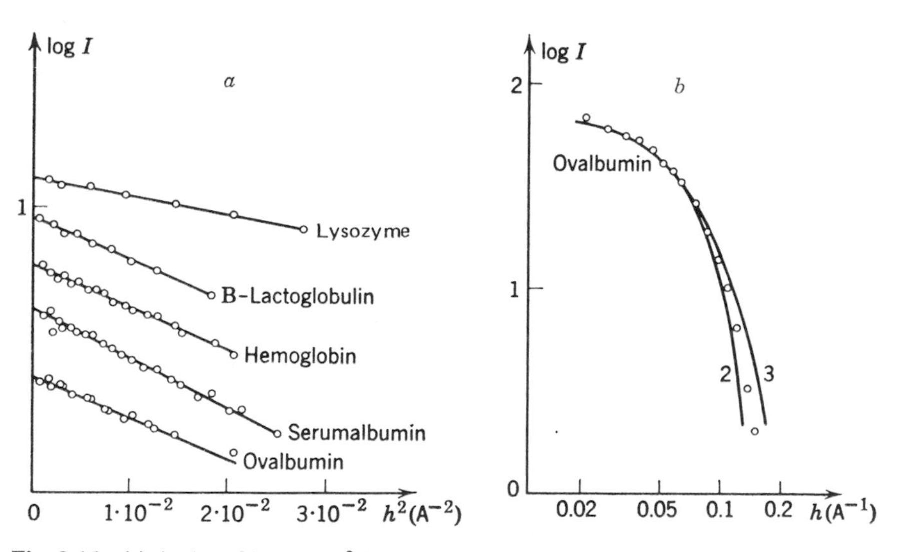

Fig. 8-16 (*a*) A plot of log **I** vs. $\mathbf{h}^2$ for solutions of several different proteins. (*b*) Log **I** vs. log **h** plot for ovalbumin (open circle). The solid lines labeled 2 and 3 are theoretical curves for ellipsoids of axial ratios 2 and 3, respectively. [H. N. Ritland, P. Kaesberg, and W. W. Beeman, *J. Chem. Phys.*, **18**, 1237 (1950).]

shape of the scattering intensity in the central maximum can be calculated for various models and compared to experimental observations. One such analysis is shown in Fig. 8-16*b* for ovalbumin considered as a prolate ellipsoid of revolution. The uncertainty in shape determination based on comparison to models is similar to that discussed for light scattering.

In the above analysis, accurate determination of absolute intensities was not required. Although not a trivial procedure, this measurement can now be made with considerable accuracy with the result that the weight-average molecular weight can be obtained from the data as well. The exact formulation is presented in References 28 and 30.

Finally, we wish to call attention to the fact that low-angle X-ray scattering may be used in favorable cases to compute radial (and other) electron density distributions. Some elegant examples of these calculations on small RNA viruses in solution are reviewed by Anderegg in Reference 31. In order to perform the calculation for particles with a center of symmetry, the electron density as a function of radius $D(r)$ is defined such that $4\pi^2 D(r)\,dr$ is the number of electron pairs with separation between r and $r + dr$. The structure factor can then be written as

$$\overline{F^2}(h) = \int_0^\infty D(r)\,\frac{\sin hr}{hr}\,4\pi r^2\,dr \qquad (8\text{-}55)$$

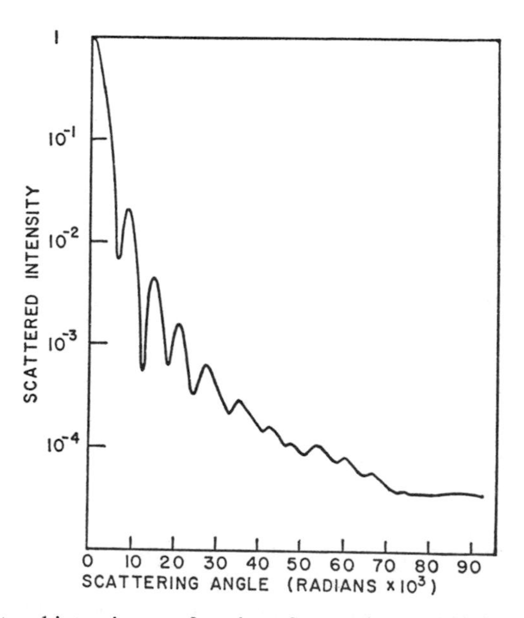

Fig. 8-17 Scattered intensity as a function of scattering angle for the protein shell of wild cucumber mosaic virus. [J. W. Anderegg, P. H. Geil, W. W. Beeman, and P. Kaesberg, *Biophys. J.*, **1,** 657 (1961).]

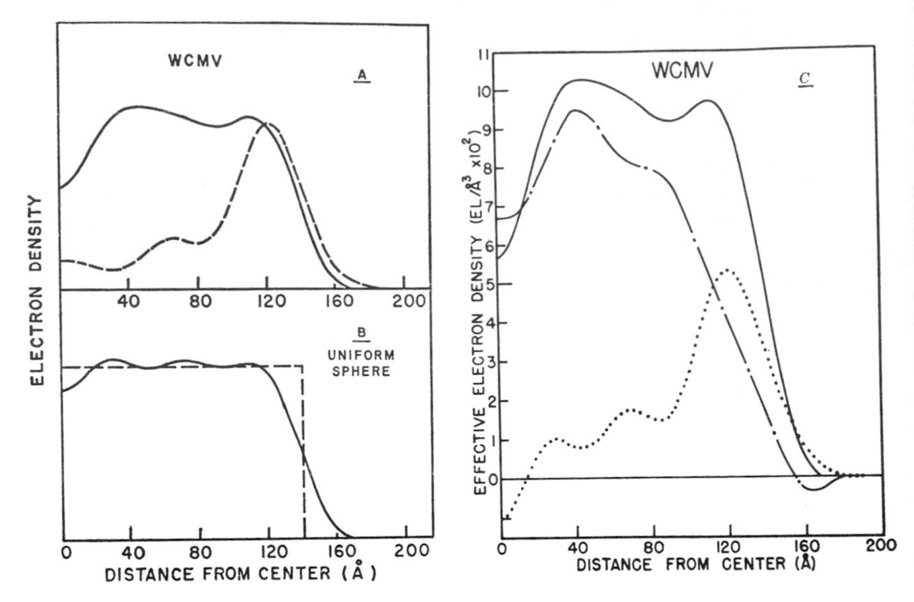

Fig. 8-18 (A) Radial density distributions for wild cucumber mosaic virus: intact virus (solid line); protein shell (broken line). Data were obtained for Bragg spacings out to 40 Å and included six maxima. (B) Theoretical scattering curve for a uniform sphere of 140 Å radius calculated as in (A). The broken line shows the true uniform density distribution of a sphere. (C) Density distributions of (A) normalized by substracting solvent contributions: intact virus (solid line); protein shell (dotted line). The difference curve (· — ·) reflects the distribution of ribonucleic acid in the intact virus. [J. W. Anderegg, P. H. Geil, W. W. Beeman, and P. Kaesberg, *Biophys. J.*, **1,** 657 (1961).]

and for spherically symmetric particles

$$F(h) = \int \rho(r) \frac{\sin hr}{hr} 4\pi r^2 \, dr \tag{8-56}$$

where $\rho(r)$ is the electron density at the radius r. The magnitude of $F(h)$ is obtained from the square root of the intensity curve. The phase for each observed maximum (i.e., the sign, + or −) is guessed, or the *form* of $F(h)$ is predicted by calculation for a model (for example, alternating phase signs are obtained for a hollow sphere model). The electron density is then calculated by Fourier transformation of Eq. 8-56.

In order to carry out such analyses it is necessary to obtain scattering curves with at least several maxima. An experimental curve for the protein shell of wild cucumber mosaic virus is shown in Fig. 8-17 and resultant electron density maps are shown in Fig. 8-18. The general features of a hollow protein sphere filled with nucleic acid in the virus are clearly indicated.

REFERENCES

1. T. Svedberg and K. O. Pedersen, *The Ultracentrifuge*, Oxford University Press, Oxford, 1940.
2. H. K. Schachman, *Ultracentrifugation in Biochemistry*, Academic Press, New York, 1959.
3. H. Fujita, *Mathematical Theory of Sedimentation Analysis*, Academic Press, New York, 1962.
4. J. H. Coates, "Ultracentrifugal analysis," in *Physical Principles and Techniques of Protein Chemistry*, Part B., S. J. Leach, Ed., Academic Press, New York, 1970.
5. G. N. Lewis and M. Randall, *Thermodynamics*, McGraw-Hill, New York, 1923.
6. K. E. Van Holde and R. L. Baldwin, "Rapid attainment of sedimentation equilibrium," *J. Phys. Chem.*, **62,** 734 (1958).
7. E. G. Richards and H. K. Schachman, "Ultracentrifuge studies with Rayleigh interference optics. I. General applications," *J. Phys. Chem.*, **63,** 1578 (1959).
8. F. E. LaBar, "Ultracentrifuge studies with Rayleigh interference optics. I. General application," *Proc. Nat. Acad. Sci., U. S.*, **54,** 31 (1965).
9. R. T. Simpson and J. L. Bethune, "Ultracentrifuge time-lapse photography. Determination of molecular weights," *Biochemistry*, **9,** 2745 (1970).
10. P. A. Charlwood, "Alternate procedure for equilibrium determinations in the ultracentrifuge," *J. Polym. Sci., Part C*, **16**(3), 1717 (1967).
11. D. A. Yphantis, "Equilibrium ultracentrifugation of dilute solutions," *Biochemistry*, **3,** 297 (1964).
12. D. C. Teller, T. A. Horbett, E. G. Richards, and H. K. Schachman, "Ultracentrifuge studies with Rayleigh interference optics. III. Computational methods applied to high-speed sedimentation equilibrium experiments," *Ann. N. Y. Acad. Sci.*, **164,** 66 (1969).
13. D. E. Roark and D. A. Yphantis, "Studies of self-associating systems by equilibrium ultracentrifugation," *Ann. N. Y. Acad. Sci.*, **164,** 245 (1969).
14. K. E. Van Holde, G. P. Rossetti, and R. D. Dyson, "Sedimentation equilibrium of low-molecular-weight associating solutes," *Ann. N. Y. Acad. Sci.*, **164,** 279 (1969).
15. E. T. Adams, Jr., "On the significance of average molecular weights from sedimentation equilibrium experiments," *Proc. Nat. Acad. Sci., U. S.*, **51,** 509 (1964).
16. R. H. Haschemeyer and W. F. Bowers, "Exponential analysis of concentration or concentration difference data for discrete molecular weight distributions in sedimentation equilibrium," *Biochemistry*, **9,** 435 (1970).
17. Th. G. Scholte, "Determination of the molecular weight distribution of polymers from sedimentation-diffusion equilibria," *Ann. N. Y. Acad. Sci.*, **164,** 156 (1969).
18. J. Vinograd and J. E. Hearst, "Equilibrium sedimentation of macromolecules and viruses in a density gradient," *Prog. Chem. Org. Natural Products*, **20,** 272 (1962).
19. J. Vinograd, "Sedimentation equilibrium in a buoyant density gradient," in *Methods in Enzymology*, Vol. 6, S. P. Colowick and N. O. Kaplan, Eds., Academic Press, New York, 1963, p. 854.
20. M. S. N. Rao and G. Kegeles, "An ultracentrifuge study of the polymerization of α-chymotrypsin," *J. Amer. Chem. Soc.*, **80,** 5724 (1958).

21. E. P. Geiduschek and A. Holtzer, "Application of light scattering to biological systems," *Adv. Biol. Med. Phys.*, **6,** 431 (1958).
22. C. Tanford, *Physical Chemistry of Macromolecules*, Wiley, New York, 1961.
23. K. A. Stacey, *Light-Scattering in Physical Chemistry*, Academic Press, New York, 1956.
24. G. Oster, "Light Scattering," in *Physical Methods of Organic Chemistry*, Vol. I, A. Weissberger, Ed., Wiley-Interscience, New York, 1960, p. 2107; S. N. Timasheff and R. Townend, "Light scattering," in *Physical Principles and Techniques of Protein Chemistry*, Part B., S. J. Leach, Ed., Academic Press, New York, 1970.
25. N. H. Frank, *Introduction to Electricity and Optics*, 2nd ed., McGraw-Hill, New York, 1950.
26. J. Valasek, *Introduction to Theoretical and Experimental Optics*, Wiley, New York, 1949.
27. W. Kauzmann, *Quantum Chemistry*, Academic Press, New York, 1957.
28. A. Guinier and G. Fournet, *Small-Angle Scattering of X-Rays*, Wiley, New York, 1955.
29. E. F. Casassa and H. Eisenberg, "Thermodynamic analysis of multicomponent solutions," *Adv. Protein Chem.*, **19,** 287 (1964).
30. O. Kratky, "X-ray small angle scattering with substances of biological interest in dilute solution," *Prog. Biophys.*, **13,** 105 (1963).
31. H. Brumberger, Ed., *Small Angle X-Ray Scattering*, Gordon and Breach, New York, 1967.
32. D. M. Engelman and P. B. Moore, "A new method for the determination of biological quaternary structure by neutron scattering," *Proc. Nat. Acad. Sci. U. S.*, **69,** 1997 (1972).

IX Optical Methods

There are many different ways in which electromagnetic radiation can interact with matter. In this chapter we restrict the discussion to those optical methods based upon absorption of radiation (and associated dispersive phenomena) that are used in studying protein conformations, primarily in solution. The part of the spectrum of interest is that ranging from ultraviolet on the short wavelength side to infrared (Fig. 9-1). Absorption of radiation in general involves the disappearance of some of the incident light energy at particular wavelengths and its conversion in most cases to heat. The phenomenon is best understood quantum mechanically, in terms of the excitation of molecules to discrete higher energy states due to the absorption of light in units of quanta. Absorption in the ultraviolet and visible region involves transitions between electronic energy levels whereas infrared absorption occurs for transitions between energy levels of molecular vibrations. After excitation, return to the ground state occurs through re-emission of light of a longer wavelength (fluorescence or phosphorescence) or, more commonly,

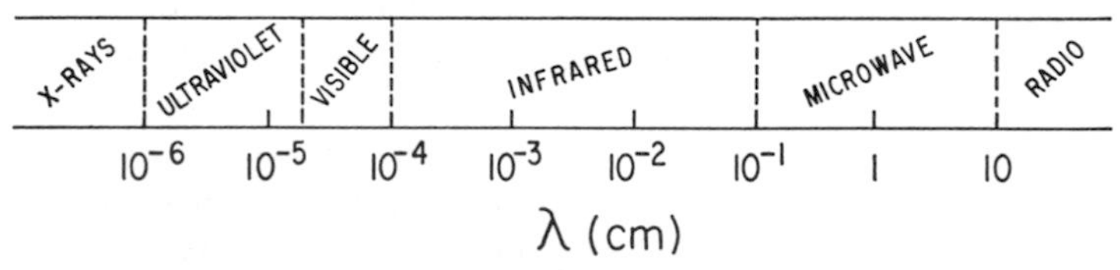

Fig. 9-1 Wavelengths in centimeters associated with various regions of the electromagnetic spectrum. Wavelengths in the ultraviolet region are expressed in nanometers (nm), for example, 10^{-5} cm in the ultraviolet region equals 100 nm. The visible region includes the range of about 380 nm (violet) to 760 nm (red) or 3800 to 7600 Å. Radiation in the infrared region is described by wavelength in cm or microns or by wave numbers ($\bar{\nu} = 1/\lambda$) in units of cm^{-1}.

through processes not involving emission of photons (the electronic energy eventually being converted to thermal energy).

Absorption properties are influenced by symmetry characteristics of molecular structures. If the irradiated substance is isotropic (same optical properties in all directions), its absorption of light can be described by a single parameter that is a function of wavelength. If the substance is anisotropic, its absorption bands may show different magnitudes depending on direction (linear dichroism). For optically active molecules, absorption differs for right- and left-circularly polarized light (circular dichroism). Table 9-1 presents the various absorption phenomena associated with electronic transitions of molecules.

Table 9-1 Absorption and dispersion phenomena arising from electronic transitions of molecules.[29]

Absorption	Dispersion
Molar (molecular) extinction (ε)	Polarizability or refractive index (n)
Linear dichroism ($\varepsilon_{\parallel} - \varepsilon_{\perp}$)	Birefringence ($n_{\parallel} - n_{\perp}$)
Circular dichroism ($\varepsilon_l - \varepsilon_r$)	Circular birefringence ($n_l - n_r$) (optical rotation)

As shown in Table 9-1, the electromagnetic interactions responsible for each type of absorption also produce characteristic dispersion phenomena. In the first case the transitions responsible for ordinary absorption as measured by an extinction coefficient also cause refraction or bending of the light beam at wavelengths different from that of the absorption band. Each material is then characterized by a refractive index that is a direct function of the polarizability of the electron cloud about the atoms of the substance. Refraction can perhaps be understood best from the point of view of the classical electron theory of optics. The process is viewed as the re-emission of light at the same frequency by electrons set into forced oscillation by the incident field. The reduced velocity of the light (compared with that in a vacuum) follows from the retardation in phase of the re-emission. The study of linear dichroism and its associated linear birefringence (difference in refractive index in two directions, e.g., parallel and perpendicular) require oriented specimens that are not usually available for proteins. However, linear birefringence is of importance in the study of hydrodynamic and electrical properties of macromolecules under an applied external field (cf. flow birefringence and transient electric birefringence). Circular dichroism and circular birefringence (optical rotation) are of considerable value in the study of protein conformation, and are discussed in the second part of this chapter.

The reader is referred to the books by Kauzmann[1] and Caldwell and Eyring[37] for a thorough treatment of the classical and quantum theories that describe these phenomena.

1 ABSORPTION AND FLUORESCENCE SPECTROSCOPY

1-1 Ultraviolet Absorption[2–7]

As indicated above, ultraviolet irradiation can induce transitions of molecules from their ground or lowest energy state to higher electronic levels. For simplicity it is common to relate an electronic transition to a particular chromophore such as the phenolic moiety of tyrosine. However, environmental factors influence these properties and it is important to remember that a measured absorption refers to a transition between the ground state of the entire system (chromophore plus environment) to an excited state of the chromophore in that environment.

Most electronic transitions occur in the far ultraviolet region and cannot be measured with conventional instrumentation (the lower wavelength limit is approximately 180 nm); however, a great deal of information about proteins can be obtained from ultraviolet absorption of the peptide bond (185–210 nm region) and the aromatic amino acids (260–280 nm). The measurement of ultraviolet absorption is based upon the familiar law of photometry

$$I = I_0 e^{-\alpha b}$$

where I_0 is the intensity of the incident light, I is the intensity of the transmitted light, α is the absorption coefficient of the material, and b is the thickness of the absorbing layer. The ratio I/I_0 is the *transmittance;* however, it is more useful to express absorption measurements in terms of *absorbance* (A):

$$A = \log_{10} (I_0/I)$$

where I_0 is in practice the intensity of light transmitted by a suitable reference material or solution. Absorbance is usually obtained directly from the spectrophotometer after adjustment of the instrument for I_0. Other names that have been used for the same quantity are absorbancy (A), optical density (O.D.), and extinction (E). Another important unit is *molar absorptivity* (a_m):

$$a_m = A/bc$$

where b is the path length (centimeters) in the spectrophotometer cell and c is the concentration in moles/liter. Other terms for this quantity are molar absorbancy index (a_m) and molar (or molecular) extinction coefficient (ε_m). The expressions $A_{1\,\text{cm}}^{1\%}$ and $E_{1\,\text{cm}}^{1\%}$, which frequently appear in biochemical literature, refer to the absorbance of a 1% (by weight) solution in a cell with a 1-centimeter optical path. Wavelengths in the ultraviolet region are given in nanometers (nm) now used in place of the older term millimicrons (mμ). Ångstroms (Å) are also sometimes used for this part of the spectrum (10 Å = 1 nm).

Most proteins in solution show an absorption maximum at about 280 nm, and absorbance measurements at this wavelength provide a convenient measure of protein concentration when the absorptivity of the protein per mole or per unit weight is known. The 280-nm absorption is due principally to tryptophan and tyrosine residues in the protein. Phenylalanine absorption occurs at slightly shorter wavelengths (about 255 nm). The ultraviolet spectra of these amino acids in solution at pH 6 are shown in Fig. 9-2. Other chromophores that contribute to a lesser extent in this region are disulfide groups (λ_{max} = 250 nm, a_m = 300) and ionized sulfhydryl groups (λ_{max} = 235 nm, a_m = 3000). The latter have been discussed in relation to the dissociation of cysteine (cf. Chapter 2). Molar absorptivities for a variety of proteins at these and other wavelengths have been tabulated.[8]

For proteins that lack cystine it is possible to determine tryptophan and

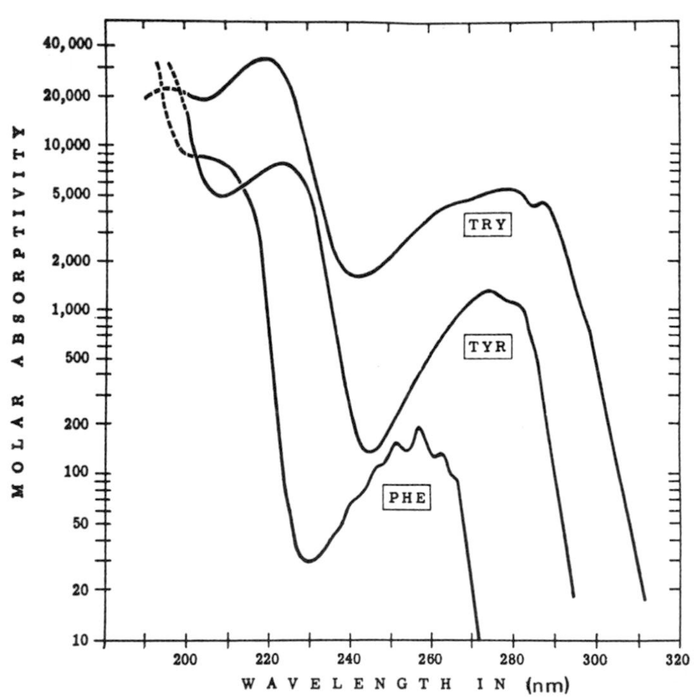

Fig. 9-2 Absorption spectra of the aromatic amino acids tryptophan, tyrosine, and phenylalanine at pH 6.[3]

tyrosine content of the protein from absorbance measurements (in 6 *M* guanidine hydrochloride, 0.02 *M* phosphate buffer, pH 6.5), according to the equations:[9]

$$A_{288} = 4815\ M_{\text{trp}} + 385\ M_{\text{tyr}}$$
$$A_{280} = 5690\ M_{\text{trp}} + 1280\ M_{\text{tyr}}$$

When cystine is present, only tryptophan can be determined by this method; tyrosine is determined independently from the difference in 295 nm absorbance at pH 12 and pH 7 (molar difference absorptivity = 2480 in 6 *M* guanidine hydrochloride), which reflects the ionization of tyrosine.

A number of factors can influence the spectral characteristics of the aromatic residues in proteins, producing a shift in the wavelength of maximum absorption or a change in the area under the curve (oscillator strength). One of the most striking is the change in the spectrum of tyrosine that occurs with ionization of the phenolic group (cf. Chapter 2). This property is used, as described above, for quantitation of tyrosine and for spectrophotometric titration of tyrosine residues in proteins. Another factor that can directly affect the chromophoric group is its involvement in hydrogen bonding, as well as environmental conditions such as polarizability of the surrounding medium, proximity of charged groups, or "coupling" with other chromophores. Changes in absorption of the aromatic amino acid residues are sometimes used to follow reactions involving these groups in particular or in observing alterations in secondary and tertiary structure of the protein which affect the spectrum of these residues.

Environmental perturbations include those identified by the terms hyperchromism and hypochromism, which refer to an increase or decrease, respectively, in absorptivity at a particular wavelength. This phenomenon is interpreted in terms of the coupling of electronic transitions of unequal energy such that absorption intensity (or oscillator strength) is shifted from one to the other. The result is increased absorption of the incident beam by one transition at the expense of the other. Another type of environmental perturbation is that resulting from the coupling of electronic transitions of the same energy which occurs when identical chromophores are in suitable proximity and orientation to each other. This phenomenon, known as exciton splitting, causes a single energy level to be split into two or more levels, depending on the number of interacting equivalent chromophores. This type of spectral modification can be expected for chromophores in ordered arrays such as peptide groups in an α-helix or β-structure in proteins and polypeptides.

Because changes in absorptivity due to environmental conditions are often rather small, it is customary to measure differences directly in a double-beam spectrophotometer by comparing the protein under study with the

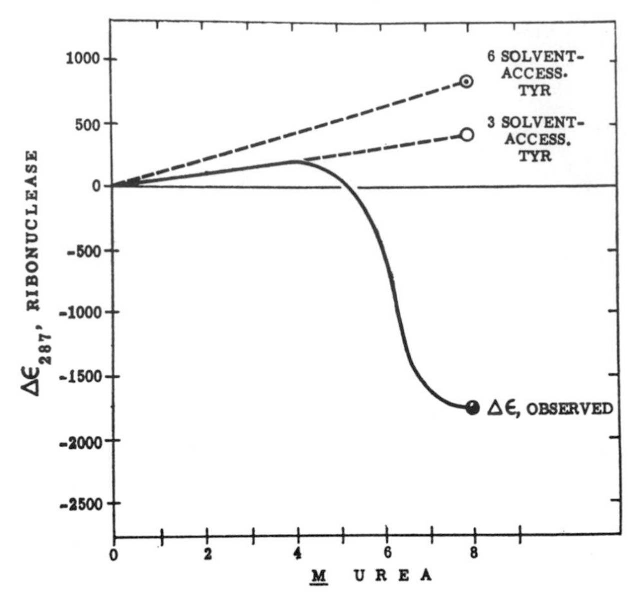

Fig. 9-3 The effect of urea concentration on the molar absorptivity of ribonuclease at 287 nm. Absorbance was measured versus a reference solution of ribonuclease containing no urea; experimental data given by the solid line. The lower dashed line is a linear extrapolation of the experimental curve, showing three tyrosine residues accessible to solvent. The upper dashed line gives experimental data for oxidized ribonuclease (six solvent-accessible tyrosine residues). [From Reference 3, based on data of C. C. Bigelow and I. I. Geschwind, *Comp. rend. trav. lab. Carlsberg. sér. chim.*, **31,** 283 (1960).]

same protein in some reference standard state. Difference spectra obtained using various solvents (solvent perturbation method) permit deductions about the accessibility of chromophores in a protein, such as tyrosine and tryptophan, to the solvent. This is illustrated in the study of the urea denaturation of ribonuclease shown in Fig. 9-3. Extrapolation of the results at low urea concentrations indicates that three of the six tyrosine residues in the protein are exposed to the solvent. If all six had been accessible to the solvent, the extrapolation would be like that shown for oxidized ribonuclease. Above 4 *M* urea the difference spectrum reflects the denaturation of the protein and the transfer of the three buried tyrosine residues from a protein environment to the aqueous urea environment. Solvent exposure studies have also been carried out by thermal perturbation methods (direct comparison of absorption of the same protein solution at two different temperatures).

Changes in spectra frequently can be correlated with alterations in other properties such as optical rotation. A classical application of these two techniques has been the determination of "melting curves" of the nucleic acids. In proteins, however, the relationship may be complex. In the heat

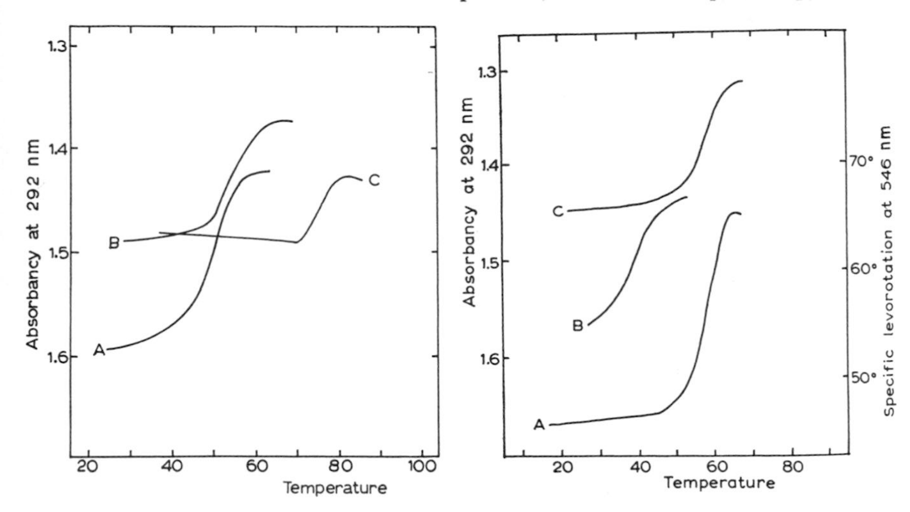

Fig. 9-4 Thermal denaturation of lysozyme followed by absorbance and optical rotation. *Left:* Spectrophotometric measurements in 8 *M* urea, 0.1 *M* KCl (Curve *A*); 0.1 *M* KCl, pH 2.2 (Curve *B*); 0.1 *M* KCl (Curve *C*). *Right:* Spectrophotometric and polarimetric measurements in dioxane solutions. Specific rotation in 20% dioxane, 0.1 *M* KCl (Curve *A*); absorbance in 40% dioxane, 0.1 *M* KCl (Curve *B*); absorbance in 20% dioxane, 0.1 *M* KCl (Curve *C*). [From J. G. Foss, *Biochim. Biophys. Acta*, **47**, 569 (1961).]

denaturation of lysozyme (Fig. 9-4) a decrease in absorbance at 292 nm provides a convenient measure of the thermally induced conformational transition of the molecule. In lysozyme the correspondence in transition temperature as measured by absorbance, optical rotation, or viscosity is fairly close, although in other proteins significant differences in temperature profile or in other dependencies such as pH may be found for the various conformation-dependent properties. For lysozyme, low pH or the presence of 8 *M* urea causes a large downward shift in the temperature denaturation profile. A similar result is obtained with dioxane, an organic solvent that would be expected to interfere with apolar interactions within the native structure. Comparable profiles for the conformational transition in 20% dioxane are shown by absorbance and optical rotation (Fig. 9-4). In favorable cases, spectrophotometric data of this type, usually obtained as difference spectra, may be used to determine approximate thermodynamic parameters.[10]

The other major region of ultraviolet absorption by proteins is centered around 190 nm and is due to transitions of the peptide group. These have been identified as an $n \rightarrow \pi^*$ transition at about 210 nm (representing the excitation of an electron from a nonbonding orbital to an excited antibonding π orbital) and a $\pi \rightarrow \pi^*$ transition (from a bonding to an antibonding π

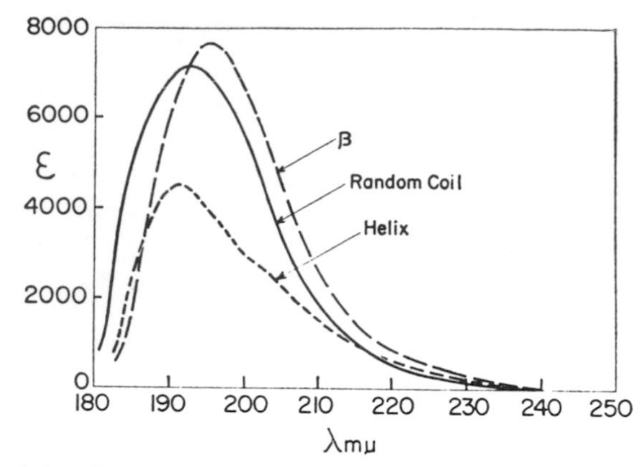

Fig. 9-5 Ultraviolet absorption spectra of poly-L-lysine hydrochloride in aqueous solution: random coil, pH 6.0, 25°C; helix, pH 10.8, 25°C; β-form, pH 10.8, 52°C. [From K. Rosenheck and P. Doty, *Proc. Nat. Acad. Sci., U.S.*, **47,** 1775 (1961).]

orbital) at 185 nm. All of these transitions contribute to the characteristic appearance of the peptide group absorption band. This band is strongly influenced by the conformation of the polypeptide chain. Figure 9-5 shows ultraviolet spectra of poly-L-lysine in three forms: random coil, α-helix, and β-pleated sheet. The decrease in absorptivity (hypochromic effect) occurring upon formation of the α-helix can be interpreted in terms of exciton splitting due to interaction of perpendicular transition dipoles of the α-helical amide bands.[11] The result is that the amide band (at 190 nm in the random coil) is split into two components with a large net hypochromicity in this wavelength region. The β-structure, shown in Fig. 9-5, gives a small shift in λ_{max} and a small hyperchromic effect.

The large hypochromic effect associated with α-helix formation permits study of the helix-random coil conversion in polypeptides by simple measurement of absorbance changes in the lower ultraviolet (190 to 220 nm), and it has been suggested that helical content in proteins might also be determined in this way. However, because of other factors affecting the absorptivity of peptide groups in native proteins as well as the possible contribution of other conformations to the hypochromicity, the application of this method is subject to uncertainty. A similar difficulty exists in the analysis of optical rotatory dispersion data for such processes.

1-2 Fluorescence[12-14]

Studies of fluorescent emission have proved useful for a number of problems in protein structure and interactions. In most cases, investigators have taken advantage of the intrinsic fluorescent chromophores in proteins (the aromatic side chains of tyrosine, tryptophan, and phenylalanine) or naturally bound chromophores such as flavin adenine dinucleotide (FAD). Another more recent approach has involved the introduction of nonbiological fluorescent molecules (extrinsic chromophores) into specific sites in proteins and synthetic polypeptides. These fluorescent "probes" may be noncovalently bound or may be attached through covalent linkages.

Fluorescence occurs when some of the energy absorbed in a transition between atomic or molecular energy levels is re-emitted as light. The type of fluorescence discussed here is that associated with electronic transitions in molecules, in which the excitation is produced by visible or ultraviolet light. As indicated earlier, transitions of electrons to higher energy levels produced by the absorption of electromagnetic radiation are usually followed by radiationless return to the ground state. However, in certain types of chromophores such radiationless "internal conversion" brings the electrons only down to a lower level excited state. In most cases this energy level, like the ground level, is a singlet state, meaning that electron spins are paired, that is, for every electron having a spin (about its own axis) of $+\frac{1}{2}$, there is another with spin of $-\frac{1}{2}$. The excited singlet state has an extremely short lifetime (10^{-9} to 10^{-8} sec) and passes immediately to the ground state with the emission of a photon of light. In almost all cases the emitted light is of lower energy than the absorbed light because of energy reduction by internal conversion, and thus the fluorescence spectrum is shifted to higher wavelengths compared to the absorption spectrum. Figure 9-6 illustrates the relation between absorption and fluorescence for a simple case.

Figure 9-7 shows the fluorescence emission spectra of the three aromatic amino acids occurring in proteins. These can be compared with the ultraviolet absorption spectra of these amino acids (Fig. 9-2). The symbol $F(\lambda)$ represents the light signal measured in a detecting photomultiplier tube

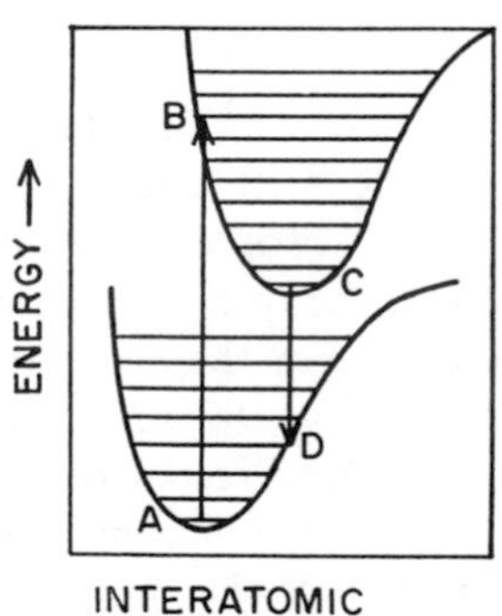

Fig. 9-6 Absorption and fluorescence transitions for a hypothetical diatomic molecule. The absorption transition causes excitation of molecules from A, the lowest vibrational level of the ground electronic state, to vibrational level B of the excited electronic state. During the lifetime of the excited state, transfer of energy through collisions causes the excited molecules to fall to their lowest vibrational levels (level C). Re-emission of radiation (fluorescence) occurs to reach level D of the ground electronic state, followed by internal conversion (radiationless energy transfer) to the lowest vibrational level.[1]

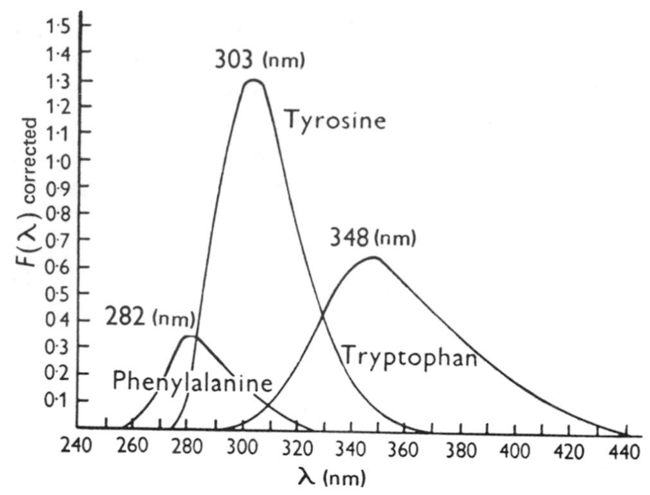

Fig. 9-7 Fluorescence emission spectra of phenylalanine, tyrosine, and tryptophan in neutral aqueous solution. Maxima occur at 282, 303, and 348 nm, respectively. [From F. W. J. Teale and G. Weber, *Biochem. J.*, **65,** 476 (1957).]

corrected for differences of sensitivity with wavelength. In the case of tyrosine a correction has been made because of the overlap of fluorescence and absorption spectra of this molecule. The correction is often necessary depending on the geometry of the measurement because of the reabsorption of that part of the fluorescent emission that falls within the absorption spectrum. The result is that fluorescent signals will be relatively reduced (attenuated) in regions of overlap and increased at higher wavelengths, and thus the experimentally observed spectrum will differ from the true molecular spectrum.

Quantitation of fluorescence is done by calculation of the quantum yield:

$$Q = \frac{\text{quanta emitted}}{\text{quanta absorbed}}$$

where quanta are the standard units of light energy, $h\nu$. In practice this is obtained by comparing the corrected integrated fluorescence with a reference of known Q and equal absorbance. If every molecule that absorbs a photon of light emits a photon upon return to the ground state, the quantum yield will be unity. This maximum is approached by some dyes. For the aromatic amino acids the quantum yields based on the data of Fig. 9-7 are 21% (tyrosine), 20% (tryptophan), and 4% (phenylalanine). More recent determinations are somewhat lower (e.g., 14% for tryptophan[15]). Interactions with molecules in the environment of the chromophore can reduce the quantum yield through external conversion, that is, processes that cause the molecule to return to the ground state without light emission. This is referred to as "fluorescence quenching." Quenching is a critical factor in liquid scintillation counting for example, which depends on the fluorescence induced in certain compounds by the β-emission of such radioactive isotopes as ^{14}C, ^{3}H, or ^{32}P. In that case however, the term is used more loosely to include other sources of signal attenuation (e.g., self absorption) as well.

Although the greatest number of biological applications of fluorescence concern problems outside the scope of this book (see, for example, the book by Udenfriend[12]), the method is becoming increasingly important for proteins. The native fluorescence spectrum of proteins, due to their aromatic amino acid content, is quite markedly dependent on environmental quenching factors and upon the possibilities for energy transfer between chromophores. This is particularly true for the tryptophan spectrum, which thus provides a highly specific conformational probe for proteins. For example, fluorescence studies may be used to investigate the role of tryptophan residues in enzyme active sites. Figure 9-8 shows the effect on the fluorescence of lysozyme produced by the binding of a substrate of the enzyme. The remarkable dequenching of tryptophan fluorescence (increase in quantum yield) and its pH

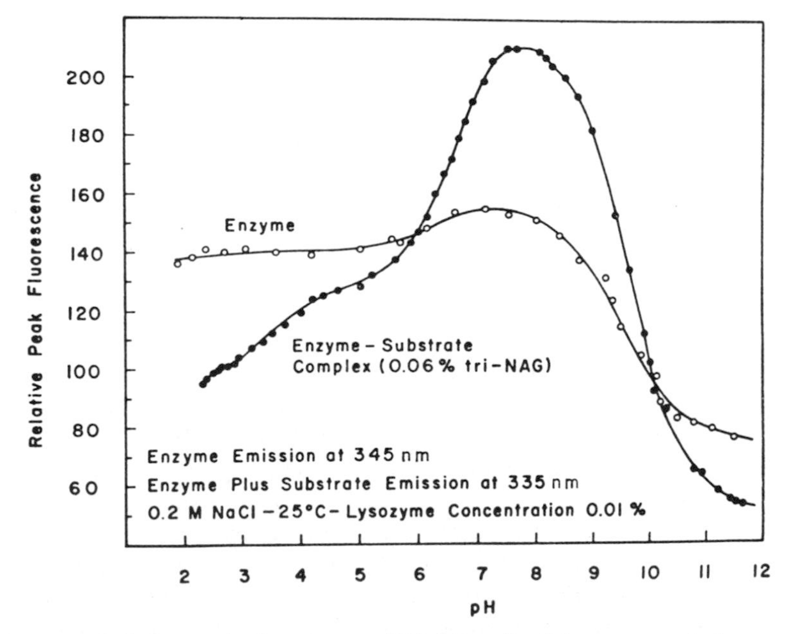

Fig. 9-8 Relative peak fluorescence (320–340 nm) of lysozyme and the complex of lysozyme with tri-*N*-acetyl-D-glucosamine as a function of pH. [From S. S. Lehrer and G. D. Fasman, *Biochem. Biophys. Res. Commun.*, **23,** 133 (1966); *J. Biol. Chem.*, **242,** 4644 (1967).]

dependence was interpreted in terms of the involvement of three tryptophan residues and two carboxyl groups in the binding region.

The study of fluorescence changes has also been used to follow protein denaturation processes and other smaller conformational changes (see Reference 14 for a review of this work). Solvent perturbation effects, as studied by absorption difference spectroscopy, can also be examined by the more sensitive fluorescence methods to obtain information on chromophore accessibility to solvent. Study of phosphorescence is also possible.[16] Because of its specificity, fluorescence can also be used as a rapid quantitative assay for proteins in the presence of many other materials, including nucleic acids.

Energy Transfer. Many biological systems seem to be able to transmit energy between chromophores by a mechanism like that of vibrational transfer between tuning forks of the same pitch. The transfer occurs when an acceptor that absorbs light of a particular wavelength lies close enough to a donor that can potentially emit light of that wavelength upon excitation. If the acceptor is fluorescent, the process can be followed by detection of the fluorescence characteristic of the acceptor after excitation at the wavelength absorbed by the donor. Energy transfer between donors and acceptors, which has been treated theoretically by Förster,[17] is considered to take place by a radiationless mechanism and can be effective over a distance as great as 50 Å. This type of energy transfer has been proposed, for example, to explain the lack of fluorescence of hemoglobin and myoglobin. This mechanism can also occur in small molecule interactions as in the quenching of tryptophan fluorescence in transferrin (a plasma iron-binding globulin) associated with the binding of ferric or cupric ions. Here, fluorescence quenching at 325 nm can be directly correlated with increased absorption by transitions associated with metal–protein interactions.[18]

The study of energy transfer has been suggested as a possible tool for measuring distances between specific sites on macromolecules. If a suitable donor is present or can be incorporated at one site and an acceptor at another, the efficiency of energy transfer to the acceptor after excitation of the donor is a direct function of the distance between them. Stryer and Haugland prepared a series of proline polymers of one to twelve units with an α-naphthyl group as donor at one end of the molecule and a dansyl group as acceptor at the other. As shown in Fig. 9-9, these groups satisfy the spectral criteria for energy transfer measurement: both are highly fluorescent, the donor can be selectively excited due to its absorption at 287 nm where the acceptor shows a minimum, and the donor's emission at about 350 nm overlaps the acceptor's absorption band. The authors then determined the extent of energy transfer between the two groups for the polymers of various lengths by measurement of fluorescence produced as a function of excitation wavelength. Figure 9-10 shows the efficiency of energy transfer plotted vs. the molecular lengths in Å

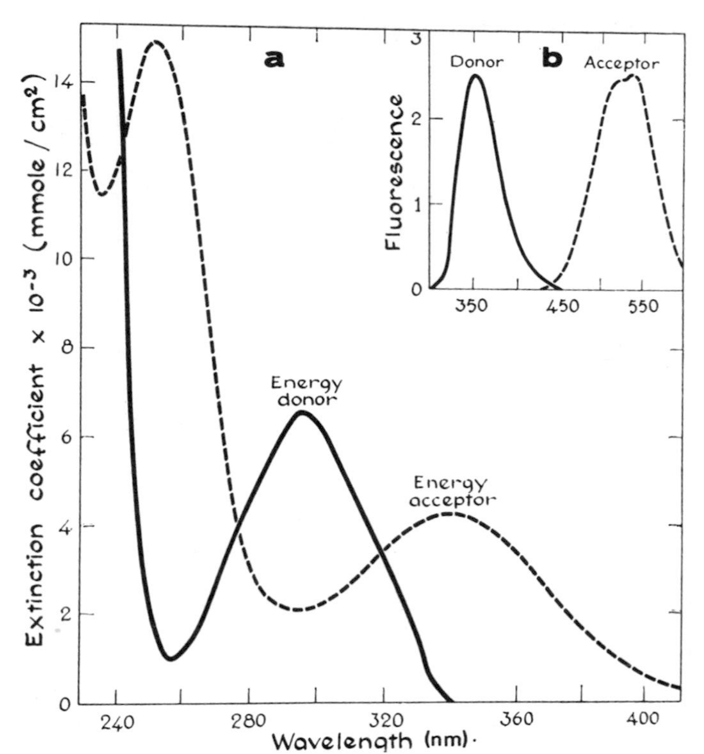

Fig. 9-9 Absorption and fluorescence spectra of an energy donor, the α-naphthyl group (solid lines), and of an energy acceptor, the dansyl group (broken lines), used for study of energy transfer. [From L. Stryer and R. P. Haugland, *Proc. Nat. Acad. Sci., U.S.*, **58,** 719 (1967).]

of the proline polymers. The sensitivity of the method in the 0 to 50 Å range of molecular dimensions is apparent. Short of complete structural analysis by X-ray methods, such distance information can only be obtained in special cases by electric birefringence, a technique requiring very complicated apparatus, or by the use of bifunctional reagents (cf. Chapter 5). The fluorescence technique is far easier to apply in practice, although its use depends upon the presence (or insertion) of suitable donor and acceptor groups in the macromolecule of interest and upon determination of a standard curve for these compounds, such as that of Fig. 9-10. These factors and others, such as the role of orientation of the donor–acceptor pair, are discussed by Stryer.[19] The method has been applied to study the interaction of a dansylated peptide substrate with carboxypeptidase (in the native zinc form and in a cobalt-substituted form). Energy transfer due to the overlap of dansyl absorption with tryptophan fluorescence was observed from the quenching of

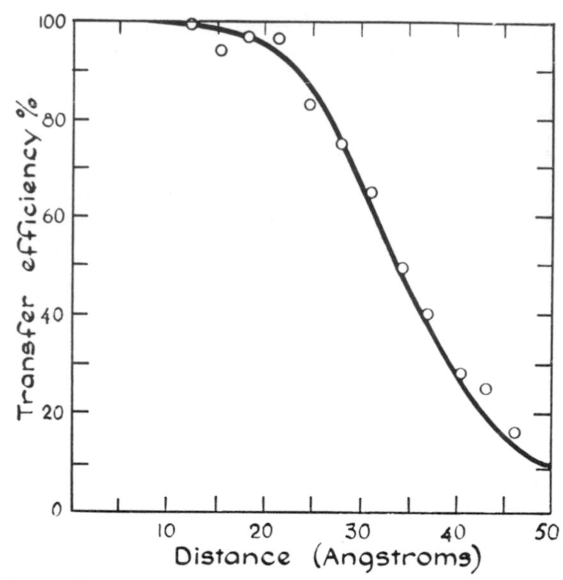

Fig. 9-10 Efficiency of energy transfer as a function of distance in the dansyl-(L-prolyl)$_n$-α-naphthyl polymers with $n = 1$ to 12 prolyl residues. [From L. Stryer and R. P. Haugland, *Proc. Nat. Acad. Sci., U.S.*, **58,** 719 (1967).]

tryptophan fluorescence and the appearance of dansyl fluorescence when substrate was bound to the zinc enzyme. In the cobalt enzyme the latter was also quenched because of overlap of a cobalt absorption band with the dansyl emission, and an estimation of the distance between substrate and metal ion sites could be made.[20]

Fluorescence Polarization. The extent of depolarization in the fluorescence produced by chromophores on proteins after excitation with polarized light has been used to follow Brownian motion of these macromolecules. From the resultant rotational diffusion coefficient, one is able to obtain information on size and shape parameters for the macromolecule in solution.

In this method the sample is irradiated (usually with polarized light) at the wavelength of absorption. The probability of absorption by any particular molecule will depend upon the orientation of the electric dipole of the chromophore (representing the strength and direction of the electronic oscillation) with respect to the direction of polarization of the electric field of the incident light. The largest contribution to the total absorption by the solution comes from those molecules with the closest alignment of electric dipole with the field direction. If no change occurs in this alignment during the lifetime of the excited state (about 10 nsec), the resultant fluorescence will show the same extent of polarization as that represented by the absorption distribution. Thus a maximal value for the fluorescence polarization is obtained. However, in solutions Brownian motion leads to rotational diffusion which randomizes the directions of the chromophores during this interval and produces depolarization additional to that caused by the initial random distribution of chromophores. Because of their rapid rotational diffusion, most small molecules show complete depolarization. In the case of chromophores that are rigidly bound to macromolecules, however, the extent of depolarization depends on the rate of rotational diffusion of the macromolecules. The characteristic times for this process are of the order of 100 to 1000 nsec, or large enough so that only partial depolarization due to diffusion can occur during the lifetime of the excited state. The rotational diffusion coefficient of the protein can then be calculated by comparison of the resultant polarization with that which would be obtained if no diffusion took place. This is done in practice by measurement of polarization (from the fluorescence observed parallel and perpendicular to the incident plane of polarization) as a function of viscosity, and extrapolation of the results to infinite viscosity. This method is probably the simplest available to obtain rotational diffusion coefficients of proteins. It is sensitive to small quantities of material, is simple to use, and does not suffer the experimental limitations common to the hydrodynamic methods. A source of error caused by energy transfer among bound ligands can be eliminated by the use of appropriate equations.[21]

One problem that has existed in the standard method of fluorescence polarization is that the validity of the measurement depends upon the requirement that the chromophore be rigidly fixed in its matrix and not undergo independent rotational motion. This problem can be circumvented by use of a new polarization method that is specifically designed to probe the independent rotational motion of chromophores.[19] In this technique the chromophore is excited with a light pulse in the nanosecond range. The polarization of the emission is then measured as a function of time in the same range. This yields the true rotational relaxation time of the chromophore. If the result is the same as the relaxation time for rotational diffusion of the macromolecule, one may conclude that the chromophore site is rigid, at least over that period of time. Although requiring more complicated equipment, this modification has the advantage that only one experiment need be done (viscosity dependence is not required), and there is a greater possibility of obtaining simultaneous data on rotational diffusion about more than one axis in anisometric macromolecules. The method has been used to investigate the hydrodynamic properties of immunoglobulin G through study of the complex of a fluorescent hapten, ε-dansyl-L-lysine, with anti-dansyl antibody. The results indicate that a good deal of flexibility exists in the antibody molecule, the antigen-binding segments (F_{ab} portions) being capable of free rotation over an angular range of 33°.[22]

1-3 Infrared Spectroscopy[23-33]

Infrared absorption spectroscopy has long served as a powerful tool in the study of the basic structure and properties of molecules. It has been extensively applied in biochemistry for identification and for structural and conformational analysis of small molecules such as carbohydrates, steroids, coenzymes, carotenoids, and metal complexes. A particularly useful reference on infrared applications to biochemical and organic chemical problems is the book by Rao.[26]

The use of infrared spectra for the study of proteins has been somewhat limited because of the experimental requirement for a medium or solvent that is transparent to the radiation and because of the enormous complexity of protein spectra. However, with recent theoretical advances and the increasing use of difference spectra, it is likely that infrared spectroscopy will become much more important for all types of macromolecules.

A rigorous theoretical discussion of infrared spectra based on quantum mechanics can be found in the book by Wilson, Decius, and Cross.[23] Infrared absorption measures transitions that are possible between energy levels representing vibrational modes of molecules. Vibrational modes refer to the movement of nuclei with respect to one another in polyatomic molecules. For a molecule of n atoms there are a total of $3n$ degrees of freedom, of which 3 represent translations, 3 are rotations, and the remaining $3n - 6$ are vibrations. In the special case of linear molecules there are only two independent rotations, and thus $3n - 5$ vibrational degrees of freedom. For example, a diatomic molecule has just one vibrational mode; a linear triatomic molecule such as CO_2 has four vibrational modes, whereas a nonlinear triatomic molecule such as water has three.

All $3n$ degrees of freedom are quantized, that is, only certain discrete energy levels are allowed. The translations, however, can be adequately described in most cases by the equations of classical physics, since the energy states are sufficiently close together to produce a continuum. Transitions between rotational energy levels involve absorption of light at discrete wavelengths and, in the case of small molecules in the vapor state, can be measured by microwave spectroscopy. Data for these transitions yield information on bond distances and angles. Transitions between vibrational energy levels are studied by infrared spectroscopy since the energy of infrared radiation ($E = h\nu$, where h is Planck's constant and ν is the frequency of the radiation) is of the same order of magnitude as the spacing between vibrational energy levels. Infrared radiation is usually specified in terms of its wavelength in microns μ ($1\ \mu = 10^4$ Å $= 10^{-4}$ cm) or in wave numbers, $\bar{\nu} = 1/\lambda$ in cm^{-1}. The infrared includes the wavelength range of 0.8 to 200 μ or, in wave numbers, 50 to 12,500 cm^{-1}.

A simple diatomic molecule may be thought of as two balls connected by a "Hooke's law spring" with a strength determined by the force constant of the bond between them. The system is capable of vibration at certain discrete frequencies only. These frequencies define a set of energy levels, the lowest of which is termed the ground state. Under the influence of radiation of infrared frequency molecules are excited to higher vibrational energy levels (the principal transition is from the ground state to the first "excited" state), and the absorption of the incident radiation can be measured. For example, in HCl vapor the major transition, the so-called stretching vibration produces an infrared absorption band at 2855 cm^{-1}.

In a triatomic molecule, such as H_2O, the two hydrogens may still be thought of as being connected to the oxygen by springs, but the vibrations of the two springs are not independent. Instead, H_2O has $3n - 6$ or 3 normal modes of vibration, each with a characteristic normal (or fundamental) frequency. A normal mode is a vibration in which all atoms move at the same frequency and are in phase, that is, they pass through their equilibrium positions simultaneously and reach their maximum displacements at the same time. The three normal mode vibrations of water are illustrated in Fig. 9-11.

Fig. 9-11 Normal modes of vibration of the water molecule in mass-weighted coordinates. To reflect actual relative motions in space, the arrows representing displacements of the oxygen atom should be only one-fourth as long as shown here.[23]

All observable vibrational transitions of the water molecule can be accounted for in terms of these three fundamental types of motion or as a linear combination of them. As the number of atoms in the molecule increases, the normal mode vibrations become more numerous and more complicated; they are often described by such colorful terms as scissoring, twisting, rocking, or wagging. The resulting infrared spectrum, as illustrated in Fig. 9-12 for acrylonitrile, represents a complicated "fingerprint" of the vibrational motions of the molecule, and is of great importance in identification and comparison of substances.

The analysis of infrared spectra is greatly facilitated by the fact that certain groupings in molecules give rise to characteristic "group frequencies," which in most cases are not strongly affected by neighboring atoms. This occurs when the displacement of atoms outside the group is negligible for that particular normal mode. The C—H stretching vibration, for example, can

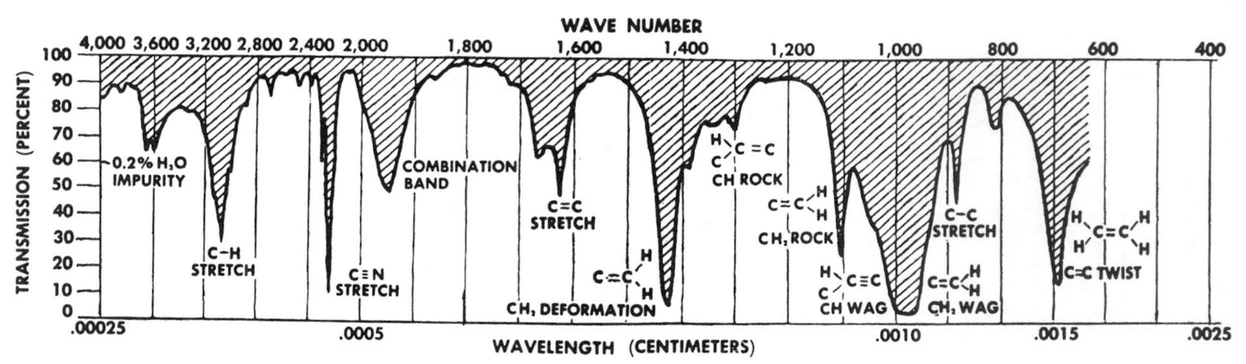

Fig. 9-12 Infrared spectrum of acrylonitrile (CH_2CHN). The curve represents the amount of energy transmitted by a sample of material as a function of wavelength of the incident radiation. The observed bands correspond to the natural frequencies of molecular vibration where energy is strongly absorbed. (From B. Crawford, Jr., "Chemical analysis by infrared," Scientific American, October 1953.)

be characteristically located at about 3100 cm^{-1} because its high frequency is not affected by the slow vibrations of other substituents on the carbon atom. Similarly, the carbonyl group stretching frequency, occurring in the range of 1600 to 1800 cm^{-1}, is readily identifiable in many compounds. On the other hand, since group frequencies can be affected to varying degrees by environmental conditions, care must always be taken in the assignment of infrared bands to particular groups.

An important application of infrared spectroscopy to protein structure has been in the detection of hydrogen bonding, although studies have been limited largely to model compounds and polypeptides in nonaqueous solvents. It is found that a shift in the characteristic frequencies of the C=O and N—H vibrations occurs when these groups are involved in hydrogen bonding, and the effect can be readily quantitated. For example, when a simple compound such as *N*-methylacetamide is in equilibrium between a hydrogen bonded and free form, two absorption bands are observed, the equilibrium constant can be determined as a function of temperature, and thus the thermodynamics of the process established. This compound has been studied as a model for the peptide amide group of proteins. From these results Klotz and Franzen were able to conclude that in aqueous solution interamide hydrogen bond formation is not energetically favored, thus casting doubt on the importance of interpeptide hydrogen bonds in the stabilization of protein configurations in aqueous solution (cf. Chapter 6).

Shifts in group frequencies produced by isotopic substitution have been used to follow isotopic exchange reactions (cf. hydrogen exchange, Chapter 11). This phenomenon, called an isotope effect, is to be distinguished from the term kinetic isotope effect which refers to the alteration in rate of a chemical reaction caused by isotopic substitution. Vibrational isotope effects can be helpful in assigning infrared bands to particular groups. For example, the identification of an N–H associated absorption band at 6000 to 6700 cm^{-1} in ribonuclease was made by obtaining a difference spectrum between the fully deuterated protein in D_2O and the native protein in D_2O.[34]

Although the infrared spectrum of a protein harbors a great deal of information about protein structure and interactions, most of it cannot be deciphered. Some help can be obtained from infrared dichroism, in which one measures the absorption of radiation polarized with its electric vector parallel or perpendicular to the axis of an oriented protein specimen. The degree of absorption by each group of atoms responsible for the vibrational transition observed depends on the orientation of its *transition moment* with respect to the electric vector of the incident radiation. Within any particular group of atoms the transition moment is a complicated function of the electronic environment. In a diatomic molecule the transition moment lies along the bond axis, and this is often assumed to be approximately true for diatomic groups that exhibit characteristic frequencies in polyatomic molecules. When a protein specimen is oriented, certain groups such as the carbonyl groups in an α-helical segment may all lie with their transition moments in the same direction. Absorption will depend upon the relation of this orientation to the direction of the electric vector of the incident light. The result ranges from no absorption when the electric vector is perpendicular to the transition moment to a maximum absorption when they are parallel. By measuring absorption with light polarized parallel and perpendicular to the axis of the sample (e.g., fiber axis or one axis of a crystal), one determines the dichroic ratio (absorbance ⊥/absorbance ∥) at each wavelength of interest. If this number is greater than one, the transition is said to exhibit perpendicular dichroism (⊥); if less than unity, parallel (∥) dichroism.

Infrared dichroism has proved to be valuable in evaluating the contribution of the α-helix and the β-pleated sheet conformation in proteins for which geometrically oriented specimens can be prepared. In the α-helix the interpeptide hydrogen bond (cf. Fig. 1-2) is nearly parallel to the helical axis; in the β-structure (cf. Fig. 14-4) the hydrogen bonds are perpendicular to the direction of the chains (corresponding to the fiber axis in oriented specimens). Therefore, if the C=O stretching vibration occurred as the principal component of an observed vibrational band, it would show ∥ dichroism in the α-configuration and ⊥ dichroism in the β-configuration. The same result would be expected for the N—H stretching vibration, whereas the N—H deformation mode, in which motion is principally in components perpendicular to the hydrogen bond axis, should exhibit opposite properties. This is exactly what Ambrose and Elliott found when they analyzed the dichroic properties of oriented α- and β-structures in fibrous proteins and polypeptides (Table 9-2). The agreement of these findings with predictions

Table 9-2 Dichroism of the principal infrared absorption bands of the peptide group.[a]

	Frequency (cm^{-1})		Dichroism	
Assignment	α	β	α	β
N—H stretch	3300	3300	∥	⊥
C=O stretch (amide I)	1660	1640	∥	⊥
N—H deformation (amide II)	1545	1525	⊥	∥

[a] From E. J. Ambrose and A. Elliott, *Proc. Roy Soc.*, **A205,** 47 (1951).

based on the Pauling and Corey α-helix hastened the early acceptance of this model. A more recent study of α- and β-structural components using infrared spectral data has been reported for keratin.[35]

New experimental and theoretical developments have greatly increased the potential of infrared investigations of proteins. The use of difference spectra in studying protein reactions is likely to prove as valuable in the infrared as it has been in the ultraviolet region. It permits the selective observation of certain molecular vibrations in the presence of high background absorption, and should be particularly useful in studying interactions of proteins with small molecules such as enzyme substrates or inhibitors. It has been shown that good results are obtained when the ligand contains a triple bond, because in that case its absorption band is well separated from most of the diffuse protein absorption.[36]

Theoretical advances in infrared spectroscopy have included the analysis of band splitting and frequency shifts due to "coupling" of vibrational energy levels. More is being learned, too, about the nature of the normal modes in proteins. For example, it had long been assumed that the 1650 to 1700 cm^{-1} absorption in proteins (called the amide I band) was a pure carbonyl stretching vibration with its transition moment along the C═O bond. This band is now recognized to be a normal mode with about 80% of its potential energy associated with the C═O stretch and the rest equally distributed between a C—N stretching and an N—H bending motion; the direction of the transition moment is 17° to the C═O bond axis. Similar information on other normal modes has confirmed the relative validity of earlier assignments, and has been theoretically developed in the coupled oscillator models of Miyazawa.[27,32]

2 OPTICAL ROTATORY DISPERSION AND CIRCULAR DICHROISM[1,37–42]

The phenomenon of optical rotation in amino acids and the methods of measurement have been discussed earlier. Since optical rotation is an additive property, one might expect that proteins, on a molar basis, would exhibit optical rotation equal to the sum of the contributions from the amino acid residues. In native proteins, however, this is not true because the fixed and specific three-dimensional structure of the molecule produces new potential fields and possibilities for interaction that strongly influence rotatory properties. Because of the sensitive dependence of optical rotation on protein conformation, the measurement of this property has become quite common in the study of protein structure, even though analysis of the results in an absolute sense is difficult. For comparative purposes, however, it has proved to be a useful way of following gross conformational changes in proteins, such as denaturation. In most cases the observation of a change in optical rotation can be correlated with the occurrence of a conformational change. The converse, however, is not always true; slight conformational alterations may not be detected. The method has the advantage that measurements can be made quickly without destroying the sample and with little restriction on conditions such as temperature and solvent.

In the study of proteins and other macromolecules, optical rotation measurements may be made over a broad range of wavelength (e.g., 185 to 600 nm) and the resulting spectrum is referred to as the *optical rotatory dispersion* (ORD). The magnitude of the rotation at a particular wavelength is usually expressed in terms of the reduced mean residue rotation $[m']_\lambda$ defined as:

$$[m']_\lambda = \frac{3}{n^2 + 2} \frac{\text{MRW}}{100} [\alpha]_\lambda \qquad (9\text{-}1)$$

The Lorentz correction factor $3/(n^2 + 2)$, where n is the refractive index of the solvent, has been introduced in an attempt to account for the effect of solvent polarizability on $[\alpha]_\lambda$. MRW is the mean residue weight of the protein and depends on the amino acid composition; it is usually taken as 115 for proteins of unknown amino acid composition. $[\alpha]_\lambda$ is the specific rotation as defined previously (Chapter 2). This quantity is used to permit comparison of proteins with differing molecular weights and amino acid composition. In aqueous solvents ($n = 1.33$), $[m']_\lambda$ is about $0.9[\alpha]_\lambda$ for proteins.

Although ORD data contain, in principle, a great deal of information on macromolecular structure, interpretation is limited by problems in theoretical understanding of the phenomenon. In early work, much attention was given to the correlation of experimental data with particular structural models, such as the α-helix, in order to predict just how much of a protein might exist in that conformation. Analysis of optical rotation was first attempted for spectral regions away from any absorption frequency; these regions reflect contributions from all of the absorption bands. Later, attention was shifted to the region of the absorption bands themselves where the most striking changes in rotation occur (*Cotton effects*). In the Cotton effect region the rotation due to the optical activity of the absorbing chromophore is maximal, although residuals from all other absorption bands will contribute a background of variable magnitude. In both cases the basis of the approach was largely empirical. A serious limitation was a lack of knowledge of other types of conformations that might occur in proteins and their contribution to ORD.

More recently a greater emphasis has been placed on the measurement of circular dichroism, the absorptive corollary of optical rotation. It has the advantage that the optical activity due to a particular type of chromophore can be better isolated from that of other absorbing groups.

2-1 Theoretical Considerations of Optical Activity[1,37]

In order to give some idea of the nature and complexity of optical activity and to help the student avoid oversimplification in regard to studies on macromolecules, some aspects of the theoretical background should be considered. A more complete presentation can be found in the book by Kauzmann.[1] Equation 9-2 represents the basic quantum mechanical description of optical rotation:

$$[\alpha]_\lambda = \frac{n^2 + 2}{3} \frac{9600\pi N}{hcM} \sum_{\substack{\text{all } i \\ \text{transitions}}} \frac{\lambda_i^{\,2}(R_i)}{\lambda^2 - \lambda_i^{\,2}} \tag{9-2}$$

where n is the index of refraction, N is Avogadro's number, h is Planck's constant, c is the velocity of light, M is the molecular weight, λ is the wavelength of observation, λ_i is the wavelength associated with an electronic transition i, and R_i is the rotational strength of that transition. The summation is carried out over all electronic transitions that occur in the molecule. The term R_i is defined by

$$R_i = \text{Im}\,[\bar{m}_i \cdot \bar{\mu}_i] \tag{9-3}$$

where the symbol Im means that one takes only the imaginary part of the dot product in the parenthesis (i.e., that part multiplied by the factor $\sqrt{-1}$); this is simply a mathematical device to account for the difference in phase of $\bar{m}_i$ and $\bar{\mu}_i$. The symbols $\bar{m}_i$ and $\bar{\mu}_i$ are vectors representing the electric and magnetic moments of the ith transition, and their dot product is given by

$$\bar{m}_i \cdot \bar{\mu}_i = m_i \mu_i \cos\theta \tag{9-4}$$

or simply the product of the magnitude of the vectors times the cosine of the angle between them.

Although somewhat forbidding, these equations allow one to see a number of properties of optical rotation. First, any electronic transition in a molecule that has a nonvanishing R_i will contribute to the measured rotation (Eq. 9-2). The extent of its contribution will depend on the distance of its wavelength λ_i from the wavelength λ where the measurement is being made. As λ becomes very large with respect to all the transition wavelengths λ_i of a molecule, $[\alpha]$ obviously approaches zero. The same is true when λ is very small compared with λ_i; as λ approaches zero ($\lambda \rightarrow 0$), the summation term in Eq. 9-2 reduces to the sum of the R_i which can be shown theoretically to be zero.[1] Thus optical rotatory power goes to zero at both ends of the spectrum. The wavelength dependence in Eq. 9-2 also illustrates the dispersive nature of optical rotation, that is, the effect of the interaction of the incident light with the molecule at a particular transition wavelength λ_i is dispersed over a broad wavelength range.

Second, it is shown (Eq. 9-4) that the magnitude of R_i depends on the magnitude of the electric moment for the transition. This is the factor that determines the strength of the absorption observed at that wavelength. However, the strengths of absorption and optical rotation are not always directly related. Because of the dependence of R_i on μ_i and the angle between the vectors (Eq. 9-4), it turns out that weak absorption bands may make as large a contribution to the optical rotation as strong absorption bands. Thus the determination of the frequencies at which optical activity occurs may help to locate weak absorption bands.

It is clear, in addition, that any transition for which R_i is zero will not contribute to [α]. This will occur when $\bar{m}_i$ or $\bar{\mu}_i$ is zero (Eq. 9-3) or when the two vectors are perpendicular such that $\cos\theta = 0$ (Eq. 9-4). A particularly important case described by Kauzmann is that of a molecule that is identical to its mirror image. This will of course be the case for any arrangement of atoms that possesses a plane of symmetry. It can be shown that in this case the rotatory strengths of the two mirror image molecules are equal in magnitude but opposite in sign, that is, for a particular transition, $R_i = -R_i$. This is only possible if $R_i = 0$. There is no mirror image of a protein, however, that is identical with the original molecule, since the protein is composed initially of L-amino acids lacking the required symmetry elements and is, in addition, folded into a tertiary structure that imparts additional asymmetry. Thus, in general, all chromophores will lie in regions that do not possess an identical mirror image, their electronic transitions will be acted upon by the asymmetric potential fields about them, and optical activity will result.

Unfortunately, it becomes more difficult to show theoretically how the nonvanishing R_i actually comes about when the mirror image condition is satisfied. There have been several different theoretical approaches. In the one-electron theory[43] the transition of a single electron in a chromophore is treated from the point of view of quantum theory. If isolated from other atoms, the chromophoric group (e.g. C=O) is identical with its mirror image, and thus the R_i for its transitions must vanish. However, if placed in an asymmetric environment, the wave functions of the ground and excited states of the chromophore are perturbed in such a way that the rotational strength no longer vanishes. In the case of the 2950 Å absorption of C=O in camphor, it can be shown that induction by the asymmetric environment of a small electric moment parallel to the large magnetic moment of this transition could account for the observed rotational strength.[1]

In a different type of approach, Kirkwood[44] proposed the theoretical treatment of optical rotation in terms of the interaction of the electric moment of a transition with small electric moments induced on other groups by that transition. This theory has the advantage that rotation can then be expressed in terms of polarizabilities and geometry. It also shows more clearly the role of geometry in the interactions that give rise to optical activity, of particular importance in the study of protein conformational changes with ORD. However, it does not account for the large contributions of transitions that have small electric moments but large magnetic moments.

Although Eq. 9-2 adequately describes the contributions to the optical rotation measured at $\lambda \neq \lambda_i$, it is not suitable in wavelength regions where $\lambda \to \lambda_i$ because it predicts that $[\alpha] \to \infty$, which is not observed experimentally. A modified equation that takes into account dissipative forces that shift the phase of the electronic oscillations must be used. This modified equation predicts that when λ reaches λ_i, the contribution of the ith transition changes sign and is zero at λ_i (the *Cotton effect*).

Figure 9-13 shows the optical rotatory dispersion that might be observed for a single electronic transition, together with the absorption spectrum for that transition. The maximum and minimum of the Cotton effect occur near the inflection points of the absorption curve. The Cotton effect has been

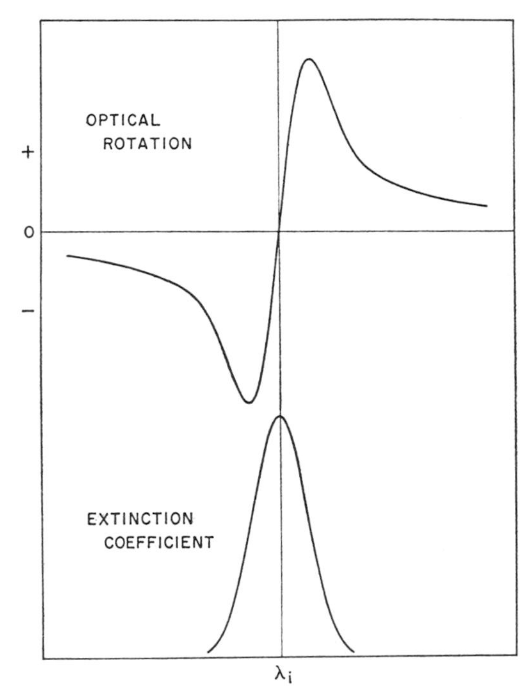

Fig. 9-13 Idealized Cotton effect at an isolated optically active absorption band with its maximum at λ_i. [From P. Urnes and P. Doty, *Advan. Protein. Chem.*, **16,** 401 (1961).]

arbitrarily shown as *positive* (i.e., with its maximum on the high wavelength side of the absorption peak. If the maximum is on the low wavelength side, the Cotton effect is said to be *negative*).

Since the Cotton effect of the *i*th transition in a real molecule is superimposed on contributions to the rotation from other transitions, a clear maximum and minimum, as shown in Fig. 9-13, is not always observed. If the rotational strength of the *i*th transition is very small compared to the dispersion contributed at λ_i by all the other transitions, it may appear as only a minor "wiggle" on a smooth dispersion curve. This is often observed at the absorption maxima of coenzymes or other ligands when they are bound to proteins. The effect is due to perturbation of the chromophore by the asymmetric electronic environment of the protein. Figure 9-14 shows the

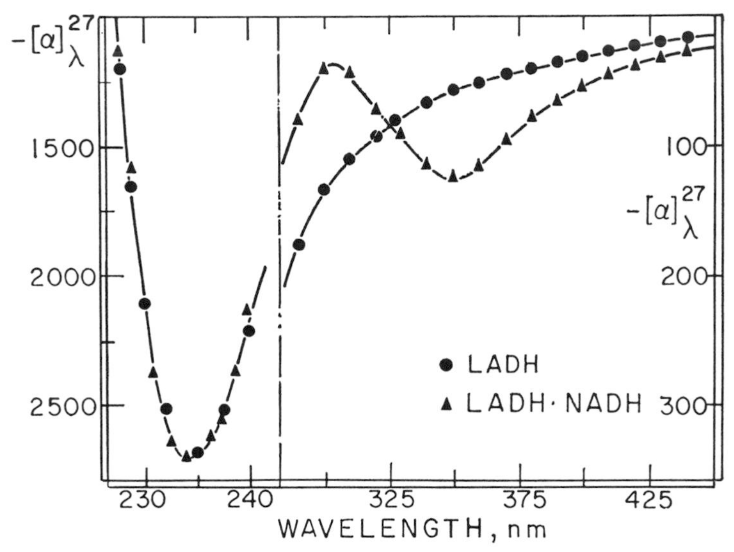

Fig. 9-14 Extrinsic Cotton effect in the optical rotatory dispersion of liver alcohol dehydrogenase (LADH) produced by the binding of the coenzyme nicotinamide adenine dinucleotide (NADH). The intrinsic Cotton effect of the enzyme at 234 nm is not affected by the coenzyme. [From D. D. Ulmer and B. L. Vallee, *Advan. Enzymol.*, **27**, 37 (1965).]

extrinsic Cotton effect produced by NADH (nicotinamide adenine dinucleotide) when bound to liver alcohol dehydrogenase. The region around the absorption maximum of the coenzyme (340 nm) has been magnified in comparison with the region of the protein Cotton effect. If the ORD curve of the apoenzyme (protein alone) were subtracted from that of the complex, the contributions in the 340 nm region of all other transitions would be eliminated, and the resulting Cotton effect would appear as in Fig. 9-13.

Another general characteristic of optical rotatory dispersion is that in wavelength regions sufficiently far removed from all the λ_i, Eq. 9-2, comprising a summation over many terms, can be approximated by a single equation containing factors that represent average values of the R_i and λ_i. This empirical relation, called the Drude equation, expresses the reduced mean residue rotation of Eq. 9-1 in the form

$$[m']_\lambda = \frac{a_c \lambda_c^2}{\lambda^2 - \lambda_c^2} \tag{9-5}$$

where λ_c and a_c are constants to be determined from the experimental data. It is found that the values of these constants are fairly uniform for unfolded proteins or random polypeptide chains in solution, about $-600°$ for a_c and 220 nm for λ_c. Many native proteins also obey a one-term Drude expression although a_c and λ_c values vary considerably.

2-2 Empirical Analysis of ORD Data

It was noted early in the investigation of the optical rotatory properties of proteins that the value of the specific rotation at the sodium D-line $[\alpha]_D$ tended to increase when the protein was denatured by reagents such as urea or acid. Similarly, it was found that values of λ_c obtained with the Drude approximation (Eq. 9-5) changed with protein denaturation. An empirical correlation of high λ_c (greater than 230 nm) with a highly organized structure and low λ_c (less than 230 nm) with a disorganized structure has been made for many proteins. Similar groupings for $[\alpha]_D$ also were made.

Theoretical considerations by Moffitt[45] led him to suggest the following equation to describe the optical rotation of proteins containing α-helical sections:

$$[m']_\lambda = \frac{a_0 \lambda_0^2}{(\lambda^2 - \lambda_0^2)} + \frac{b_0 \lambda_0^4}{(\lambda^2 - \lambda_0^2)^2} \tag{9-6}$$

where b_0 is particularly sensitive to amount of helical structure in the molecule and a_0 is a function of both residue and helical contributions. The value of λ_0 is chosen to give the best straight line when $[m']_\lambda(\lambda^2 - \lambda_0^2)$ is plotted vs. $1/(\lambda^2 - \lambda_0^2)$; b_0 is obtained from the slope and a_0 from the intercept. (It was found that $\lambda_0 = 212$ nm in the wavelength region 350–600 nm and $\lambda_0 = 216$ nm in the 240–280 nm region give a good fit for most proteins and polypeptides.)

Studies on synthetic polypeptides have demonstrated that the value of b_0 for the right-handed α-helix is usually around $-630°$, whereas b_0 falls to near zero when the polypeptide is in the random-coil form. Left-handed α-helices are expected to have $b_0 = +630°$, and this has indeed been observed for some polypeptides (e.g., poly-γ-benzyl-L-aspartate and poly-D-amino acids). The assumption that a linear interpolation may be made between $b_0 = 0$ and $b_0 = -630°$ to determine the percent α-helix seems to fit the data for model polypeptides fairly well. The successful use of b_0 for the calculation of α-helical content in proteins has been verified for two proteins, hemoglobin and myoglobin. Both of these proteins have about 75% helical content as determined by X-ray crystallography, in agreement with the estimation made from b_0. However, structure determinations on other proteins such as cytochrome *c* and carbonic anhydrase show poorer agreement between α-helix values predicted from optical rotation and those found in the crystal.

2-3 Analysis of Cotton Effects

The Cotton effects that occur in the ORD curves of proteins are of value in structural studies now that instruments are available for measurements in the ultraviolet. Figure 9-15 shows a typical ORD curve of a protein. A small Cotton effect is observed at the wavelength of absorption of the aromatic amino acids (about 280 nm); however, the region of greatest optical activity is that corresponding to the transitions of the peptide group. Polypeptides in the α-helical conformation show similar optical rotatory behavior. Typically, a minimum in the optical rotatory dispersion curve is observed at about 233 nm (negative Cotton effect); the rotation then passes through zero, and rises to a maximum positive value at about 198 nm (positive Cotton effect). Values of $[m]_{233} = -12{,}700°$ and $[m]_{190} = +80{,}000°$ are found for 100% helical polypeptides. The percent helix in model compounds can be calculated by making a linear interpolation between the limits of $[m]_{233} = -1800°$ for 0% helix and the value for 100% helix. For left-handed helices the sign of the Cotton effect is reversed, that is, $[m]_{233} = +12{,}700°$. In general, there has been good correlation between the values of b_0 and the magnitude of the helical content as calculated from the Cotton effect. The Cotton effect measurements offer a distinct advantage in cases where b_0 values may be influenced by prosthetic groups and other optically active chromophores absorbing at higher wavelengths.

As indicated above, the optical rotatory dispersion of proteins in the 190 to 233 nm region represents the superposition of two Cotton effects, one identified by the minimum observed in the range of 220 to 233 nm and the other by the maximum at 190 to 200 nm. The two Cotton bands may be correlated with the $n\pi^*$ and $\pi\pi^*$ transitions, respectively, of the peptide group, discussed earlier. Their contribution to the observed ORD varies for different structures. For example, poly-L-proline shows quite different optical rotatory behavior in this region of the spectrum in its two helical forms. Polyproline I exhibits a strong positive Cotton band (probably

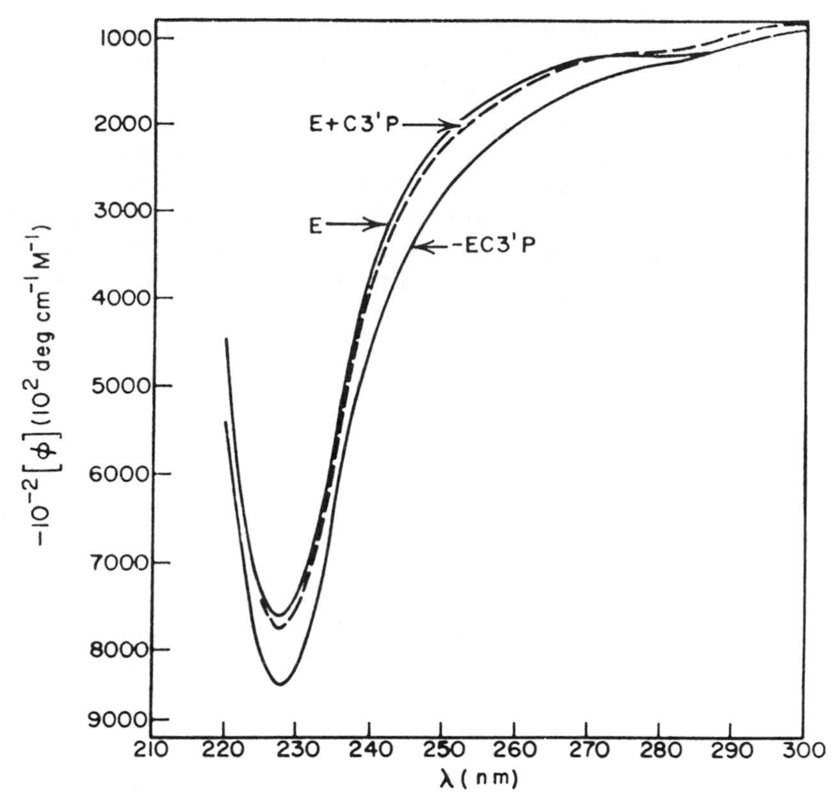

Fig. 9-15 Ultraviolet rotatory dispersion of ribonuclease (E) and its complex with cytidine 3′-phosphate (EC3′P). The broken line is the sum of ORD for enzyme and ligand measured separately. Data are given as molar rotation $[\phi]$ where $[\phi] = 100\ \alpha/lc$. α is the observed rotation in degrees, l is the path length in centimeters, and c is the concentration in moles/liter. [From R. E. Cathou, G. G. Hammes and P. R. Schimmel, *Biochemistry*, **4**, 2687 (1965).]

$\pi \rightarrow \pi^*$) centered at about 217 nm whereas form II, which arises by mutarotation, shows a weaker negative Cotton band at 202 nm. This change parallels a shift in absorption maximum from 210 to 202 nm.[46]

Theoretical treatments of peptide group optical activity have been carried out for the α-helix and, more recently, for β-polypeptide structures.[47] Differences in the optical activity of different backbone structures can be used in the analysis of protein ORD data provided one can reasonably limit the conformational possibilities that contribute to the ORD to only a few; for example, the ORD of a protein may be analyzed in terms of three reference configurations (α-helix, β-form, and random coil).[48] The possible contribution of other types of backbone configurations to the ORD is not yet known. As more complete protein structures become available from X-ray studies, better understanding of the origin of ORD in proteins and better empirical correlations should be possible.

Cotton effects have been observed in the binding of dyes such as acridine orange to polypeptides and proteins. It has been shown, in agreement with theory, that the Cotton effect of a dye bound to the helical form of poly-L-glutamic acid is of opposite sign to that observed when it is bound to the mirror image molecule, poly-D-glutamic acid (Fig. 9-16). A weak Cotton effect is observed when the polymer is in the random coil form.[49] Proteins

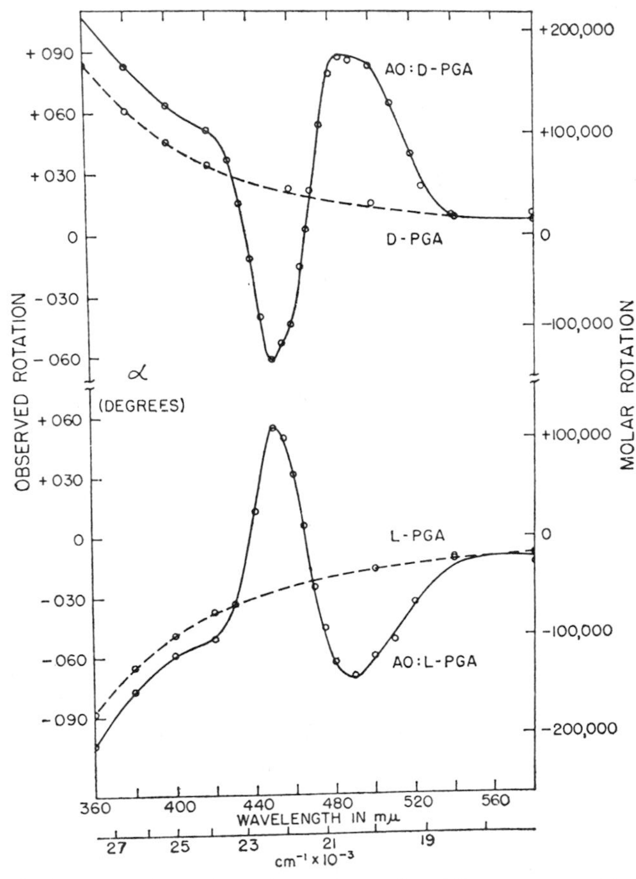

Fig. 9-16 Optical rotatory dispersion of complexes of acridine orange and helical poly-L-glutamic acid (AO:L-PGA), and of acridine orange with poly-D-glutamic acid (AO:D-PGA), a helix of opposite screw-sense to L-PGA. Broken lines show the ORD of the polymers in this wavelength region in the absence of bound dye. [From L. Stryer and E. R. Blout, *J. Amer. Chem. Soc.*, **83**, 1411 (1961).]

that contain prosthetic groups or coenzymes bound in an appropriate asymmetric environment also exhibit Cotton effects near the absorption maxima of those chromophores. Such Cotton effects are highly dependent on the conformation of the protein in the vicinity of the chromophore and may be used to advantage in studying binding reactions and conformational changes about a chromophore. In this area the use of difference spectropolarimetry should prove valuable in picking up small conformational changes.[50]

2-4 Circular Dichroism

Circular dichroism (CD) represents the absorptive counterpart of optical rotation. The relationship of the absorption and dispersion phenomena associated with electronic transitions was given previously (Table 9-1). As indicated there, circular dichroism refers to the unequal absorptivity for right and left circularly polarized light (i.e., $\varepsilon_l - \varepsilon_r \neq 0$). In most cases of interest in relation to proteins, one wishes to assess the perturbations produced in a suitable chromophore, which may itself be symmetric, by an asymmetric environment.

In the measurement of circular dichroism, linearly polarized light at the wavelength of an optically active transition emerges from the sample elliptically polarized due to the unequal absorption of the left- and right-handed components. The difference in absorbance $(A_l - A_r)$ is used to calculate the dichroism of the sample, expressed as *molecular ellipticity*, as follows:

$$[\theta]_\lambda = 2.303\left(\frac{4500}{\pi}\right)(\varepsilon_l - \varepsilon_r) \text{ degrees}$$

where ε_l and ε_r represent the extinction coefficients of the sample for left and right circularly polarized light. Physically, ellipticity represents the angle whose tangent is the ratio of the minor to the major axis of the emergent elliptically polarized light. An isolated ellipticity band is characterized by the wavelength of maximum ellipticity, the value of ellipticity $[\theta]_\lambda$, and by an expression of the band width.

Although the basic theoretical problems in interpreting circular dichroism results are the same as for ORD, this method has the advantage that the positive and negative ellipticity bands appearing at the wavelengths of optically active transitions are generally more readily resolved than the corresponding ORD bands. Figures 9-17*a* and 9-17*b* illustrate the CD spectra of poly-L-lysine in three different conformations and its optical rotatory dispersion under the same conditions. The distinct ellipticity bands of the α-helical conformation (at 190.5, 207, and 221 nm) and of the β-pleated sheet conformation (at 195 and 217 nm) may be compared with the less well

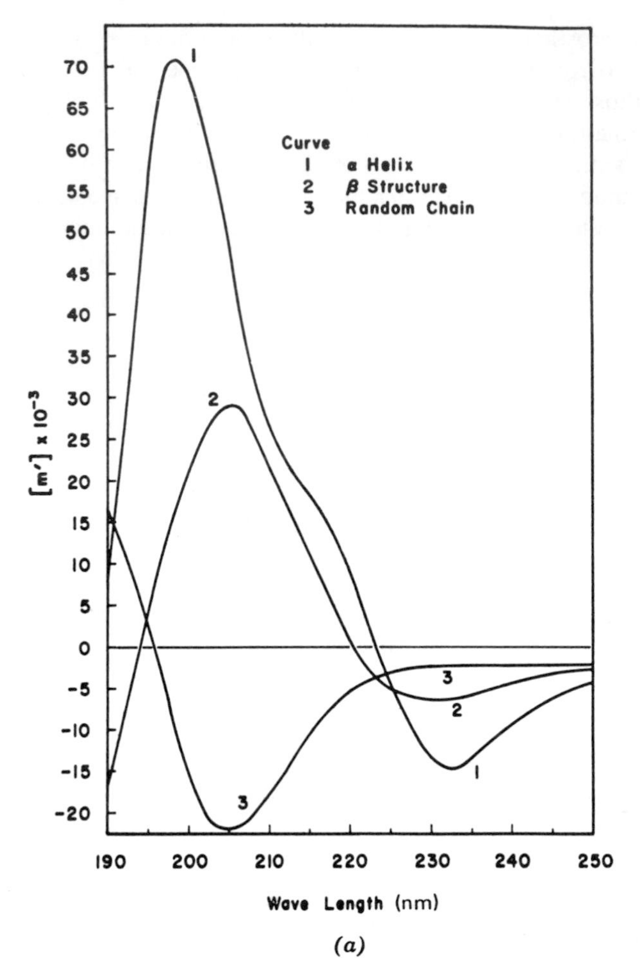

(a)

Fig. 9-17a The optical rotatory dispersion of poly-L-lysine in the α-helical, β, and random coil conformations. Absorption spectra of these forms are given in Fig. 9-5. [From N. Greenfield, B. Davidson, and G. D. Fasman, *Biochemistry*, **6,** 1630 (1967).]

resolved Cotton effects seen in the ORD data. Theoretical analysis of CD in synthetic polypeptides has yielded good agreement with experimental findings. Figure 9-18 shows the experimental CD of poly-L-alanine (α-helical conformation) and a theoretical curve based on the calculated rotational strengths of the $n\pi$ and $\pi\pi^*$ transitions of the peptide groups.

The study of circular dichroism has been applied to the problem of general structural analysis for a variety of proteins; like ORD, it is difficult to obtain absolute information but the method is valuable in comparative

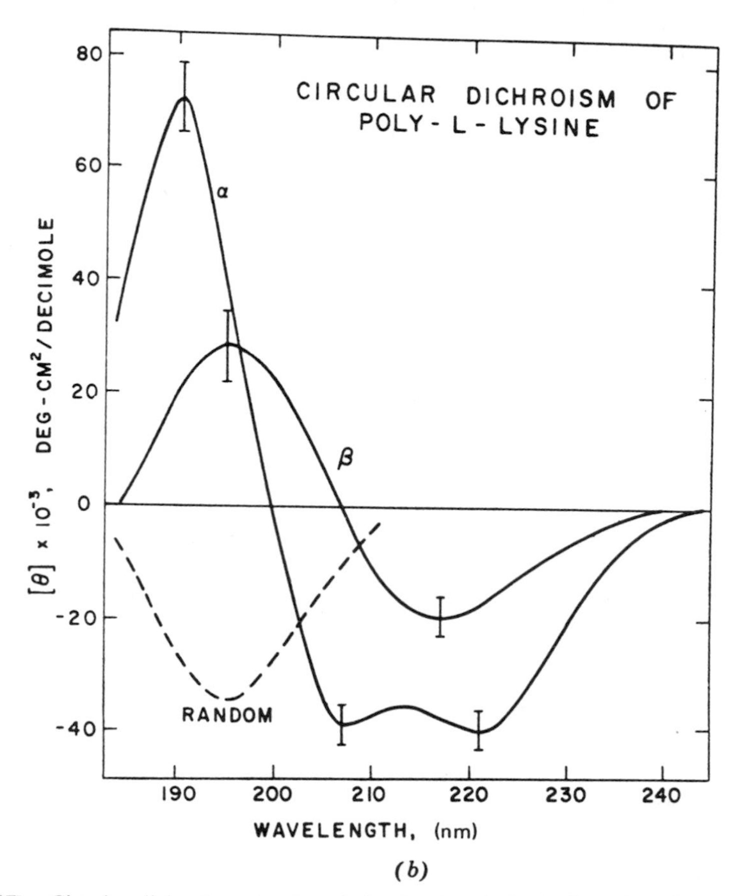

(b)

Fig. 9-17b Circular dichroism of poly-L-lysine in the α-helical, β, and random coil conformations. [From R. Townend, T. F. Kumosinski, S. N. Timasheff, G. D. Fasman, and B. Davidson, *Biochem. Biophys. Res. Commun.*, **23,** 163 (1966).]

studies. In particular it provides a sensitive probe for conformational changes occurring upon interaction of strongly absorbing ligands with protein molecules or for the detection of subtle environmental differences for ligands in related molecules. Figure 9-19 shows the effects of oxygenation and of oxidation (to methemoglobin) on the CD spectrum of hemoglobin in relation to the absorption spectra of these forms in the visible, Soret (about 420 nm), and peptide absorption regions. Significant changes in the globin part of the ellipticity spectrum occur upon removal of the heme group in hemoproteins; in metmyoglobin these have been attributed to an alteration in the number of amino acid residues in the α-helical segments of the molecule.[51] A further interesting development has been the measurement of magnetic circular

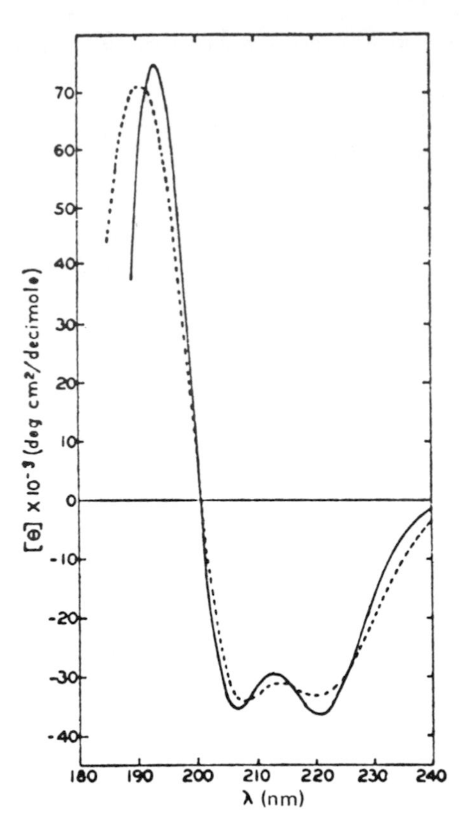

Fig. 9-18 Circular dichroism of poly-L-alanine in the α-helical conformation (solid line) and a theoretical curve (broken line) based on rotational strengths of peptide group transitions. [From R. W. Woody, *J. Chem. Phys.*, **49,** 4797 (1968).]

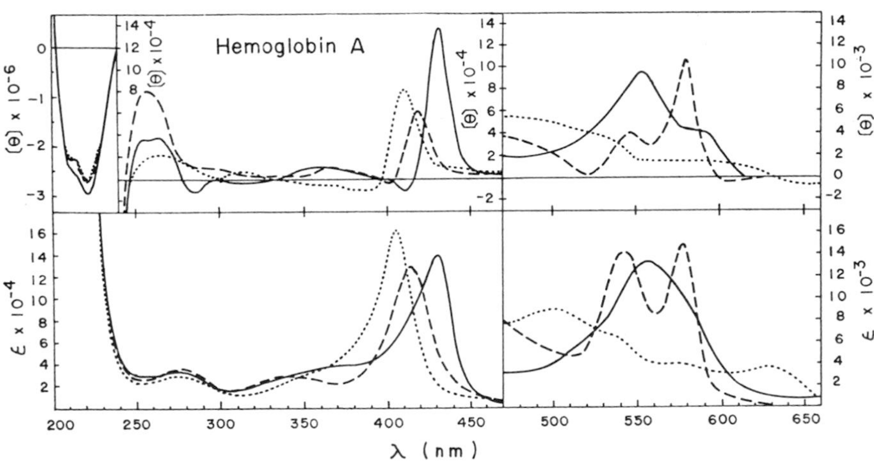

Fig. 9-19 Circular dichroism and absorption spectra of native human hemoglobin: deoxygenated hemoglobin (solid line); oxygenated hemoglobin (dashed line); methemoglobin (dotted line). Ellipticity and extinction (absorptivity) are given on a molar heme basis. [From Y. Sugita, M. Nagai and Y. Yoshimasa, *J. Biol. Chem.*, **246,** 383 (1971).]

dichroism in proteins. In conjunction with circular dichroism, this new technique provides additional data on the nature of optically active transitions.[52]

REFERENCES

1. W. Kauzmann, *Quantum Chemistry*, Academic Press, New York, 1957.
2. G. Weber and F. W. J. Teale, "Interaction of proteins with radiation," in *The Proteins*, 2nd ed., Vol. 3, H. Neurath, Ed., Academic Press, New York, 1965.

Ultraviolet Spectroscopy

3. D. B. Wetlaufer, "Ultraviolet spectra of proteins and amino acids," *Adv. Protein Chem.*, **17,** 303 (1962).
4. G. H. Beaven, "The ultraviolet absorption spectra of proteins and related compounds," *Adv. Spectrosc.*, **2,** 331 (1961).
5. W. B. Gratzer, "Ultraviolet absorption spectra of polypeptides," in *Poly-α-amino Acids*, G. D. Fasman, Ed., Marcel Dekker, Inc., New York, 1967.
6. J. W. Donovan, "Ultraviolet absorption," in *Physical Principles and Techniques of Protein Chemistry*, Part A, S. J. Leach, Ed., Academic Press, New York, 1969.
7. T. T. Herskovits, "Difference spectroscopy," *Methods in Enzymology*, Vol. 11, C. H. W. Hirs, Ed., Academic Press, New York, 1967, p. 748.
8. D. M. Kirschenbaum, "Molar absorptivity and $A_{1cm}^{1\%}$ values for proteins at selected wavelengths of the ultraviolet and visible region. V," *Int. J. Protein Research* **4,** 63 (1972). Earlier compilations also are cited here.
9. H. Edelhoch, "Spectroscopic determination of tryptophan and tyrosine in proteins," *Biochemistry*, **6,** 1948 (1967).
10. J. F. Brandts, "The thermodynamics of protein denaturation. I. The denaturation of chymotrypsinogen," *J. Amer. Chem. Soc.*, **86,** 4291 (1964).
11. I. Tinoco, Jr., A. Halpern, and W. T. Simpson, "The relation between conformation and light absorption in polypeptides and proteins." in *Polyamino Acids, Polypeptides and Proteins*, M. A. Stahmann, Ed., University of Wisconsin Press, Madison, Wisc., 1962.

Fluorescence

12. S. Udenfriend, *Fluorescence Assay in Biology and Medicine*, Academic Press, New York, Vol. 1 (1962) and Vol. 2 (1969).
13. S. V. Konev, *Fluorescence and Phosphorescence of Proteins and Nucleic Acids*, Plenum Press, New York, 1967.
14. R. F. Chen, H. Edelhoch, and R. F. Steiner, "Fluorescence of proteins," in *Physical Principles and Techniques of Protein Chemistry*, Part A., S. J. Leach, Ed., Academic Press, New York, 1969.

15. J. Eisinger and G. Navon, "Fluorescence quenching and isotope effect of tryptophan," *J. Chem. Phys.*, **50,** 2069 (1969).
16. R. M. Purkey and W. C. Galley, "Phosphorescence studies of environmental heterogeneity for tryptophyl residues in proteins," *Biochemistry*, **9,** 3569 (1970).
17. Th. Förster, "Transfer mechanisms of electronic excitation," *Discussions Faraday Soc.*, **27,** 7 (1959). Also see O. Sinanoglu, Ed., *Modern Quantum Chemistry* (Istanbul Lectures, Part III), Academic Press, New York, 1965.
18. S. S. Lehrer, "Fluorescence and absorption studies of the binding of copper and iron to transferrin," *J. Biol. Chem.*, **244,** 3613 (1969).
19. L. Stryer, "Fluorescence spectroscopy of proteins," *Science*, **162,** 526 (1968).
20. S. A. Latt, D. S. Auld, and B. L. Vallee, "Surveyor substrates: Energy-transfer gauges of active center topography during catalysis," *Proc. Nat. Acad. Sci., U. S.*, **67,** 1383 (1970).
21. G. Weber and S. R. Anderson, "The effects of energy transfer and rotational diffusion upon the fluorescence polarization of macromolecules," *Biochemistry*, **8,** 371 (1969).
22. J. Yguerabide, H. F. Epstein, and L. Stryer, "Segmental flexibility in an antibody molecule," *J. Mol. Biol.*, **51,** 573 (1970).

Infrared Spectroscopy

23. E. B. Wilson, Jr., J. C. Decius, and P. C. Cross, *Molecular Vibrations*, McGraw-Hill, New York, 1955.
24. S. Mizushima, *Structure of Molecules and Internal Rotation*, Academic Press, New York, 1954.
25. M. Davies, Ed., *Infrared Spectroscopy and Molecular Structure*, Elsevier, Amsterdam, 1963.
26. C. N. Rao, *Chemical Applications of Infrared Spectroscopy*, Academic Press, New York, 1963.
27. T. Miyazawa, "Characteristic amide bands and conformations of polypeptides," in *Polyamino Acids, Polypeptides and Proteins*, M. A. Stahmann, Ed., University of Wisconsin Press, Madison, Wisc., 1962.
28. W. P. Jencks, "Infrared measurements in aqueous media," *Methods in Enzymology*, Vol. 6, S. P. Colowick and N. O. Kaplan, Eds., Academic Press, New York, 1963, p. 914.
29. J. A. Schellman and C. Schellman, "The conformation of polypeptide chains in proteins," in *The Proteins*, 2nd ed., Vol. 2, H. Neurath, Ed., Academic Press, New York, 1964.
30. N. B. Colthup, L. H. Daly, and S. E. Wiberley, *Introduction to Infrared and Raman Spectroscopy*, Academic Press, New York, 1964.
31. S. Timasheff and M. J. Gorbunoff, "Conformation of proteins," *Ann. Rev. Biochem.*, **36,** 13 (1967).
32. T. Miyazawa, "Infrared spectra and helical conformations," in *Poly-α-amino Acids*, G. D. Fasman, Ed., Marcel Dekker, Inc., New York, 1967.
33. R. D. B. Fraser and E. Suzuki, "Infrared methods," in: *Physical Principles and Techniques of Protein Chemistry*, Part B, S. J. Leach, Ed., Academic Press, New York, 1970.

34. J. Hermans, Jr. and H. A. Scheraga, "Structural studies of ribonuclease. IV. The near infrared absorption of the hydrogen-bonded peptide NH group," *J. Amer. Chem. Soc.*, **82,** 5156 (1960).

35. C. B. Baddiel, "Structure and reactions of human hair keratin: An analysis by infrared spectroscopy," *J. Mol. Biol.*, **38,** 181 (1968).

36. J. O. Alben and W. S. Caughey, "An infrared study of bound carbon monoxide in the human red blood cell, isolated hemoglobin, and heme carbonyls," *Biochemistry*, **7,** 175 (1968).

Optical Rotatory Dispersion and Circular Dichroism

37. D. J. Caldwell and H. Eyring, *The Theory of Optical Activity*, Wiley-Interscience, New York, 1971.

38. S. Beychok, "Circular dichroism of biological macromolecules," *Science*, **154,** 1288 (1966).

39. W. F. H. M. Mommaerts, "Ultraviolet circular dichroism in nucleic acid structural analysis," in *Methods in Enzymology*, Vol. 12, Part B, L. Grossman and K. Moldave, Eds., Academic Press, New York, 1968.

40. B. Jirgensons, *Optical Rotatory Dispersion of Proteins and Other Macromolecules*, Springer-Verlag, New York, 1969.

41. J. T. Yang, "Optical rotatory dispersion," in: *Poly-α-amino Acids*, G. D. Fasman, Ed., Marcel Dekker, Inc., New York, 1967.

42. S. Beychok, "Circular dichroism of poly-α-amino acids and proteins," in *Poly-α-amino Acids*, G. D. Fasman, Ed., Marcel Dekker, Inc., New York, 1967.

43. E. U. Condon, W. Altar, and H. Eyring, "One-electron rotatory power," *J. Chem. Phys.*, **5,** 753 (1937).

44. J. G. Kirkwood, "On the theory of optical rotatory power," *J. Chem. Phys.*, **5,** 479 (1937).

45. W. Moffitt, "Optical rotatory dispersion of helical polymers," *J. Chem. Phys.*, **25,** 467 (1956).

46. F. A. Bovey and F. P. Hood, "The optical rotatory properties of poly-L-proline," *J. Amer. Chem. Soc.*, **88,** 2326 (1966).

47. R. W. Woody, "Optical properties of polypeptides in the β-conformation," *Biopolymers*, **8,** 669 (1969).

48. M. E. Magar, "On the analysis of the optical rotatory dispersion of proteins," *Biochemistry*, **7,** 617 (1968).

49. B. C. Myhr and J. G. Foss, "Polyglutamic acid-acridine orange complexes. Cotton effects in the random coil region," *Biopolymers*, **4,** 949 (1966).

50. B. J. Adkins and J. T. Yang, "Difference spectropolarimetry as a probe for small conformational changes," *Biochemistry*, **7,** 266 (1968).

51. E. Breslow, S. Beychok, K. Hardman, and F. R. N. Gurd, "Relative conformations of sperm whale metmyoglobin and apomyoglobin in solution," *J. Biol. Chem.*, **240,** 304 (1965).

52. G. Tollin, "Magnetic circular dichroism and circular dichroism of riboflavin and its analogs," *Biochemistry*, **7,** 1720 (1968).

X Methods Based on Charge or Charge Distribution

1 MULTIPLE EQUILIBRIA[1–6]

A protein molecule may have many combining sites for interacting reversibly with small ions or molecules. In the systems most often studied, this interaction is with small ions; titration studies with H^+ are particularly informative. In these cases the charge properties of the protein molecule come into play and multiple equilibria are involved. The theory and experimental techniques for the study of these systems will be discussed here and applied in particular to hydrogen ion equilibria. The approach is equally applicable to systems of uncharged molecules.

1-1 General Equations

If 1 mole of protein P may combine with n moles of a small molecule or ion A, the following equations hold:

$$\begin{aligned} P + A &= PA \\ PA + A &= PA_2 \\ &\cdots\cdots \\ PA_{i-1} + A &= PA_i \\ &\cdots\cdots \\ PA_{n-1} + A &= PA_n \end{aligned} \tag{10-1}$$

If the equilibrium constant for the ith association is designated k_i, then the following equations can be written relating the activities of the various components (in parentheses):

$$
\begin{aligned}
(\mathrm{PA}) &= k_1(\mathrm{P})(\mathrm{A}) \\
(\mathrm{PA}_2) &= k_2(\mathrm{PA})(\mathrm{A}) = k_1k_2(\mathrm{P})(\mathrm{A})^2 \\
&\cdots\cdots\cdots\cdots \\
(\mathrm{PA}_i) &= k_i(\mathrm{PA}_{i-1})(\mathrm{A}) = (k_1k_2\cdots k_i)(\mathrm{P})(\mathrm{A})^i \\
&\cdots\cdots\cdots\cdots \\
(\mathrm{PA}_n) &= k_n(\mathrm{PA}_{n-1})(\mathrm{A}) = (k_1k_2\cdots k_i\cdots k_n)(\mathrm{P})(\mathrm{A})^n
\end{aligned}
\tag{10-2}
$$

Binding experiments are generally conducted by adding a known amount of A to a given quantity of protein and then determining the parameter $\bar{\nu}$ defined by

$$\bar{\nu} = \frac{\text{moles bound A}}{\text{moles total protein}} \tag{10-3}$$

Many of the methods which have been used to determine $\bar{\nu}$ are discussed in detail by Steinhardt and Reynolds.[5] Equilibrium dialysis, gel filtration, use of specific ion electrodes (e.g., determination of pH), ultracentrifugation, and spectrophotometry are applicable to specific systems for the determination of $\bar{\nu}$.

Since the ith species of complex PA_i contains i moles of bound A per mole of protein,

$$\bar{\nu} = \frac{(\mathrm{PA}) + 2(\mathrm{PA}_2) + \cdots + i(\mathrm{PA}_i) + \cdots + n(\mathrm{PA}_n)}{(\mathrm{P}) + (\mathrm{PA}) + (\mathrm{PA}_2) + \cdots + (\mathrm{PA}_i) + \cdots + (\mathrm{PA}_n)} \tag{10-4}$$

The concentration of the ith species may be replaced with $i(k_1k_2\cdots k_i)(\mathrm{P})(\mathrm{A})^i$ by virtue of Eq. 10-2 to give

$$\bar{\nu} = \frac{k_1(\mathrm{A}) + 2k_1k_2(\mathrm{A})^2 + \cdots + i(k_1k_2\cdots k_i)(\mathrm{A})^i + \cdots + n(k_1k_2\cdots k_n)(\mathrm{A})^n}{1 + k_1(\mathrm{A}) + k_1k_2(\mathrm{A})^2 + \cdots + (k_1k_2\cdots k_i)(\mathrm{A})^i + \cdots + (k_1k_2\cdots k_n)(\mathrm{A})^n} \tag{10-5}$$

Note that the protein concentration has been canceled out of the numerator and denominator. Thus $\bar{\nu}$ is independent of protein concentration provided that the protein does not change its state of aggregation upon binding A.

Protein chemists sometimes prefer to consider the equilibria in terms of a set of dissociation constants K, given by

$$K_1 = \frac{(\mathrm{PA}_{n-1})(\mathrm{A})}{(\mathrm{PA}_n)}, \qquad K_2 = \frac{(\mathrm{PA}_{n-2})(\mathrm{A})}{(\mathrm{PA}_{n-1})}, \ldots, K_n = \frac{(\mathrm{P})(\mathrm{A})}{(\mathrm{PA})}$$

where (A) is the concentration of free A. The last dissociation constant is numerically equal to the reciprocal of the first association constant and so forth. The quantity $\bar{r}$ may be defined as

$$\bar{r} = n - \bar{\nu} \tag{10-6}$$

and the analogous equation to Eq. 10-5 for dissociation becomes

$$\bar{r} = \frac{K_1/(\mathrm{A}) + 2K_1K_2/(\mathrm{A})^2 + \cdots + nK_1K_2\cdots K_n/(\mathrm{A})^n}{1 + K_1/(\mathrm{A}) + K_1K_2/(\mathrm{A})^2 + \cdots + K_1K_2\cdots K_n/(\mathrm{A})} \tag{10-7}$$

Several general features of Eqs. 10-5 and 10-7 are worth noting at this point.

1. The values of $\bar{\nu}$ and $\bar{r}$ represent an average property of multiple species of the complex.
2. The resultant thermodynamic parameters obtained by this approach also are averaged quantities. For example, the dissociation $\mathrm{PA}_i \rightleftharpoons \mathrm{PA}_{i-1} + \mathrm{A}$ involves, in general, a set of molecular species PA_i that differ in occupied binding sites. Each of these may dissociate via i different pathways, for which ΔG may differ significantly.
3. Although the unknown constants appearing in the equations may be evaluated by numerical methods for simple cases, the experimental data are seldom accurate enough to permit such analysis when n is large as in many protein problems.

It follows that if binding studies are to be useful, certain simplifying assumptions must usually be made.

1-2 Multiple Equilibria Involving Equivalent Combining Sites

In most cases it is physically realistic to suppose that the interaction of a protein with a small molecule or ion is the same for chemically similar (equivalent) binding sites. For example, the carboxylate groups of glutamic acid might be expected to have similar affinities for protons even though they are in slightly different chemical environments. Exactly equivalent combining sites are anticipated for the binding of substrate or inhibitors to multisubunited enzymes with the proper symmetry. For n equivalent noninteracting sites. the addition of A to PA_i is equally likely to take place on each of the $(n - i)$ available sites. We may define (but not always measure) the equilibrium constant for addition of A to *a particular* site in terms of an intrinsic equilibrium constant k_0. The thermodynamic equilibrium constants for equivalent noninteracting sites may then be expressed in terms of k_0 by taking into account the proper statistical factor. For the first association for example, there are n sites of affinity k_0 so that k_1 is nk_0. It can be shown that the ith thermodynamic equilibrium constant is related to the intrinsic equilibrium constant by

$$k_i = \frac{n - i + 1}{i}\, k_0 \tag{10-8}$$

The term $(n - i + 1)/i$ is often called the statistical factor and is seen to be plausible from the following argument. The complex PA_i has i times the chance of dissociating to give A than would be the case for a species with only one A bound. Similarly, in the association of A with PA_{i-1} to yield PA_i, PA_{i-1} has $i - 1$ out of n sites occupied so that $n - (i - 1)$ sites are available to A. Thus the association of A with PA_{i-1} to give PA_i is $n - i + 1$ times greater than for a single site, while the dissociation of PA_i is i times more likely. The net result is that k_i must be $(n - i + 1)/i$ times the value for a single site, as given by Eq. 10-8. Further discussion of these concepts with several pertinent examples is given by Klotz.[1]

Substitution of Eq. 10-8 into Eq. 10-5 gives a result that may be greatly simplified by employing the binomial theorem.[1] The result, which may also be derived on more intuitive grounds,[2] is

$$\bar{\nu} = \frac{nk_0(\text{A})}{1 + k_0(\text{A})} \tag{10-9}$$

Rearrangements of this equation suggest straightforward ways to determine k_0 and n from experimental data. The reciprocal form of Eq. 10-9 is

$$\frac{1}{\bar{\nu}} = \frac{1}{nk_0}\frac{1}{(\text{A})} + \frac{1}{n} \tag{10-10}$$

thus a plot of $1/\bar{\nu}$ vs. $1/(\text{A})$ gives a straight line with slope $1/nk_0$ and intercept $1/n$. Alternatively, the linear form

$$\frac{\bar{\nu}}{(\text{A})} = k_0 n - k_0 \bar{\nu} \tag{10-11}$$

may be used to obtain $-k_0$ and nk_0 from the slope and intercept respectively of a plot of $\bar{\nu}/(\text{A})$ vs. $\bar{\nu}$. Equations 10-10 and 10-11 form the basis for the Klotz plot and Scatchard plot, respectively.

Experimental data for the binding of cupric ions to bovine serum albumin are shown in Fig. 10-1. A convenient representation of the primary data is a plot of $\bar{\nu}$ against the log of free $[Cu^{2+}]$ as shown in Fig. 10-1*a*. Data for the binding of Ca^{2+} to casein, plotted according to Eq. 10-10 (Fig. 10-1*b*) or 10-11 (Fig. 10-1*c*) yield a straight line from which it is calculated that 16 moles of Ca^{2+} are bound to each mole of protein. The Scatchard plot emphasizes a common occurrence in this type of study, namely that precise experimental data can often not be obtained near saturating conditions, since the bound A is determined as the small difference between 2 large numbers, total A − free A. This results in a great deal of uncertainty in the extrapolation for n. A similar uncertainty exists in the Klotz plot since the extrapolated value of $1/n$ is a small decimal fraction. The linearity of the plot,

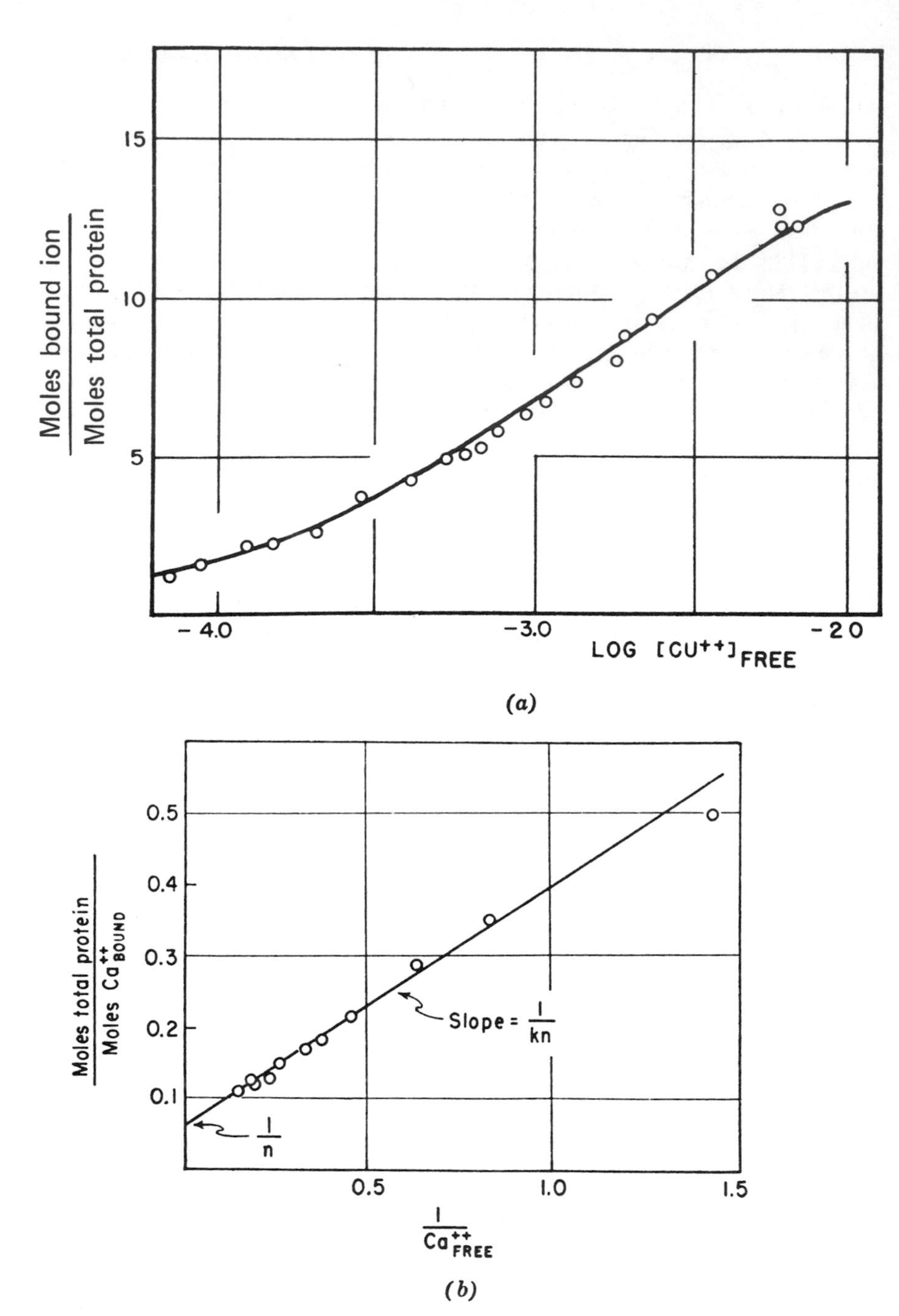

Fig. 10-1 (*a*) Binding data for the interaction of cupric ions with bovine serum albumin at 25°C and pH 4.83. [I. M. Klotz and H. G. Curme, *J. Amer. Chem. Soc.*, **70,** 939 (1948).] (*b*) A "Klotz plot" for the binding of calcium to casein.

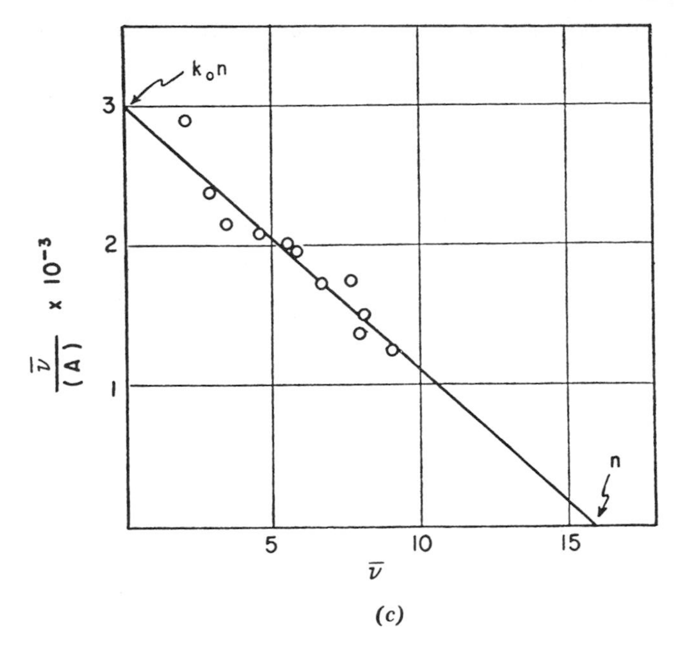

(*c*)

Fig. 10-1 (continued) (*c*) A Scatchard plot of the data shown in (*b*). [(*b*) and (*c*) are from Klotz,[1].]

however, demonstrates that, within experimental error, the assumption of identical sites with no interaction between them was valid.

When the plots according to Eqs. 10-10 and 10-11 are curved, either the binding sites are not in fact equivalent or interaction between binding sites must be suspected. Cooperative binding in which the binding of A *increases* the affinity for further binding of A, gives rise to a curved Scatchard plot in which the slope becomes more negative with increasing $\bar{\nu}$. Thus a curve with initial positive slope may pass through a maximum as the slope becomes negative. This behavior is characteristic of substrate binding by allosteric proteins, as in the oxygen binding of hemoglobin. More commonly, negative interactions occur such that the binding of A decreases the affinity for A. The slope of the Scatchard plot therefore becomes less negative with increasing $\bar{\nu}$, a result also expected for proteins with nonequivalent binding sites (i.e., the stronger binding sites are occupied first). Negative effects may result from steric hindrance, conformational changes, or because of electrostatic interactions. The latter occurs whenever the bound molecule is charged, since its binding will exert a net repulsive force toward further binding. The effect will be marked whenever large number of ions are bound (as in H^+ titration) or if the binding sites are in close proximity to each other.

The free energy of association for equivalent sites that are subject to interaction may be given as the sum of an intrinsic value when $\bar{\nu} = 0$ and an interaction term

$$\Delta G^0 = \Delta G^0_{\text{int}} + RT\phi(\bar{\nu})$$

or (10-12)

$$k = k_0 \exp [-\phi(\bar{\nu})]$$

where $\phi(\bar{\nu})$ is an arbitrary function. Substitution into Eq. 10-9 gives

$$\bar{\nu}/(n - \bar{\nu}) = k_0 \exp [-\phi(\bar{\nu})](\text{A}) \tag{10-13}$$

or in logarithmic form

$$\log (\bar{\nu}/n - \bar{\nu}) - \log (\text{A}) = \log k_0 - 0.434\phi(\bar{\nu}) \tag{10-14}$$

The left-hand side of Eq. 10-14 may be plotted (if n is known) against $\bar{\nu}$ to give a horizontal line for no interaction, upward curvature for cooperative interaction, or downward curvature for negative interaction. The use of an arbitrary function, however, is not of great utility in describing molecular processes. Consequently, models must be adopted in order to develop a useful form for this function.

Several models for ion binding have been reviewed.[3,4] A basic model is that formulated by Linderstrøm-Lang using the Debye-Hückel approximation. The theory assumes that charge is evenly distributed on the surface of a conducting sphere. Consequently, the work required to bring an ion to the surface will be proportional to the average *net* charge, $\bar{Z}$. It is convenient to replace $\phi(\bar{\nu})$ with $\phi(\bar{Z})$ and establish the reference as $\bar{Z} = 0$ rather than $\bar{\nu} = 0$. The result is

$$\bar{\nu} = \frac{nk_0 \exp (-2wz_i\bar{Z})(\text{A})}{1 + k_0 \exp (-2wz_i\bar{Z})(\text{A})} \tag{10-15}$$

where z_i is the charge of the ion being bound. The parameter w must be calculated in terms of an appropriate model for the protein (e.g., size and shape) and values of the ionic strength and temperature. The value of w decreases, but does not vanish, in the presence of increasing amounts of supporting electrolyte. Alternatively, w may be ascertained from the experimental binding curve as noted below. By analogy to Eq. 10-13, we may write

$$\bar{\nu}/(n - \bar{\nu}) = k_0 \exp (-2wz_i\bar{Z})(\text{A}) \tag{10-16}$$

which may be expressed in Scatchard form as

$$\frac{\bar{\nu} \exp (2wz_i\bar{Z})}{(\text{A})} = k_0(n - \bar{\nu}) \tag{10-17}$$

For electrostatic interactions then, a plot of $\bar{\nu} \exp (2wz_i\bar{Z})/(\mathrm{A})$ vs. $\bar{\nu}$ should be a straight line if the theory is valid, if w can be calculated, and if $\bar{Z}$ is known (e.g., from titration studies or because the protein is initially isoionic). The equivalent to Eq. 10-14 for electrostatic interactions is

$$\log (\bar{\nu}/n - \bar{\nu}) - \log (\mathrm{A}) = \log k_0 - 0.868wz_i\bar{Z} \qquad (10\text{-}18)$$

A plot of the left side of Eq. 10-18 vs. $\bar{Z}$ has a slope of $-0.868wz_i$ and only the difference in $\bar{Z}$ (obtained from $\bar{Z} = \bar{\nu}z_i$) is required. Obtaining a straight line by this procedure with a reasonable value for w indicates that a model for the protein with identical sites and electrostatic interactions among charges uniformly distributed over the surface approximates the binding behavior of the real protein within experimental error. Other models may also fit the data or may be required if this simple model fails.[2]

1-3 Hydrogen Ion Equilibria

The interaction of H^+ with macroions is accompanied by negative interactions that are normally due solely to electrostatic forces. Consequently, Eqs. 10-15 and 10-17 can be used to describe the binding. Historically, however, it has been customary to represent hydrogen ion equilibria in terms of dissociation rather than association. The equation analogous to Eq. 10-15 for dissociation from a single set of equivalent binding sites is

$$\bar{r} = \frac{nK_0 \exp (2w\bar{Z})/(\mathrm{A})}{1 + K_0 \exp (2w\bar{Z})/(\mathrm{A})} \qquad (10\text{-}19)$$

This may be expressed in logarithmic form as

$$\log (\alpha/1 - \alpha) - \mathrm{pH} = \mathrm{p}K_0 - 0.868w\bar{Z} \qquad (10\text{-}20)$$

where $\alpha = \bar{r}/n$ and $\mathrm{p}K_0 = -\log K_0$. The negative log of hydrogen ion concentration is replaced with the experimentally observed quantity, pH, which in fact measures more nearly the activity than the concentration of hydrogen ions.

The simplest type of titration that can be performed on a polyion is one where all the titratable groups are of one kind. An excellent example is discussed in detail by Tanford[2] for the titration of polymethacrylic acid, and only a few conclusions are summarized here. The titration at several salt concentrations is shown in Fig. 10-2 as α vs. pH. The titration is broader than expected for carboxyl groups, and the midpoints of the curves ($\alpha = 0.5$) representing apparent pK's, by analogy to the Henderson-Hasselbalch equation, are above pH 5. Such values are considerably higher than that expected for isolated carboxyl groups (about 4.75), even at high ionic strength. At low ionic strength the apparent pK is greater than 7, due to the marked

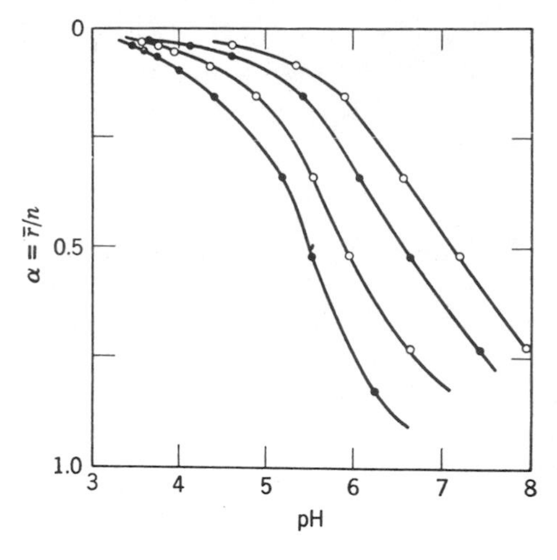

Fig. 10-2 Titration curves for polymethacrylic acid in various initial concentrations of KCl. The KCl concentrations were 0.001, 0.01, 0.1, and 1.0 *M* from the top to the bottom curves, respectively. [From Tanford,[2] based on data of R. Arnold and J. Th. G. Overbeek, *Rec. trav. chim.* **69,** 192 (1950).]

electrostatic attraction of the negatively charged polyion for the positively charged hydrogen ion. A plot of the data according to Eq. 10-20 is shown in Fig. 10-3, where α is used on the abscissa rather than $\bar{Z}$, since in this case $\bar{Z} = \alpha n$. Two important features of this plot should be noted. First, according to Eq. 10-20, the intrinsic pK is obtained when no electrostatic interactions are present, that is, when α is zero. A value of 4.85 is thus obtained at high ionic strength, in reasonable accord with expectations. Second, it should be noted that the curves are not straight, indicating that w is not constant with α, in this case because the linear polyelectrolyte expands with increasing charge. The interested student is referred to Tanford[2] for details of such

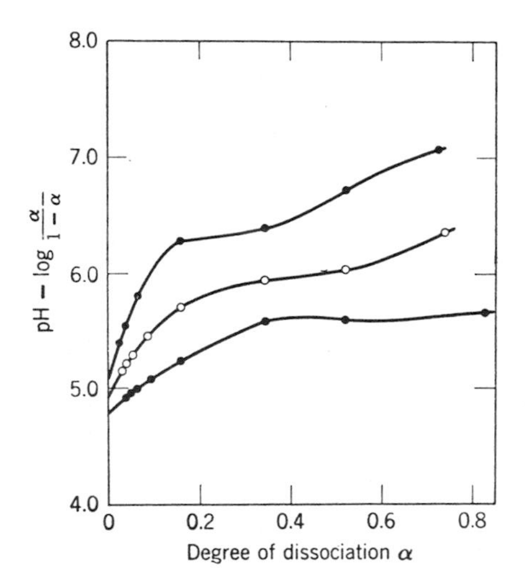

Fig. 10-3 A plot of the data of Fig. 10-2 according to Eq. 10-20. The top, middle, and lower curve are for 0.01, 0.1, and 1.0 *M* KCl, respectively. [From Tanford,[2] based on data of R. Arnold and J. Th. G. Overbeek, *Rec. trav. chim.* **69,** 192 (1950).]

calculations and for a theoretical description of the behavior of charged linear polymers. A similar analysis to reflect conformational changes is sometimes possible for the titration behavior of proteins (e.g., see the results for serum albumin discussed by Tanford[2]).

1-4 Hydrogen Ion Titration of Proteins

Proteins contain a number of groups that are capable of interacting with H^+. These are listed in Table 10-1 along with the anticipated values of the intrinsic

Table 10-1 Protein groups capable of association-dissociation reactions with H^+ and their intrinsic pK's.[4]

Titratable group	pK_{int} expected from data on small molecules
α-COOH	3.75
Side-chain COOH	4.6
Imidazole	7.0
α-NH_2	7.8
Sulfhydryl	8.8
Phenolic	9.6
Side-chain NH_2	10.2
Guanidyl	>12

pK_0 as obtained from studies on model systems. In view of the different intrinsic pK's present in a protein and the presence of marked electrostatic interactions, it is not surprising that the titration curve of a protein (e.g., plot of $\bar{r}$ vs. pH) is broad and somewhat difficult to interpret with precision.

A typical titration curve for a protein is shown in Fig. 10-4. In addition to the usual ordinate scale based on the experimental data, two other useful scales with different points of reference are shown. The $\bar{r}$ scale sets "0" at the end point on the acid side where all the carboxyl groups have presumably been titrated. The $\bar{Z}_h$ scale sets "0" at the isoionic pH of the protein. This value is determined by measuring the pH of the protein solution after passing it through a mixed-bed ion exchange resin to eliminate all ions in solution except the hydrogen and hydroxyl ions produced by the dissociation of water and the protein.

The full titration curve of a protein, as in Fig. 10-4, can be approximately broken down into three sigmoid curves. The range of pH 1.5 to 6 or slightly above corresponds to the ionization of side-chain carboxyl groups. Between pH 6 and 8.5 the titration is attributed to histidine residues and any

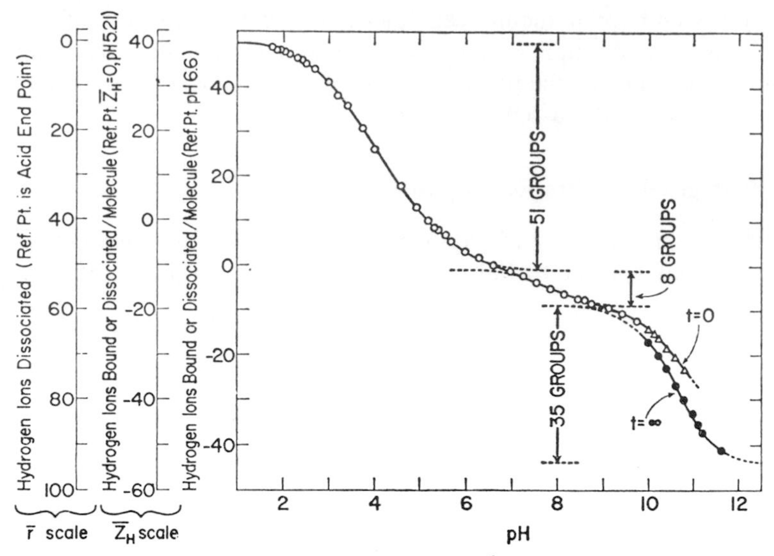

Fig. 10-4 The titration curve of β-lactoglobulin.[4] Details in text.

terminal α-amino groups present in the protein. The third range, from pH 8.5 up, includes ε-amino groups of lysine, phenolic hydroxyl groups of tyrosine, and sulfhydryl groups of cysteine. It is in general not possible to observe the titration of arginine since the pK of the guanidino group is too high (about 12). Several aspects of protein titration curves are of particular interest.

1. The number of groups in the acidic, neutral, and basic range can be counted and compared to analytical amino acid analyses or may be compared before and after certain treatments. In Fig. 10-4, for example, 51 titratable groups appear in the acid range (pH 1.5–6.0) while 8 groups are found in the neutral range. Amino acid analysis of β-lactoglobulin, however, shows 53 carboxyl groups that would be expected to titrate in the acid range whereas the number of groups titratable in the neutral region (histidine + N–terminal α-amino groups) is 6. This disparity disappears in the titration of the denatured protein where 53 groups are found in the carboxyl region and only 6 in the neutral region. These results show that 2 carboxyl groups of the native molecule are abnormally titrating in the neutral region. Correlation of the titration data with studies of viscosity and optical rotation in β-lactoglobulin has led to the suggestion that the anomalously titrating carboxyl groups are buried in a hydrophobic interior but come to the surface as a result of a conformational rearrangement that occurs near pH 7.5.

It is frequently possible to use titration data to determine the total number of groups titrating in the basic region (phenolic, sulfhydryl, and lysine ε-amino groups) but usually they cannot be distinguished because of the similarity of their pK's. Since the tyrosine spectrum is changed upon ionization, the number of phenolic groups may be determined by means of spectrophotometric titration. This method is complicated by the fact that spectral changes may occur in the indole and phenolic rings due to conformational changes affecting the electrostatic environment of the rings. Some examples of proteins with normal and anomalous pK's for various groups are given in Table 10-2.

2. Titrations in proteins are frequently found to be time-dependent as illustrated by the alkaline part of the titration of β-lactoglobulin as shown in Fig. 10-4. There is a considerable discrepancy between the $t = 0$ data obtained at the time of mixing and the $t = \infty$ curve representing data extrapolated to infinite time. Apparently under alkaline conditions β-lactoglobulin undergoes slow conformational changes which give rise to new environments for the titratable groups and thus different pK values. Since the effect is time-dependent, the titration study can be used to obtain kinetic data on this denaturation process. In cases where the denaturation process responsible for a time-dependent titration curve is sufficiently slow, the data can be extrapolated back to zero time to determine the number of titratable groups in the native molecule. An example of such a system is α-chymotrypsinogen, as shown in Fig. 10-5. Of the four tyrosine residues in the molecule, only two titrate normally in the native state whereas upon denaturation in 6.4 M urea all four titrate normally. Time-dependent titrations can be either reversible or irreversible. If the curve for the forward and back titrations based on data extrapolated to $t = \infty$ are identical, then by the criterion of titration (at least) the conformational change accounting for the time-dependent shift is reversible. If hysteresis remains even after extrapolation to infinite time, an irreversible denaturation process is indicated. This means that the molecule has acquired a new conformation that is "frozen-in" and cannot be returned to the native conformation. It is not obvious which form represents the lowest free energy; the new conformation may actually be thermodynamically favored in that particular experimental environment.

3. Another area in the study of proteins in which titration data have proved quite useful is in the binding of ions, coenzymes, or prosthetic groups to proteins. For example, comparison of titration data for insulin and zinc insulin, which contains one atom of zinc for every two insulin molecules, shows that the presence of zinc causes a shift in titration of two groups from the neutral region to the carboxyl region. Upon titration of these groups at low pH, the zinc ion is released from the protein. A possible interpretation is that the zinc ion is bound to two imidazole or α-amino groups, whose pK is lowered as a result.

4. Titration studies may be used to study association–dissociation equilibria in proteins provided the associated protein has a different titration curve from the monomeric protein. The advantages of a titration study over other techniques are its simplicity, the small amounts of material needed, and the rapidity with which measurements can be made in an automated pH-meter (pH-stat).

Table 10-2 Intrinsic pK values measured for various proteins.[4]

Type of group	pK
α-Carboxyl	
Insulin	3.6
Side-chain carboxyl	
Lysozyme (3 of *ca*. 16 groups)	$-\infty$
Serum albumin	4.0
Ovalbumin	4.3
Conalbumin	4.4
Corticotropin	4.6
Insulin	4.7
β-Lactoglobulin (49 of 51 groups)	4.8
β-Lactoglobulin (2 of 51 groups)	7.3
Imidazole	
Myoglobin (6 of 12 groups)	$-\infty$
Hemoglobin (*ca*. 22 of *ca*. 38 groups)	$-\infty$
Insulin	6.4
Ribonuclease	6.5
Myoglobin (6 of 12 groups)	6.6
Chymotrypsinogen	6.7
Ovalbumin	6.7
Conalbumin	6.8
Lysozyme	6.8
Serum albumin	6.9
β-Lactoglobulin	7.4
Phenolic	
Conalbumin (11 of 18 groups)	9.4
Insulin	9.6
Chymotrypsinogen (1 of 4 groups)	9.7
Corticotropin	9.8
Papain (11 of 17 groups)	9.8
Ribonuclease (3 of 6 groups)	9.9
Serum albumin	10.4
Chymotrypsinogen (1 of 4 groups)	10.6
Lysozyme	10.8
Chymotrypsinogen (2 of 4 groups)	∞
Conalbumin (7 of 18 groups)	∞
Ovalbumin	∞
Papain (6 of 17 groups)	∞
Ribonuclease (3 of 6 groups)	∞

Table 10-2 (*continued*)

Type of group	pK
Side chain amino	
Serum albumin	9.8
β-Lactoglobulin	9.9
α-Corticotropin	10.0
Ovalbumin	10.1
Ribonuclease	10.2
Lysozyme	10.4
Chymotrypsinogen (3 or 13 groups)	∞
Guanidyl	
Insulin	11.9
α-Corticotropin	~12

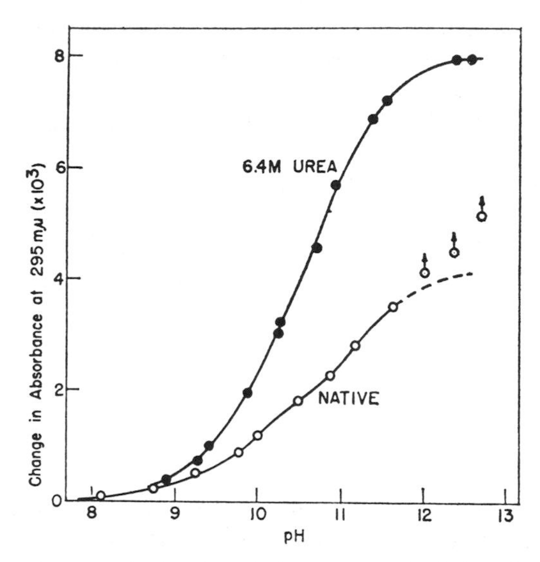

Fig. 10-5 Spectrophotometric titration of the tyrosyl side chains of α-chymotrypsinogen in 0.1 *M* KCl at 25°C (lower curve). The native protein undergoes conformation changes at alkaline pH, and the latter points (arrows) are time dependent. The upper curve is for the denatured protein in 6.4 *M* urea. Four tyrosines titrated and no time dependence was observed. [From Tanford[4].]

5. Titration methods are particularly suitable for studying reactions involving peptide bond rupture such as enzyme-catalyzed proteolysis. At pH's around neutrality where most enzymatic digestions are carried out, peptide bond breakage produces charged carboxylate ions on one side and amino groups that are only partially charged on the other. Thus there is a net release of protons that can be titrated with standard base. The titration results are often correlated with changes in other molecular properties such as molecular weight, optical rotation, or viscosity.

6. Further information that can be obtained from titration studies includes a comparison of intrinsic pK and w with expected values (e.g., by use of Eq. 10-20). For this purpose that part of the titration curve representing a single type of equivalent site must be isolated and the value of $\bar{Z}$ must be known (for calculation of pK_0, but not of w). The latter is most conveniently obtained by knowledge of the isoionic point. If the isoionic protein is obtained by passing it through an ion-exchange column, such that only H^+ or OH^- ions are bound, then

$$\bar{Z} \sum_i \mathrm{P}_i = (\mathrm{OH}^-) - (\mathrm{H}^+) \tag{10-21}$$

since electrical neutrality must be maintained. Consequently, the value of $\bar{Z}$ itself must be negligibly small if the pH is between 3 and 11 and if the protein is not too dilute. In the absence of binding of other ions, the charge may be calculated directly from the titration curve, once this zero charge reference is established. If an acid titration is performed starting at the isoionic point, and the maximum acid binding is obtained, this value corresponds to the total *cationic basic sites* present in the protein,[3] since the positive and negative charges are nearly identical at the isoionic point. This important fact often permits the number of arginine residues to be determined from titration studies, even though direct titration is not possible due to their high pK (~12.5).

2 ELECTROPHORESIS

Because protein molecules possess a net charge under most conditions (except at the isoelectric point where the net charge is 0), they will migrate in solution under the influence of an electric field. Electrophoresis is a powerful method that utilizes charge differences for the separation and the purification of proteins. Since it can detect relatively minor changes in charge, it is also useful for studies like those discussed in the previous section on titration, for example, chemical or conformational changes that alter titratable groups. An important advantage is that it can resolve components in a mixture, whereas titration results reflect the average population of the mixture.

Electrophoresis as most commonly used by the biochemist may be divided into three broad categories.

1. In *moving boundary electrophoresis* the protein dissolved in buffer is placed in a cell and buffer of the same composition is layered above it. An electric field is applied and the rate of movement of the boundaries between protein solution and buffer is followed by sensitive optical methods similar to those used in the analytical ultracentrifuge.

2. In *zone electrophoresis* the protein solution is placed at the starting position in an inert supporting medium (paper, starch gel, acrylamide gel, etc.) as a thin band or spot. Under the influence of the applied potential the proteins migrate to give distinct bands or zones. These may be located by staining, ultraviolet absorption (in some cases), or by analysis after elution of fractions.

3. In *continuous flow electrophoresis* the protein solution is applied continuously to an inert support such as paper or glass beads or allowed to flow between two closely spaced parallel plates. The sample flows down the support by gravity and an electric field applied at right angles causes migration of components away from the main stream. Fractions are collected in tubes located across the bottom of the support.

2-1 Moving Boundary Electrophoresis[2,7–9]

The Tiselius cell, normally used for moving boundary electrophoresis, is a U-shaped tube in three sections. As shown in Fig. 10-6, various chambers of the cell can be isolated by sliding the sections across one another for filling with the protein and buffer solutions and then returned to form a continuous channel with two protein-buffer boundaries. The buffer cups shown are connected to large vessels in which the electrodes are mounted and the entire assembly, on a special support, is immersed in a constant temperature bath. When temperature equilibrium has been achieved, the current is applied. The protein migrates to the cathode or anode depending upon whether its net charge is positive or negative. The rate of movement or mobility depends primarily on the magnitude of the charge on the molecule and on its size and shape. An idealized separation of three proteins A, B, and C of decreasing mobility is shown in Fig. 10-7. It should be noted that it is possible to obtain some A free of B and C and some C free of A and B. The boundaries may be visualized with a Schlieren or interference optical system giving a curve of the refractive index gradient dn/dx as a function of distance, x, from the starting position or of the refractive index n versus x, respectively.

The rate of movement of the protein in an electric field is usually

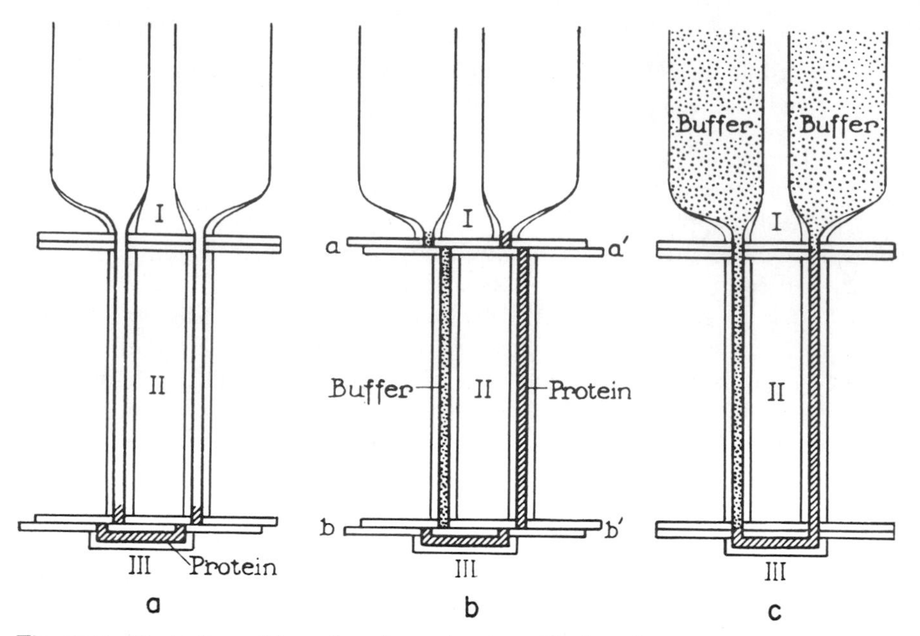

Fig. 10-6 Illustration of boundary formation in a Tiselius electrophoresis cell. (From Longsworth in Reference 7.)

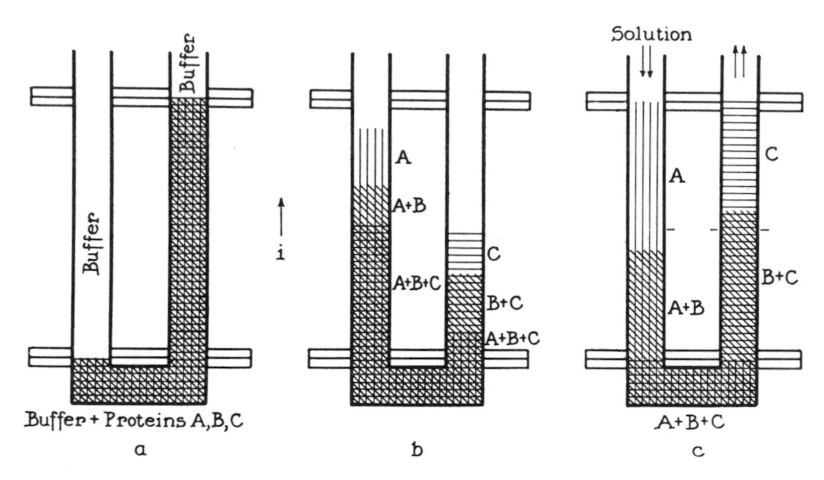

Fig. 10-7 Schematic diagram of the electrophoretic separation of a protein mixture without diffusion. In the diagram, all three proteins have the same charge (moving up in the left channel, down in the right). Protein A has the highest mobility, C the lowest. The concentration gradients are not stabilized against convection in the bottom portion of the "U" tube, so that in order to afford maximum separation, the electrophoresis apparatus has a provision for causing bulk solution flow (compensation) to achieve the separation shown on the right. (From Longsworth in Reference 7.)

expressed in terms of its mobility U, defined by the equation

$$U = \delta h/E\,\delta t = \kappa A\,\delta h/I\,\delta t \tag{10-22}$$

where δh is the distance the boundary has moved from the starting position during the time period δt and E is the applied field. E is determined experimentally as $I/\kappa A$, where I is the current flowing through the cell, κ is the conductivity of the solution, and A is the cross sectional area of the cell. The unit of mobility, velocity per unit field, is equal to (centimeters/seconds)/(volt/centimeter) or centimeter^2/volt second. Theoretical considerations indicate that the velocity with which a particle in solution moves in an electric field should be directly proportional to the charge of the particle (Q) and to the electric field E and inversely proportional to the frictional coefficient

$$\text{velocity} = QE/f \tag{10-23}$$

For spheres, we recall that according to Stokes' law, $f = 6\pi\eta r$. Actually this equation is not useful for calculating Q from a knowledge of f or vice versa using electrophoresis data because of the many other factors that influence the mobility of macromolecules. For example, charged particles are surrounded by a diffuse cloud of ions (often referred to as the *double layer*) and when this cloud is distorted under the influence of the electric field, becoming thinner in the direction of migration and thicker on the trailing side of the protein, the mobility of the molecule is reduced. A precise theory for treating this phenomenon is presently not available. Mobility is also reduced as a result of the backward flow of liquid as the protein molecules migrate in one direction. Another complication comes from the fact that the surface of the glass cell acquires a charge, which affects particle mobility.

The nature of the ionic atmosphere about a particle has a very great effect on mobility and in some cases Eq. 10-23 is completely invalid. Under conditions where the double layer becomes thin with respect to the size of the particle, mobility is found to depend on the surface charge potential of the particle and to be independent of its size and shape. It has been observed experimentally that tobacco mosaic virus (molecular weight $= 40 \times 10^6$) has the same mobility as do aggregates of its protein subunit (of much smaller molecular weight), and even more surprising, that glass beads coated with bovine serum albumin have the same mobility as the free protein.

Because of the theoretical difficulties, electrophoresis data do not give accurate values for protein charges; however, agreement with titration results to within 10 per cent has been obtained by the use of more sophisticated equations. On the other hand, electrophoresis is highly sensitive to small changes in charge between closely related proteins or to differences produced by chemical modification of a protein. Figure 10-8 shows the pH dependence of mobility for normal human hemoglobin and for the abnormal hemoglobin

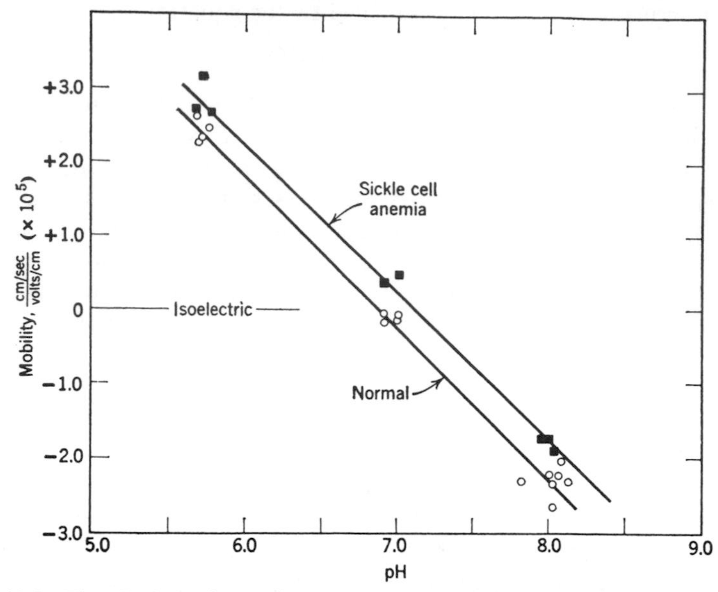

Fig. 10-8 The pH dependence of electrophoretic mobility for normal and sickle-cell hemoglobin. The normal protein has a slightly lower isoelectric point. The mobility difference is nearly constant throughout this pH range as expected for two proteins differing only in charge associated with glutamyl residues. [L. Pauling, H. A. Itano, S. J. Singer, and I. C. Wells, *Science*, **110**, 543 (1949).]

found in individuals with sickle-cell anemia. The results indicated a difference of two charges between the two species. This finding has been confirmed by amino acid analysis, which has shown that two glutamyl residues of normal hemoglobin are replaced by valyl residues in sickle-cell hemoglobin.

Electrophoresis is a powerful tool for the separation and identification of components in protein mixtures. Figure 10-9 shows the electrophoretic pattern of human blood plasma with its characteristic albumin, globulin, and fibrinogen peaks. Two additional peaks, called the δ-boundary in the ascending limb and the ε-boundary in the descending limb, are also generally observed and are found to migrate only very slowly. The δ-boundary is simply a protein concentration boundary that remains near the original boundary of the ascending limb. The ε-boundary represents a buffer concentration gradient caused by a build-up of the faster moving buffer ions between the original boundary position and the slow moving protein boundary.

In general, the Schlieren patterns in the ascending and descending limbs of the Tiselius cell are similar to one another, as shown in Fig. 10-9, but are not exact mirror images. This is because of the varying composition of the solution along the path of migration, as indicated schematically in Fig. 10-7*c*.

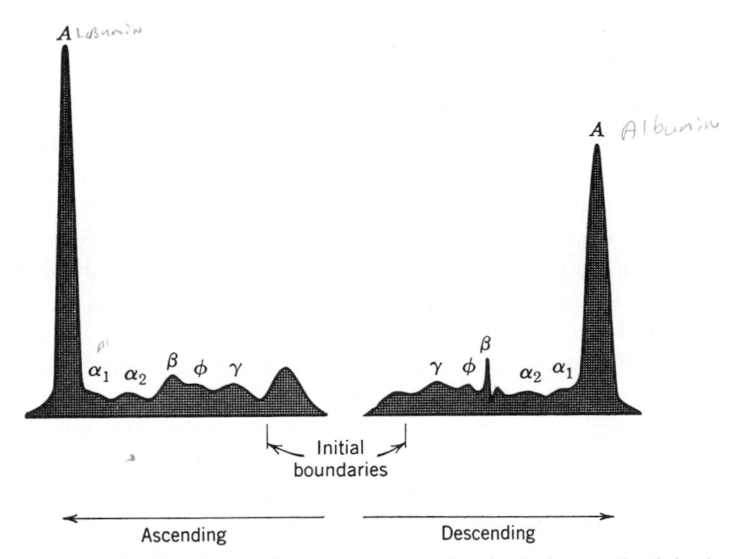

Fig. 10-9 Moving boundary electrophoretic pattern (refractive index gradient) for human blood plasma at pH 8.6 where the proteins are negatively charged. The A and ϕ peaks are albumin and fibrinogen, respectively. The other components are globulins, some of which are still identified as indicated here. [R. A. Alberty, *J. Chem. Educ.*, **25,** 619 (1948).]

Because ion mobilities at any point are strongly dependent upon the local environment, conductivity and concentration gradients of ions are set up that influence the shape and mobility of the protein boundaries and also contribute to the refractive index increment. In particular, pH gradients can occur that strongly influence boundary shape by affecting the charge of the protein molecules. Since the peaks in the two limbs are found to differ in area as well as in position and shape, it is clear that peak areas usually cannot be taken as a reliable measure of protein concentration. Accurate proportions of the various components in a mixture can be obtained by extrapolation to zero protein concentration.

The observation of a single symmetrical boundary in a protein preparation is frequently used as a criterion of homogeneity. The cautions required to interpret sedimentation patterns in this way are required here as well. A more sensitive test is to show that boundary spreading during electrophoresis at the isoelectric point (the pH at which mobility is 0) is the same as that produced by diffusion alone in the absence of the field. Another method involves reversal of the applied field during the observation of boundary spreading. Electrophoresis methods are also useful in studying the binding of ions to proteins, based on their effect on mobility or isoelectric point of the protein, and in the study of interacting systems such as in dye-binding or antigen–antibody reactions.

2-2 Zone Electrophoresis[10]

Although moving boundary electrophoresis is effective in the analysis and characterization of protein mixtures, it does not permit complete separation of the components, requires large amounts of material, and necessitates a relatively complex experimental procedure. These difficulties are circumvented by zone electrophoresis where the principle of electrophoretic mobility coupled, in some cases, with molecular sieving is used for the identification, isolation, and characterization of ionizable solute molecules. Here, the components of the mixture applied in a small spot or band on a supporting medium (such as paper or one of various types of gels) move out completely into the medium and form zones. The size of the zones depends in most cases on the size of the original spot plus diffusion broadening. The supporting medium prevents convective flow.

Zone electrophoresis has become increasingly popular in recent years. A wide variety of support media and many types of experimental apparatus, localization techniques, and procedures for isolation of individual zones have been devised. An extended discussion of the relative advantages of these techniques is not possible here, but the several systems considered below should serve as representative examples of what can be achieved. As suggested above, zone electrophoresis can be roughly subdivided into two categories, namely, media in which the support does not physically impede the movement of molecules and media in which molecular exclusion effects may be pronounced.

2-3 Zone Electrophoresis in Nonretarding Support Media[10–12]

One of the earliest and, for a time, most widely used support media for zone electrophoresis of proteins and peptides was filter paper.[11] A simple apparatus is shown in Fig. 10-10. After the paper has been impregnated with buffer, the protein sample is applied near one end and the paper is placed between two glass plates. An electric current flowing between the electrodes causes ion migration through the paper. Proteins migrate toward the pole opposite to their own charge; the distance of migration of different proteins depends on their mobilities. The protein can be located by staining (e.g., with amido black or nigrosin), and subsequently scanned with a densitometer. A densitometer tracing from the electrophoresis of serum stained with bromphenol blue is shown in Fig. 10-11. Another method of analysis that permits recovery of the separated materials involves cutting the paper into small strips and eluting the material from each strip for assay by standard analytical techniques. Many modifications of this simple apparatus have been developed for various purposes.[10]

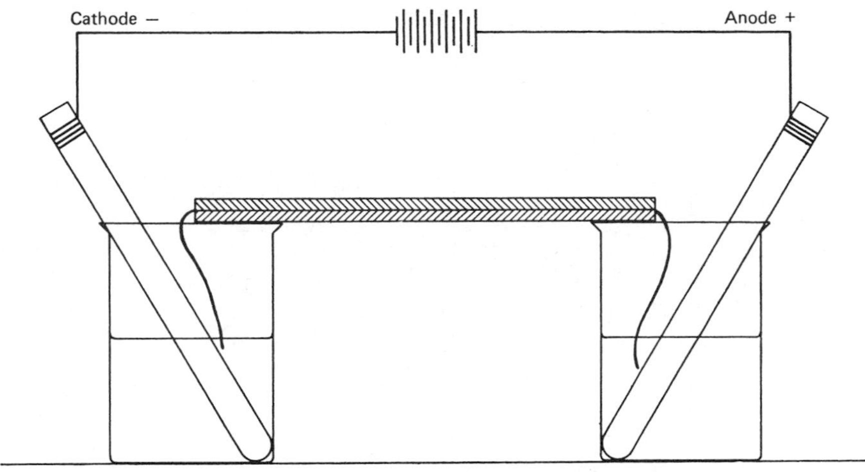

Fig. 10-10 A simple paper electrophoresis system. The filter paper strip is wetted, the sample applied, and is placed between two glass plates to minimize evaporation. The ends of the paper dip into buffer vessels containing electrodes across which a potential is applied.[10]

Certain difficulties may on occasion be encountered with paper electrophoresis. For example, some proteins are markedly bound to the support and a background smear of adsorbed protein is therefore not unusual. These problems are less severe if cellulose acetate strips are employed. A number of reliable commercial apparatuses are now available for cellulose acetate electrophoresis. The method is often preferred for routine clinical analysis of serum and other proteins. Similar results can be obtained by electrophoresis in slabs of low concentration agar or in agarose gels.

The electrophoretic patterns obtained in the above media are qualitatively similar to those expected on the basis of mobilities observed in free electrophoresis. The actual calculation of mobility from zone electrophoresis is, however, complicated by the fact that the true path length through the supporting medium is uncertain. There is the additional factor of endosmosis which causes bulk solvent flow in the support. The extremely small quantities of protein required is an obvious advantage for some studies; conversely, the procedures described are difficult to scale up for preparative purposes. Larger quantities of protein may be run in a column packed with a support such as cellulose fibers (Porath column). In general, the staining procedures employed result in somewhat different amounts of dye bound per milligram for different proteins, and precise quantitation is difficult unless prior calibration is made. An important advantage of zone electrophoresis is the possibility of carrying out tests directly on the supporting material to locate specific proteins. A great number of such localization techniques have been developed. For

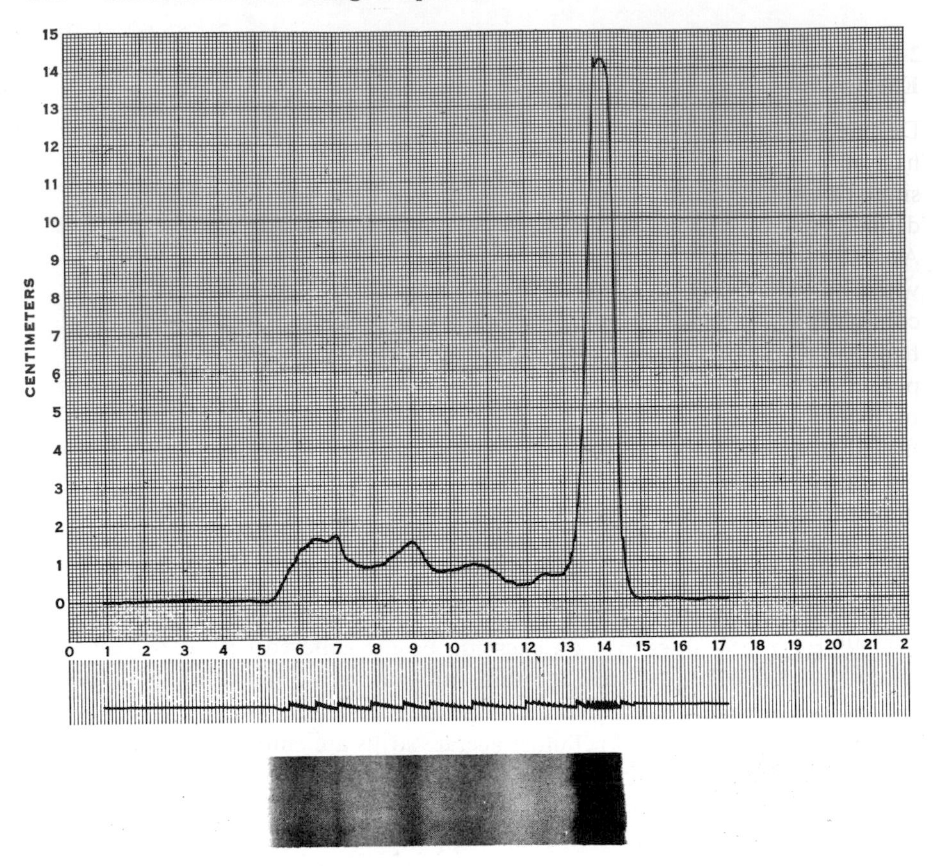

Fig. 10-11 Paper electrophoretic pattern of serum obtained with the Beckman R system. The stained strip is scanned with a densitometer to give the upper tracing. The sawtooth pattern has spikes proportional to the area between two points. (Courtesy of Beckman Instruments.)

example, certain enzymes have been located by impregnating the medium after electrophoresis with a substrate that is converted to an insoluble colored precipitate by the enzyme of interest. Thus certain enzymes can be located unambiguously even in a very impure mixture. Sometimes the procedure is done by coupling a substrate conversion to another enzymatic reaction that leads to an insoluble precipitate. Radioautography of fixed and dried strips and other biological assays are frequently employed. Specific staining of carbohydrate and lipid-containing proteins is also a useful method of localization.

2-4 Zone Electrophoresis Combined with Molecular Exclusion Effects[8,10,13]

Despite the wide applicability of simple zone electrophoresis, the technique has only moderate resolving power in protein mixtures because of the similarity in mobilities of many proteins. For example, only about five well-defined zones are observed for a mixture as complex as serum (Fig. 10-11). An extraordinarily important observation was made by Smithies in 1955[14] when he showed that serum proteins could be separated into many more components by electrophoresis in appropriately prepared gels of partially hydrolyzed starch. Smithies correctly concluded that the remarkable electrophoretic separations achieved in this system was due to hindrance of migration by the gel matrix, a factor dependent on size of the proteins. This was confirmed by Poulik and Smithies[15] by direct comparison of paper and starch gel electrophoresis. Those authors also demonstrated the value of two-dimensional electrophoresis where electrophoretic separation is carried out on filter paper after which the paper is introduced into a starch gel and electrophoresed at right angles to the initial field. The resolution of each band obtained by the filter paper method into multiple components in the second direction is illustrated in Fig. 10-12.

Experimentally starch gel electrophoresis may be done either in a horizontal or in a vertical position. Figure 10-13 is a diagram of the horizontal starch gel set-up. A solution of partially hydrolyzed starch is brought to the boiling point, poured into a tray, and allowed to solidify. It is covered with an appropiate sealer, and contact with the buffer cup containing the electrode is made by a wick of filter paper or cloth impregnated with buffer. The protein solution is introduced by soaking a small piece of filter paper and inserting it into a slot cut in the gel. When the electric current is applied, the protein migrates into the gel from the filter paper. After the run is completed the starch gel may be removed from the tray, sliced to obtain a clean surface, and stained with a dye. As in the case of unimpeded zone electrophoresis, unstained protein can be partially recovered from the gel by appropriate methods of elution, or substances can be localized by biological assays or other procedures previously discussed.

Starch gel has been superseded to a great extent by polyacrylamide gel for zone electrophoresis of proteins, primarily because this support material requires a less complicated experimental apparatus and procedure. Polyacrylamide came into common use as an electrophoretic support as a result of Raymond's introduction of a simple technique for employing a continuous buffer system in polyacrylamide gels and of Ornstein and Davis's method of electrophoresis in tubes.[13] In the latter method the sample is applied in a

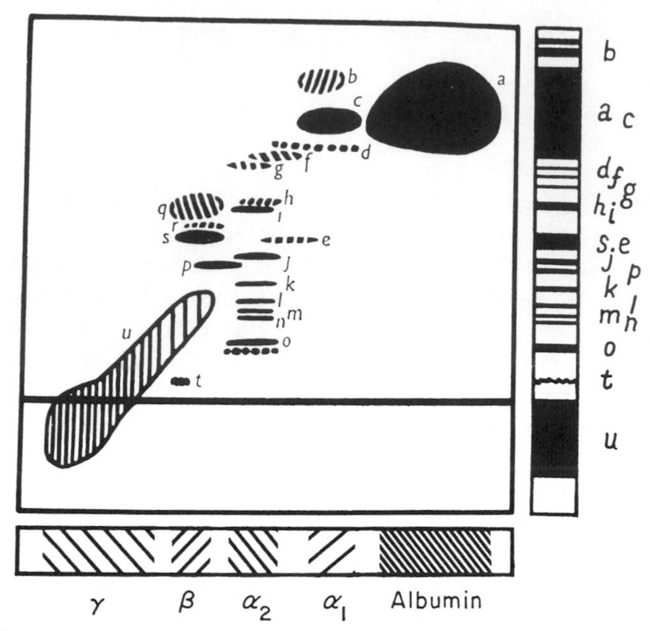

Fig. 10-12 Illustration of one type of two-dimensional electrophoresis. An abnormal serum was first separated by electrophoresis on filter paper. The paper is placed in a starch gel and the current is applied orthogonally. The final pattern is compared to the separate one-dimensional runs on filter paper (below) and starch gel (on right). The letters refer to various serum components. [From O. Smithies, *Advan. Protein Chem.*, **14,** 83 (1959).]

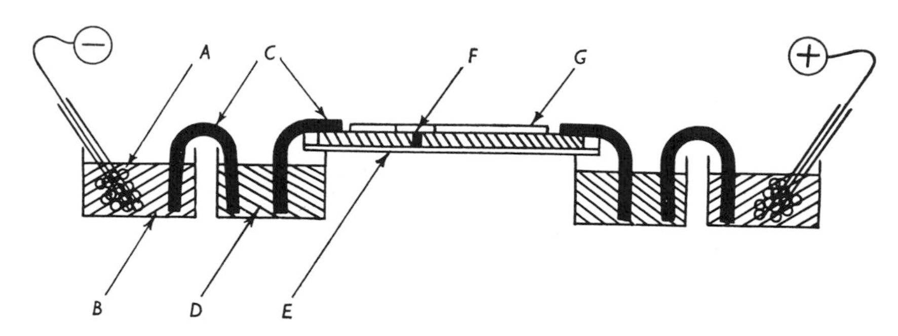

Fig. 10-13 Components of a starch-gel electrophoresis apparatus. *A*, Ag/AgCl electrode; *B*, concentrated NaCl solution in electrode compartment; *C*, filter-paper bridges soaked in bridge solution; *D*, compartment containing bridge solution; *E*, starch gel contained in plastic tray; *F*, position of sample insertion; *G*, cover to prevent loss of water during electrophoresis. [From O. Smithies, *Biochem. J.*, **68,** 630 (1958).]

gel support, which is followed by a spacer gel and finally by the running gel where separation occurs. This physical arrangement, coupled with a discontinuous buffer system, causes the protein components to "stack" as a series of very thin bands prior to entering the running gel. As a result, somewhat larger loading volumes can be used compared with nonstacking systems. Actually, some sharpening of the protein zone occurs with continuous buffer systems as well, simply due to the more rapid rate of migration of molecules in the liquid sample layered on top of the gel as compared to migration within the gel. Ornstein and Davis coined the name "disc" electrophoresis for their technique because of its dependence on a *dis*continuous buffer system, and partially because of the *dis*coid shape of the separated zones. Since continuous buffer systems are also used in gels contained in tubes, the term disc electrophoresis is best reserved for discontinuous buffer systems.

A simple, inexpensive, but serviceable apparatus for acrylamide electrophoresis is shown in Fig. 10-14. Acrylamide monomer, some crosslinking molecules, a catalyst, and an initiator are mixed with the buffer of choice and allowed to polymerize in the tubes. The upper and lower buffer reservoirs are filled and a thin layer of protein solution is placed under the buffer on top of the gels. The sample is stabilized with respect to density either by the protein itself or by addition of sucrose. After electrophoresis the gel is removed from the tubes for staining or other analysis. Further experimental details and a discussion of discontinuous gel electrophoresis may be found in the symposium on gel electrophoresis published in the *Annals of the New York Academy of Sciences*.[13]

As the pore size of a gel is decreased (by using higher concentrations of acrylamide), the distance a protein migrates under otherwise identical conditions is reduced. It is reasonable to suppose that the relative retardation of a protein as a function of gel concentration is related to molecular parameters, and many empirical relationships between molecular weight and relative retardation have been developed. These are discussed and referenced by Chrambach and Rodbard[16] who have introduced a promising theory for carrying out gel electrophoresis in a more quantitative manner. They also suggest methods for the estimation of free mobility and net charge from results obtained by gel electrophoresis conducted at a variety of gel concentrations. Under these circumstances mobility is correlated with molecular size.

Dissociation of proteins into subunits in solvents such as 8 *M* urea with an added sulfhydryl reagent is conveniently studied by gel electrophoresis.[17] If some assurance is possible that the polypeptide chains are in a nearly random-coil form in this solvent, a fairly reliable molecular weight estimate can be obtained from the retardation as a function of gel concentration. Considerable caution is required in this and the above determinations of relative mobility as a function of gel concentration, since results will depend on the

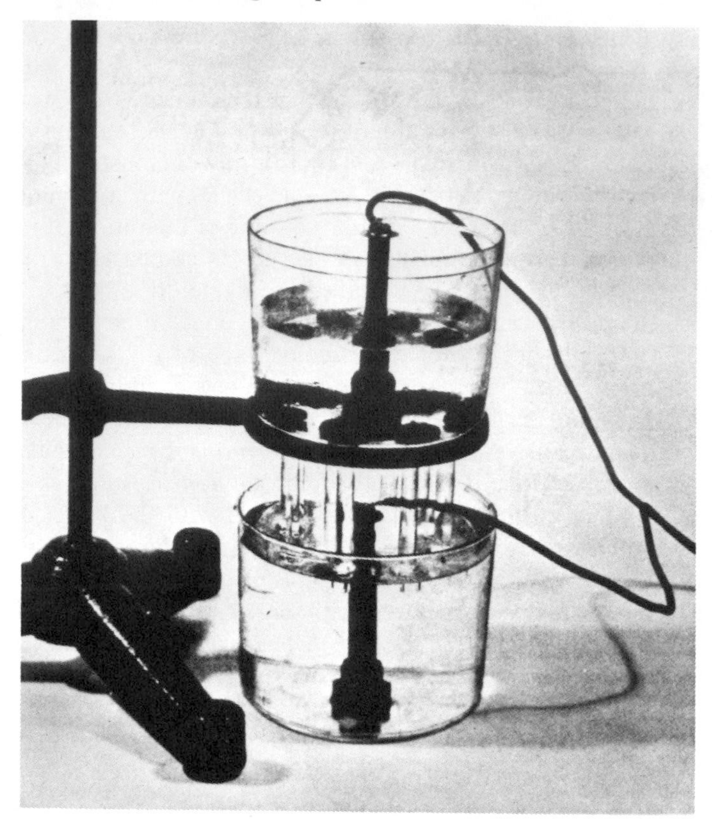

Fig. 10-14 A simple electrophoresis apparatus for polyacrylamide described by Davis in Reference 13. Holes are drilled into polystyrene refrigerator dishes and the glass tubes are held in place with rubber grommets in the upper reservoir. The graphite electrodes were obtained from flashlight batteries.

completeness of polymerization of the gel. Nearly quantitative polymerization is achieved only under very carefully controlled conditions. For this reason, many investigators prefer to estimate molecular weights by comparing the behavior of the unknown protein with that of standard proteins of known molecular weight.

The remarkable resolution of macromolecular mixtures by polyacrylamide gel electrophoresis has prompted the design of numerous preparative units, some of which are available commercially. The interested student will quickly find a detailed description of design and performance for many of these by consulting the index pages of *Analytical Biochemistry*.

In addition to the applications already discussed, gel electrophoresis is a powerful tool for following any type of reaction, chemical or enzymatic,

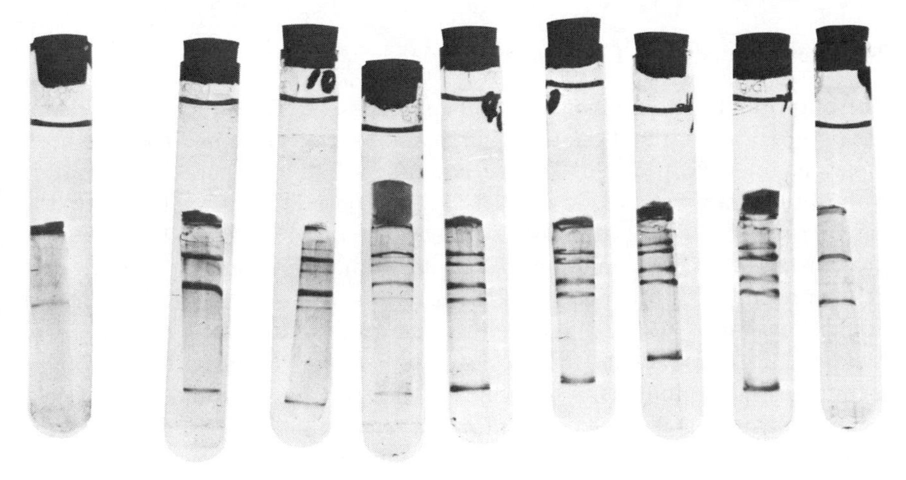

Fig. 10-15 The use of polyacrylamide electrophoresis to study the digestion of calf skin collagen by tadpole collagenase. Left to right: enzyme alone, 0 time control, 10 min, 20 min, 40 min, 2 hours, 4 hours, 8 hours, and 24 hours incubation time. Typical α, β and γ collagen bands (cf. Chapter 17) are replaced by faster moving species as the reaction proceeds. [From J. Gross and Y. Nagai, *Proc. Nat. Acad. Sci. U.S.*, **54,** 1197 (1965).]

that causes a change in the rate of migration of the material under observation, for example, changes in charge or molecular size. Very small amounts of material are required for this method, offering an obvious advantage over titration or moving boundary electrophoresis. For example, gel electrophoresis has proved to be valuable in identifying the products produced by the action of a specific collagenase on collagen (Fig. 10-15) and in revealing the three types of polypeptide chains in fibrinogen and fibrin obtained after disulfide bond rupture (Fig. 10-16). The important application of gels run in the presence of sodium dodecyl sulfate is discussed in the following section.

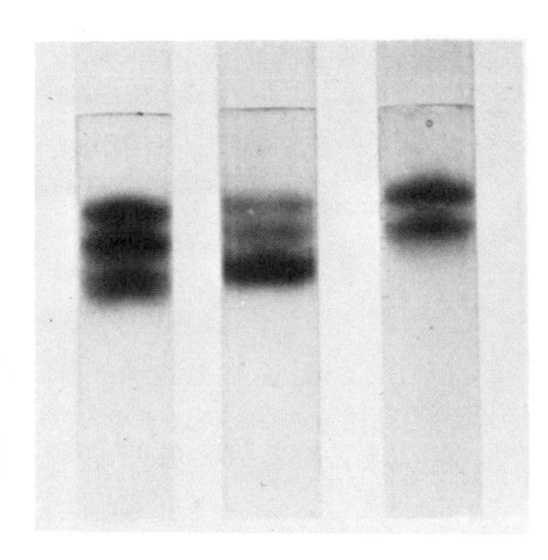

Fig. 10-16 Polyacrylamide separation of the α, β, and γ chains of bovine fibrinogen in 8 *M* urea at pH 10.5 (at left). Disulfide bridges were reduced with mercaptoethanol, then reacted with ethyleneimine in 8 *M* urea. The center gel shows a fraction enriched in γ chain; the gel at right contains only α and β chains. [Experiment by H. Castleman.]

2-5 Continuous Flow Electrophoresis[10,18]

The purpose of continuous flow electrophoresis is to facilitate the collection on a preparative scale of relatively large amounts of proteins that have been separated in an electric field. The method of continuous paper electrophoresis is shown diagrammatically in Fig. 10-17*a*. The protein solution flows onto the hanging paper and migrates downward by gravitational flow. As it does so, proteins are caused to migrate in the horizontal direction by the electric field applied at right angles to the direction of flow. At the bottom the liquid flows into test-tube collectors and can be analyzed. If the paper is stained at the end of the run, as shown in Fig. 10-17*b*, the path the proteins take under the simultaneous influence of the gravitational and electric field is revealed. Another type of continuous electrophoresis is done in free solution. In this method protein solution is introduced between two parallel glass plates that are separated by a distance so small that convective flow is impossible. Collection and analysis is similar to that described for paper electrophoresis.

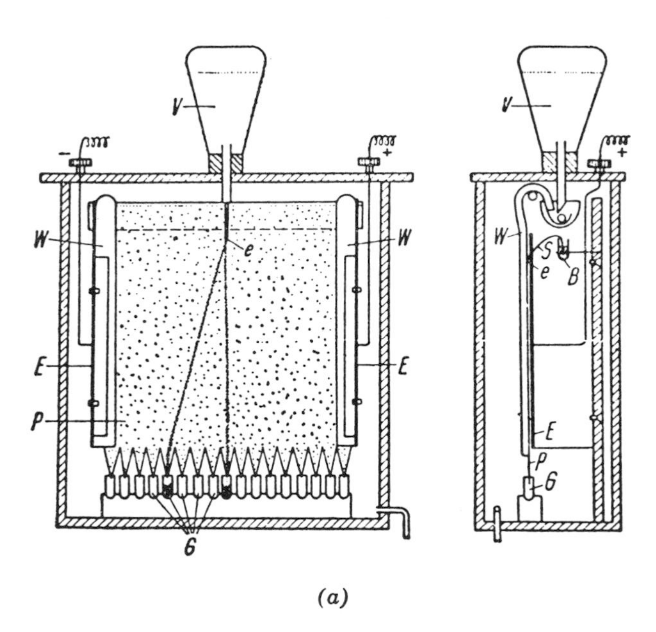

Fig. 10-17 Continuous flow electrophoresis. (a) Two views of the apparatus. (b) Examples of migration direction. [From P. Alexander and H. P. Lundgren, Eds., *A Laboratory Manual of Analytical Methods in Protein Chemistry Including Polypeptides*, Pergamon Press, New York, 1966.]

Fig. 10-17 (*continued*)

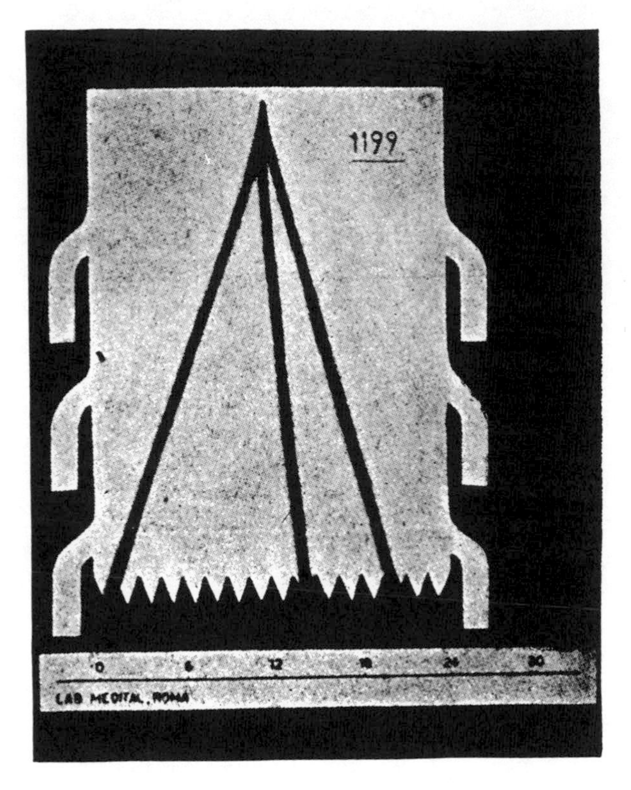

(*b*)

3 SUBUNIT STOICHIOMETRY AND MOLECULAR WEIGHTS BY GEL ELECTROPHORESIS

Sodium dodecyl sulfate (SDS) can be used under suitable conditions to dissociate oligomeric proteins into their individual polypeptide chains. The surface charge on the molecules of the SDS–protein complex is almost entirely due to the exposed sulfate ions, due to the high degree of SDS binding (e.g., about 1.4 g/g protein). Thus surface charge per unit area tends to be a constant regardless of the charge of the polypeptide chains. In addition it is found that such complexes assume the shape of a long rod of constant width where length is a function of molecular weight of the polypeptide portion.[19–21] These factors form the basis for the empirical determination of molecular weight from gel electrophoresis of SDS–protein complexes, as developed by Maizel and co-workers.[22] Their studies (confirmed by Weber and Osborn[23] on 40 different proteins) demonstrated that in gels where significant molecular exclusion occurs, a nearly linear relationship exists between electrophoretic mobility of a protein–SDS complex and the logarithm of the molecular weight of the protein (Fig. 10-18). Similar results have been obtained by other workers, and the method is gaining wide acceptance as a procedure for estimating molecular weights. Some caution, of course, is essential in interpreting the results, since factors such as deviation of partial specific volume from the average and incomplete dissociation of protein or binding of SDS will cause variations in molecular weight values determined in this way.

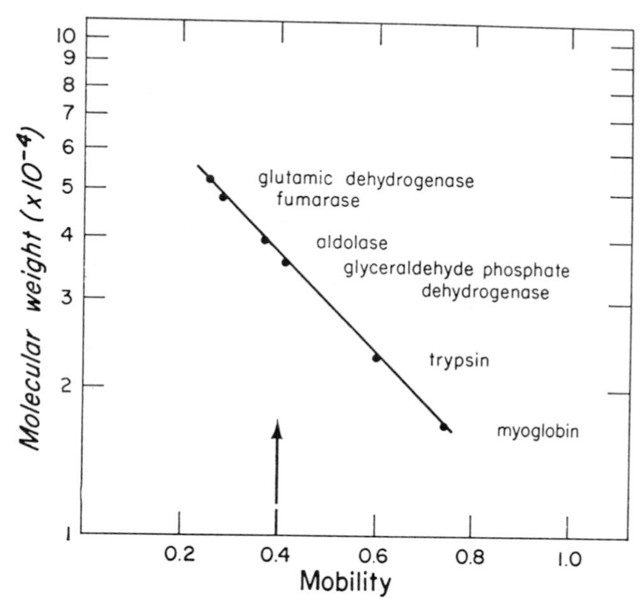

Fig. 10-18 Polyacrylamide gel electrophoretic mobility of protein–SDS complexes for several proteins run in the same pore-size gel. A linear relation between the molecular weight of the dissociated polypeptide chains and mobility is indicated. (From Weber and Osborne.[23])

A method has recently been developed by Davies and Stark[24] for determining molecular weight and number of subunits in an oligomeric protein by carrying out gel electrophoresis of the SDS–protein complex after first introducing covalent crosslinks into the protein. The bifunctional reagent dimethyl suberimidate $NH{=}C(OCH_3){-}(CH_2)_6{-}(CH_3O)C{=}NH$ was used to produce crosslinkages between lysyl residues. Most of these will form within a single subunit of the protein, but some will occur between subunits,

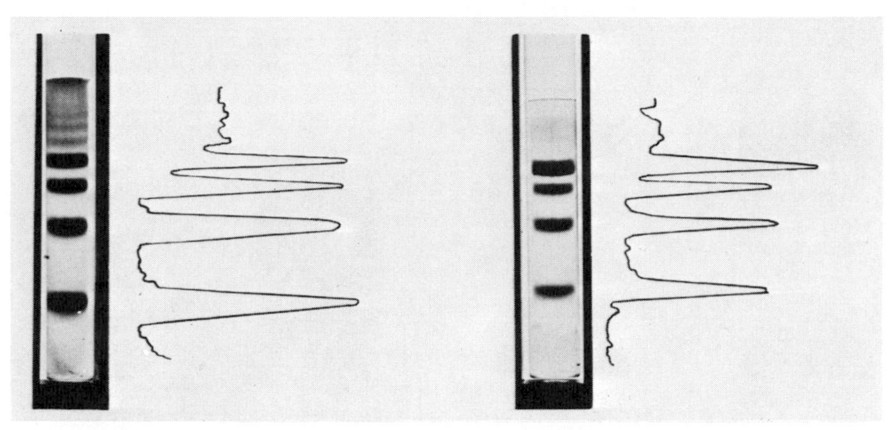

Fig. 10-19 Polyacrylamide gel electrophoresis patterns of multiple species produced in glyceraldehyde-3-phosphate dehydrogenase (left) and aldolase (right) as a result of intramolecular crosslinking with dimethyl suberimidate. Both proteins have been run as SDS-complexes. The fastest running species (lowest in pattern) is the single subunit. Above this appear dimers, trimers and tetramers, indicating a tetrameric form for both proteins. Small amounts of slower running species result from intermolecular crosslinking. [From Davies and Stark.[24]]

and the final electrophoretic pattern will reveal species with molecular weights representing all possible subunit combinations. For example, a protein with four identical polypeptide chains will produce electrophoretic bands corresponding to monomer, dimer, trimer, and tetramer (Fig. 10-19). In the absence of the crosslinking agent, only the monomer band is observed after SDS treatment. This elegantly simple procedure, or some modification of it, should greatly simplify the determination of subunit stoichiometry for most small oligomeric proteins.

Subunit stoichiometry can also be determined by gel electrophoresis of hybrid molecules obtained by reconstitution of the oligomer from mixtures of chemically modified subunits with unmodified ones[25] (Fig. 10-20) or by

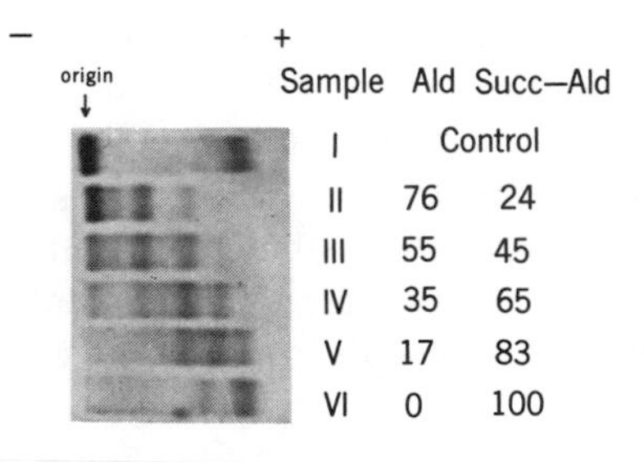

−
+
origin
Sample Ald Succ—Ald
I Control
II 76 24
III 55 45
IV 35 65
V 17 83
VI 0 100

Fig. 10-20 Determination of subunit stoichiometry by cellulose acetate electrophoresis of hybrid protein molecules produced by reconstitution of subunits from aldolase and succinyl-aldolase (a modified protein with charge different from the native form, obtained by reaction with succinic anhydride). Reconstitution mixtures prepared with various proportions of the two types of subunits show three bands representing hybrid molecules which migrate between the parent molecules. The result indicates the presence of four subunits in the reconstituted molecules. [From Meighen and Schachman.[25]]

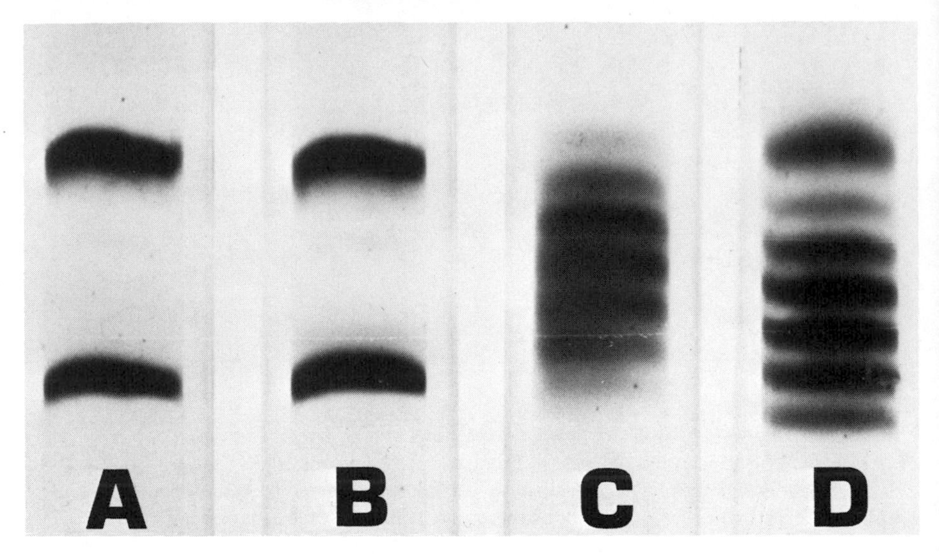

Fig. 10-21 Subunit stoichiometry by polyacrylamide gel electrophoresis of mixed hybrids obtained by dissociation and reconstitution of mixtures of aspartate-β-decarboxylase obtained from two bacterial species (*A. faecalis* and *P. dacunhae*). The holoenzymes from the two species (A) and the apoenzymes (B) exist as oligomers of sedimentation coefficient 19S and have markedly different electrophoretic mobilities. Upon dissociation to a 6S form and mixed reconstitution, a total of seven bands are observed (C and D) consisting of the two parental types and five hybrids. This clearly shows that the enzyme contains six of the "6S" subunits. Experimental conditions for the two reconstitution experiments are described by Tate and Meister.[26]

mixed reconstitution using subunits from homologous but electrophoretically distinct proteins[26] (Fig. 10-21).

4 ELECTROFOCUSING

A protein subjected to an electric field in a pH gradient will migrate electrophoretically and band at the pH at which it is isoelectric (i.e., has no net charge). A nonconvective supporting material or density gradient is required to stabilize the electrofocusing system. Unlike normal electrophoresis, the width of the final zones is independent of time and is determined by the balance of diffusion, the strength of the applied field, and the steepness of the pH gradient. Excellent protein separations have been obtained with this method since 1955,[27] but the procedure was seldom used until recently when a convenient method for establishing a stable pH gradient became commercially available.

The gradient is formed by the redistribution of carrier ampholytes (primarily composed of small polyamino-, polycarboxylic acids with a nearly continuous range of isoelectric points, typically from pH 3 to 10) by an electric field in a sucrose gradient[28] or other support such as a polyacrylamide gel.[29,30] As ampholyte molecules separate, the pH will change until all molecules come to (near) equilibrium in the area of their isoelectric point. Migration into the electrode vessels is blocked by dilute H_2SO_4 and ethanolamine at the anode and cathode, respectively. Proteins (which may initially be loaded throughout the column) will subsequently "focus" at their isoelectric pH. The protein bands are recovered from the sucrose gradient by fraction collection or may be stained in the gels as discussed for gel electrophoresis.

Isoelectric focusing affords remarkable resolution for the separation of proteins of similar properties, although only proteins that remain soluble at their isoelectric point can be studied. In most cases, proteins which differ by only one charge will be completely separated. At least 40 protein bands are easily resolved in blood serum. Although such resolution can generally be used to advantage, it should be noted that apparent heterogeneity may be of trivial significance in some cases. For example, proteins differing in amide content or sialic acid content (for glycoproteins) may reflect chemical or enzymatic change, incurred during isolation, that is of little consequence for the structural study of interest. The possibility that this is the explanation for the multiple bands observed for many highly purified proteins should soon be clarified. Highly purified proteins may also produce multiple bands because of the presence of conformational isomers, isozymes, or partially filled sites for the essentially irreversible binding of metals, coenzymes, and other small molecules. Such mechanisms may account for the microheterogeneity of L-amino acid oxidase from snake venom shown in Fig. 10-22. These results demonstrate the close correspondence between the analytical gel method (Fig. 10-22*a*) and the preparative procedure using a sucrose gradient (Fig. 10-22*b*). D-Amino acid oxidase from hog kidney gives only two bands in electrofocusing (Fig. 10-22*c*). By the use of enzymatic staining procedures, these bands were identified as the apoenzyme (protein component only) and the holoenzyme (protein plus the coenzyme flavin adenine dinucleotide). No microheterogeneity of either component was detected.

Isoelectric focusing may also be done in solvents containing uncharged dissociating agents such as 8 *M* urea. The separation of the individual polypeptide chains from dissociated oligomeric proteins may be used to advantage in the study of isozymes, proteins containing nonidentical polypeptide chains, and related problems. Another important application is in the separation of specifically cleaved peptides (e.g., by trypsin, CNBr). Figure 10-23 illustrates the separation of the CNBr peptides obtained from bovine fibrinogen.

5 ELECTRIC BIREFRINGENCE AND DIELECTRIC DISPERSION[31,32]

Two other methods of studying proteins, as well as other macromolecules, based on their charge properties are electric birefringence and dielectric dispersion. Birefringence, as indicated earlier (Table 9-1), is the dispersive counterpart of linear dichroism. Macromolecules such as proteins in solution may produce birefringence if they are preferentially aligned in one direction

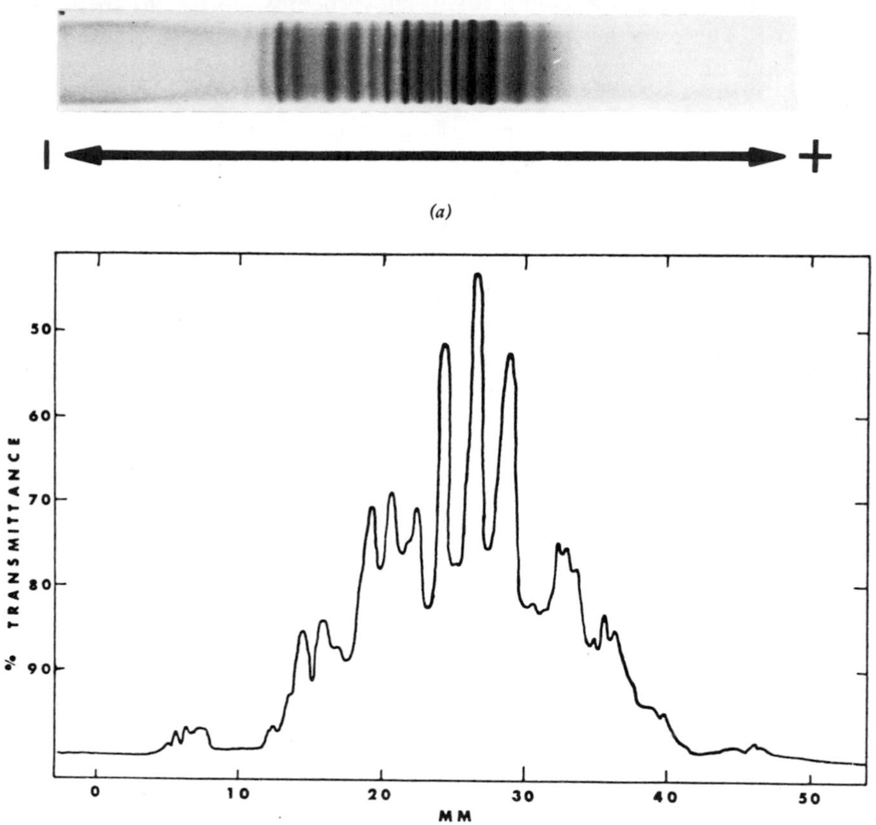

Fig. 10-22 (*a*) Demonstration of microheterogeneity of L-amino acid oxidase by electrofocusing in polyacrylamide gel: gel stained with Coomassie blue and its densitometric tracing. (*b*) Electrofocusing of L-amino acid oxidase in a sucrose gradient. Fractions were analyzed for absorbance at 275 nm (solid circles) and enzymatic activity (open circles). The pH gradient is shown by the solid line. (*c*) Gel electrofocusing of D-amino acid oxidase. Only two isozymes are indicated. Staining for enzyme activity without (1) and with (2) coenzyme shows that one of the bands has lost the cofactor. Staining for total protein (3) reveals the existence of additional minor contaminants. [From M. B. Hayes and D. Wellner, *J. Biol. Chem.*, **244,** 6636 (1969).]

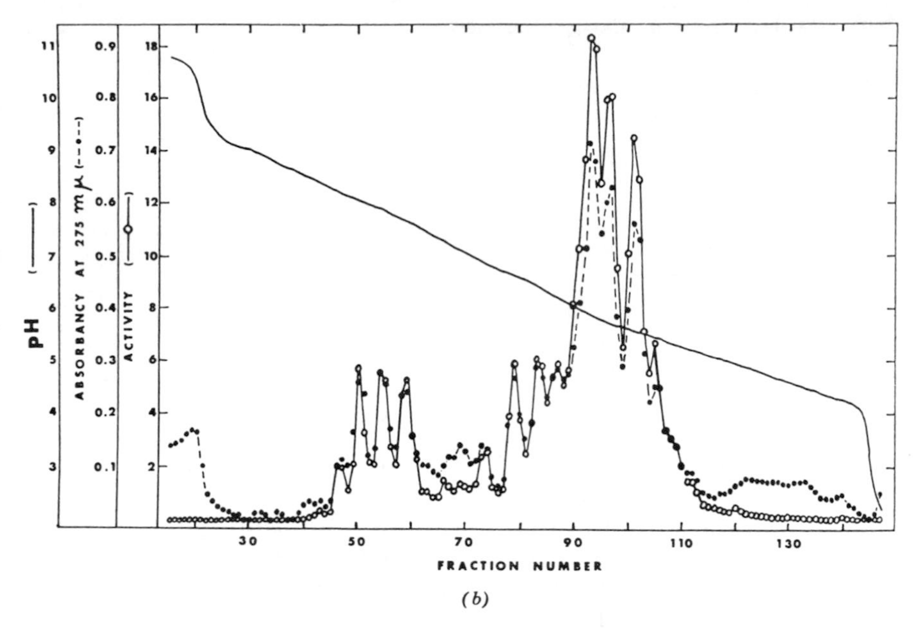

pH (——)
ABSORBANCY AT 275 mμ (--•--)
ACTIVITY (—o—)
11 10 9 8 7 6 5 4 3
0.9 0.8 0.7 0.6 0.5 0.4 0.3 0.2 0.1
18 16 14 12 10 8 6 4 2
30 50 70 90 110 130
FRACTION NUMBER

(b)

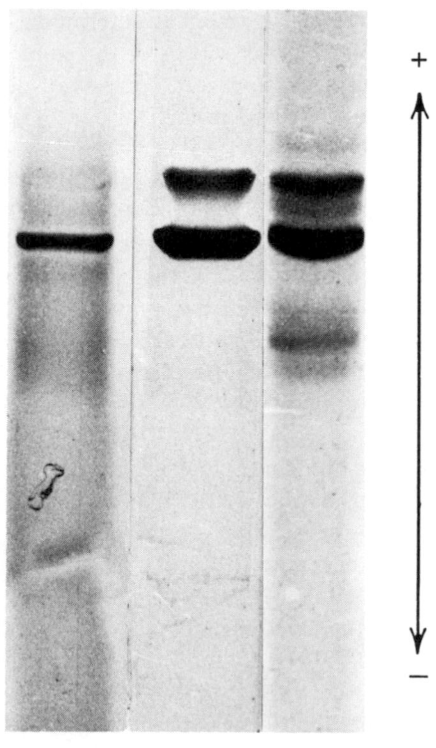

+
−
1
2
3

(c)

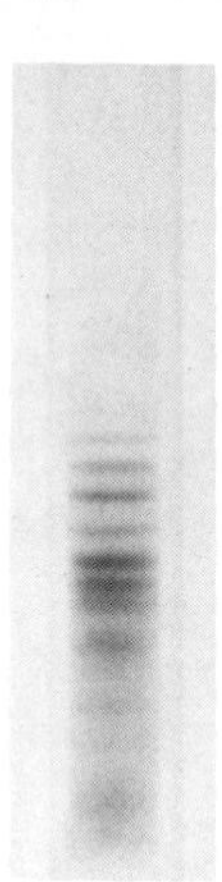

Fig. 10-23 Isoelectric focusing of cyanogen bromide peptides from bovine fibrinogen. [Experiment by H. Castleman.]

by an external force. When an electric field is used for this purpose, the degree of alignment is determined by electrical properties of the molecule.

In the method of transient electric birefringence, macromolecules are aligned in an electric field applied as a short pulse (of the order of microseconds), and the degree of alignment is determined from birefringence (the difference between the refractive indices of the solution parallel and perpendicular to the field). The measurement consists of determining the intensity of light passing through crossed polarizing lenses and falling on a photomultiplier tube where it is converted to an oscilloscope signal. The optical system is designed so that, as the molecules are aligned and the solution becomes birefringent, a tracing appears on the oscilloscope screen. In the case of proteins the effect is primarily one of *form birefringence*, that is, that produced by the alignment of nonspherical particles in a medium of different refractive index. *Intrinsic birefringence*, on the other hand, refers to a difference in refractive index between perpendicular directions within the particle itself, such as in DNA, which possesses a large negative birefringence ($n_{\parallel} < n_{\perp}$) where the parallel direction represents the axis of the double helix.

Although electric birefringence measurements provide a considerable amount of information, application of the method is limited by the requirement that the solution be relatively nonconducting. The magnitude of the birefringence gives a measure of the dipole moments in the molecule (induced and permanent) interacting with the applied field to cause alignment. The shape of the rise of the birefringence (at the onset of the electric pulse) permits one to distinguish between induced or permanent moment orientation whereas the decay of the birefringence due to rotational diffusion gives a precise measure of the rotational diffusion coefficient. This last quantity, which may also be determined from flow birefringence measurements (cf.

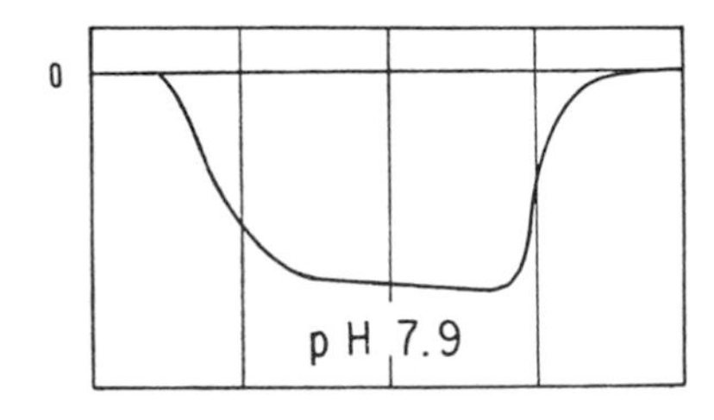

Fig. 10-24 Oscilloscope tracing depicting the transient electric birefringence of a fibrinogen solution in 0.1 *M* glycine, pH 7.9. The vertical lines are 20 μsec apart. [From A. E. V. Haschemeyer and I. Tinoco, Jr., *Biochemistry*, **1,** 996 (1962).]

Chapter 7), is highly sensitive to changes in the effective hydrodynamic length of the molecule.

Figure 10-24 illustrates the transient electric birefringence observed upon application of an electric field of the order of 5000 V/cm for 50 μsec to a solution of fibrinogen. The rotational diffusional coefficient of the molecule is readily obtained from the exponential decay of the signal; in this case using Perrin's equation (cf. Chapter 7), one obtains an effective length of 550 Å for the molecule. The slightly slower rise of the signal compared to the decay indicates the contribution of a permanent dipole interaction to the orientation. From the height of the signal one obtains an estimate of the total induced dipole and permanent dipole contributions.

The dipole moment values, which may be obtained as a function of pH by this method, give a measure of the symmetry in the distribution of charged groups over the surface of the molecule. The method is also valuable in studying reactions in which charge alteration occurs on the protein molecule such as in the release of charged peptides during activation of a protein. Figure 10-25 illustrates the striking change in electric birefringence produced during the release of the two A peptides (each with a charge of −4 electrons) from fibrinogen by the enzyme thrombin (cf. Chapter 17). The same approach has been used to study the binding of a charged hapten to its specific antibody.[33]

Dipole moments and rotational diffusion coefficients of proteins can also be obtained from the variation of the dielectric constant of their solutions

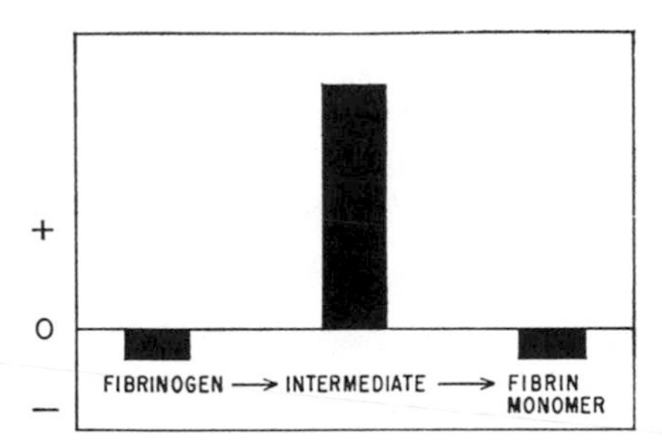

Fig. 10-25 Changes in sign and magnitude of the transient electric birefringence signal at pH 4.6 produced by the sequential release of two charged peptides from fibrinogen. [Based on data of A. E. V. Haschemeyer, *Biochemistry*, **2,** 851 (1963).]

with the frequency of the applied field (*dielectric dispersion*). When the dielectric constant of the solution is measured in a capacitor under the influence of an alternating electric field, its magnitude depends on the degree of alignment of the polar macromolecules in the field. As the frequency increases, the rotation of the molecules (by rotational diffusion) cannot keep up with the alternation of the field, and the dielectric constant changes can be used to obtain components of the rotational diffusion coefficient for both the long and short axis of the molecule. Induced dipole moments of protein molecules have also been determined, but it is now thought that they are produced primarily by polarization of the ion atmosphere about the molecules.

REFERENCES

Multiple Equilibria

1. I. M. Klotz, "Protein interactions," in *The Proteins*, Vol. 1B, H. Neurath and K. Bailey, Eds., Academic Press, New York, 1953, p. 727.
2. C. Tanford, *Physical Chemistry of Macromolecules*, Wiley, New York, 1961.
3. J. Steinhardt and S. Beychok, "Interaction of proteins with hydrogen ions and other small ions and molecules," *The Proteins*, 2nd ed., Vol. 2, Academic Press, New York, 1964, p. 140.
4. C. Tanford, "The interpretation of hydrogen ion titration curves of proteins," *Adv. Protein Chem.*, **17,** 70 (1962).
5. J. Steinhardt and J. A. Reynolds, *Multiple Equilibria in Proteins*, Academic Press, 1969.
6. J. T. Edsall and J. Wyman, *Biophysical Chemistry*, Academic Press, New York, 1958.

Electrophoresis and Electrofocusing

7. M. Bier, Ed., *Electrophoresis*, Academic Press, New York. Vol. I, 1959; Vol. II, 1967.
8. D. H. Moore, "Electrophoresis," in *Physical Techniques in Biological Research*, 2nd ed., Vol. IIA, D. H. Moore Ed., Academic Press, New York, 1968, p. 121.
9. D. J. Shaw, *Electrophoresis*, Academic Press, New York, 1969.
10. I. Smith, Ed., *Chromatographic and Electrophoretic Techniques*, 2nd ed., Vol. 2, Wiley-Interscience, New York, 1968.
11. R. Block, E. L. Durrum, and G. Zweig, *Paper Chromatography and Paper Electrophoresis*, 2nd ed., Academic Press, New York, 1958. G. Zweig and J. R. Whitaker, *Paper Chromatography and Electrophoresis*, Academic Press, New York, Vol. 1, 1967; Vol. 2, 1971.

12. H. Chin, *Cellulose Acetate Electrophoresis, Techniques and Applications*, Ann Arbor-Humphrey Science Publishers, Ann Arbor, 1970.

13. *Gel Electrophoresis* (symposium), *Ann. N.Y. Acad. Sci.*, **121,** 305–650 (1964).

14. O. Smithies, "Zone electrophoresis in starch gels: Group variations in the serum proteins of normal human adults," *Biochem. J.*, **61,** 629 (1955).

15. M. D. Poulik and O. Smithies, "Comparison and combination of the starch-gel and filter paper electrophoretic methods applied to human sera: Two dimensional electrophoresis," *Biochem. J.*, **68,** 636 (1958).

16. A. Chrambach and D. Rodbard, "Polyacrylamide gel electrophoresis," *Science*, **172,** 440 (1971).

17. M. D. Poulik, "Gel electrophoresis in buffers containing urea," *Meth. Biochem. Anal.*, **14,** 455 (1966).

18. K. Hannig, "The application of free-flow electrophoresis to the separation of macromolecules and particles of biological importance," in *Modern Separation Methods of Macromolecules and Particles*, Vol. 2, T. Gerritsen, Ed., Wiley-Interscience, New York, 1969, p. 45.

19. J. A. Reynolds and C. Tanford, "Binding of dodecyl sulfate to proteins at high binding ratios. Possible implications for the state of proteins in biological membranes," *Proc. Nat. Acad. Sci. U.S.*, **66,** 1002 (1970).

20. J. A. Reynolds and C. Tanford, "The gross conformation of protein-sodium dodecyl complexes," *J. Biol. Chem.*, **244,** 5161 (1970).

21. W. W. Fish, J. A. Reynolds, and C. Tanford, "Gel chromatography of proteins in denaturing solvents: Comparison between sodium dodecyl sulfate and guanidine hydrochloride as denaturants," *J. Biol. Chem.*, **244,** 5166 (1970).

22. A. L. Shapiro, E. Veñuela, and J. V. Maizel, Jr., "Molecular weight estimation of polypeptide chains by electrophoresis in SDS–polyacrylamide gel," *Biochem. Biophys. Res. Commun.*, **28,** 815 (1967).

23. K. Weber and M. Osborn, "The reliability of molecular weight determinations by dodecyl sulfate–polyacrylamide gel electrophoresis," *J. Biol. Chem.*, **244,** 4406 (1969).

24. G. E. Davies and G. R. Stark, "Use of dimethyl suberimidate, a cross-linking reagent, in studying the subunit structure of oligomeric proteins," *Proc. Nat. Acad. Sci., U.S.*, **66,** 651 (1970).

25. E. A. Meighen and H. K. Schachman, "Hybridization of native and chemically modified enzymes. I. Development of a general method and its application to the study of the subunit structure of aldolase," *Biochemistry*, **9,** 1163 (1970).

26. S. S. Tate and A. Meister, "Regulation and subunit structure of aspartate β-decarboxylase. Studies on the enzymes from *Alcaligenes faecalis* and *Pseudomonas dacunhae*," *Biochemistry*, **9,** 2626 (1970).

27. H. Svensson, "Zonal density gradient electrophoresis," in *Analytical Methods of Protein Chemistry*, Vol. 1, P. Alexander and R. J. Block, Eds., Pergamon Press, New York, 1960, p. 193.

28. *Science Tools* **14** (special issue devoted to isoelectric focusing), LKB-Produkter AB, Bromma, Sweden, 1967 (available from LKB Instruments Inc., Rockville, Md.).

29. H. Haglund, "Isoelectric focusing in pH gradients–A technique for fractionation and characterization of ampholytes." *Methods Biochem. Anal.*, **19,** 1 (1971).

30. D. Wellner, "Electrofocusing in gels," *Anal. Chem.*, **43,** 597 (1971).

Electric Birefringence and Dielectric Dispersion

31. S. Takashima, "Dielectric properties of proteins. I. Dielectric relaxation," in *Physical Principles and Techniques of Protein Chemistry*, Part A, S. J. Leach, Ed., Academic Press, New York, 1969.
32. K. Yoshioka and H. Watanabe, "Dielectric properties of proteins. II. Electric birefringence and dichroism," in *Physical Principles and Techniques of Protein Chemistry*, Part A, S. J. Leach, Ed., Academic Press, New York, 1969.
33. R. E. Cathou and C. T. O'Konski, "A transient electric birefringence study of the structure of specific IgG antibody," *J. Mol. Biol.*, **48,** 125 (1970).

XI Hydrogen Exchange[1–4]

Many of the hydrogen atoms in amino acids, peptides, and proteins are sufficiently labile to exchange with protons or isotopes of hydrogen ions in aqueous solutions. The rate at which exchange occurs depends on the electronegativity of the atom to which the proton is attached, the pH of the solution, and the temperature. In proteins, hydrogen atoms attached to nitrogen, sulfur, and oxygen are exchangeable. Table 11-1 shows rate constants for the exchange of the proton of the amide group in a model compound *N*-methylacetamide at room temperature as a function of pH (or pD). The measurement of hydrogen exchange is based on the change in some measureable property of the system that is sensitive to isotopic replacement. The exchange of protons with deuterium ions has been studied most extensively, although

Table 11-1 First-order rate constants of H–D exchange for the amide hydrogen in *N*-methylacetamide as a function of hydrogen (or deuterium) ion concentration in solution.[a]

Forward exchange		Back exchange	
pD	k_a (min^{-1})	pH	k_b (min^{-1})
3.23	23	3.41	7.7
3.88	6.4	4.19	1.6
4.62	1.3	4.87	0.48
5.27	0.41	5.57	0.71
5.95	0.77	—	—

[a] From S. O. Nielsen, *Biochim. Biophys. Acta*, **37**, 146 (1960).

tritium exchange is being used increasingly because of the ease in measuring the radioactive decay.

In 1955 Hvidt and Linderstrøm-Lang suggested that protons such as those in the peptide linkages of protein will have different exchange properties when they are hydrogen bonded, such as in the α-helical structure, than they have in model compounds of nonhydrogen-bonded structures. Their method for following the rate of exchange depended on the difference in density between H_2O and D_2O. After allowing exchange of a deuterated protein with water for various times, the solvent was collected by sublimation and its density accurately determined on a gradient density column. By following the rate of exchange of hydrogen atoms in insulin at various temperatures and in the presence of urea, a denaturing agent, (Fig. 11-1), they proposed the existence of several classes of hydrogen atoms in the native protein, each class having a particular exchange rate.

Since the pioneering work in the laboratory of Linderstrøm-Lang, a large number of studies of hydrogen exchange have been carried out with synthetic polypeptides and a variety of proteins. In general, small oligopeptides and polypeptides in a random coil form exchange labile hydrogens rapidly. In proteins, there is both fast and slow exchange. The rates of slow exchange are strongly dependent on conditions such as acidity and temperature and, in some cases, on the previous history of the sample. Generally, the exchange rate increases markedly with denaturation and approaches

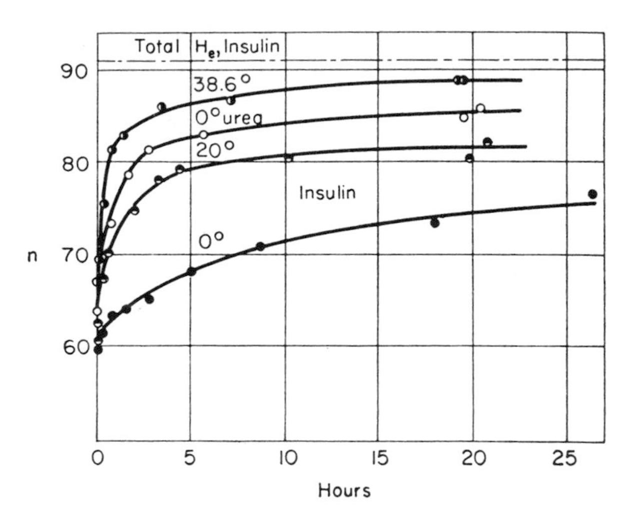

Fig. 11-1 Kinetic hydrogen exchange curves of insulin at pH 3.0 at various temperatures and in the presence of 5.2 *M* urea. The total number of labile hydrogen atoms in insulin is 91. [From A. Hvidt and K. Linderstrøm-Lang, *C. R. Trav. Lab. Carlsberg*, **29,** 385 (1955).]

the rate for random coil polypeptides. It is likely that slow exchange is associated with intrapeptide hydrogen bonding, although direct correlation is difficult to establish. Nonhydrogen-bonded peptide hydrogen atoms (and other exchangeable hydrogen atoms) that are internalized in proteins would also be expected to exchange slowly because of poor solvent accessibility (recall, however, that most internalized peptide and other exchangeable hydrogens are likely to be hydrogen bonded, from thermodynamic arguments, cf. Chapter 6). Exchange rates may also be reduced when exchangeable atoms are shielded from the external solvent by surface-bound or partially immobilized water molecules.

Although the kinetics of exchange in proteins cannot yet be completely evaluated, a number of significant findings have emerged from this type of study. One of these, of practical importance in the study of proteins, is the demonstration that subtle conformational differences may exist among samples of the same protein, depending on previous history. These differences, which may be difficult to demonstrate by other methods of study, may show up dramatically in exchange studies. Figure 11-2 shows the rates of

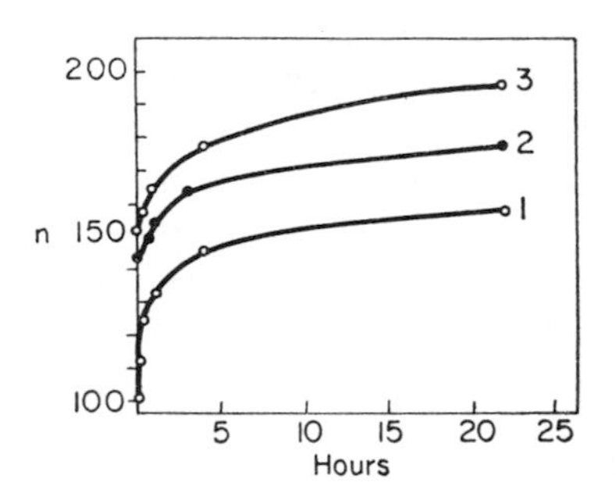

Fig. 11-2 Rates of H–D exchange between lysozyme and water at pH 3.2 and 0°C. Curves 1 and 2 refer to two different commercial samples; curve 3 is the sample shown in 2 after 1 year of storage. The ordinate gives the number of exchanged hydrogen atoms per protein molecule. The total number of labile hydrogen atoms in lysozyme is 260. [From A. Hvidt and L. Kanarek, *C. R. Trav. Lab. Carlsberg*, **33,** 463 (1963).]

hydrogen–deuterium exchange for three different samples of lysozyme. These experiments indicate that data obtained in a protein study may turn out to be valid only for that particular sample. The implications of this must be kept in mind by all investigators studying proteins in solution.

Hydrogen exchange studies have also provided insight into the dynamics of protein conformations in solution. Hvidt and Nielsen have suggested that a type of "breathing" occurs in native molecules in solution, during which atoms of the interior are transiently exposed to the solvent.[1] They find that the available kinetic data on slowly exchanging hydrogen atoms are consistent with a model in which the native molecule N is in equilibrium with a conformational isomer I:

$$N \underset{k_2}{\overset{k_1}{\rightleftharpoons}} I \xrightarrow{k_3} \text{exchange}$$

In isomer *I* the proton in question is capable of exchange with a rate constant k_3, like that found in random polypeptides. If $k_2 \ll k_3$, exchange depends on the rate constant k_1 of the conformational transition (EX_1 exchange mechanism); if $k_2 \gg k_3$, it depends on the equilibrium k_1/k_2 (EX_2 exchange mechanism).

The Hvidt-Nielsen mechanism for hydrogen exchange may be regarded as another statement of the general phenomenon of conformational equilibria in proteins, to be discussed further in Chapter 15. According to this understanding of the solution behavior of proteins, under any set of conditions, even those most favorable to the native state, a variety of partially denatured states also will exist at finite (though sometimes low) equilibrium concentrations. Hydrogen atoms that are internalized and inaccessible to solvent in the native structure may become available for exchange in one or more of these denatured states. Exchange rates will therefore depend on parameters of the conformational transitions, and it is apparent that hydrogen exchange studies could provide an extremely valuable probe for the study of conformational equilibria of proteins. Available data suggest that both EX_1 and EX_2 mechanisms may prevail in a particular protein for different regions of the molecule.[4] For example, hydrogen atoms which remain unexchanged after 20 hours at pH 5.5 in lysozyme must belong to a very stable region of the structure which is not exposed to solvent in most denatured states. Exchange of these hydrogens may be expected to be a direct function of the rate (given by k_1) of formation of the completely unfolded random coil (EX_1 mechanism). Other hydrogen atoms apparently exchange according to the EX_2 mechanism due to rates of interconversion between native and partially denatured states that are rapid compared with k_3 for exchange of a fully exposed peptide group.

A number of methods, in addition to the original one used by Hvidt and Linderstrøm-Lang, are now available for studying hydrogen exchange. Exchange with tritium may be studied with the measurement of radioactivity in a scintillation spectrometer. Some sources of error may arise, however, in the separation of protein and solvent. This method has been used effectively to study hydrogen exchange in insulin in the crystal and in solution.[5] It was found that exchange in the crystal is governed by the rate of opening of the native conformation (EX_1 mechanism), whereas reaction in solution followed the EX_2 mechanism. In another example, tritium was used to study the hydrogen exchange kinetics of chymotrypsinogen A as a function of temperature and pH. Samples were separated from labeled solvent by gel (Sephadex G25) filtration. The results indicate multiple pathways for exchange involving segmental motions of a local nature as well as full-scale cooperative unfolding.[6]

In tritiated ribonuclease two classes of hydrogen exchange differing in activation energy have been distinguished, each class consisting of a distribution of first-order rates. The results suggest that two pathways of the type previously discussed are available, one proceeding directly from the folded protein and the other from an intermediate thermally unfolded conformation.[7]

Infrared spectroscopy has been used as a more direct means for measuring deuterium exchange. Most of the slowly-exchanging labile hydrogen atoms are in peptide groups and their exchange with deuterium produces a shift in the amide II band at about 1550 cm^{-1} (cf. Chapter 9) to about 1450 cm^{-1}. Thus infrared analysis in this region of the spectrum provides a rapid measure of exchange rates, although precise quantitative interpretation is difficult. Figure 11-3 illustrates the effect of deuterium exchange on the infrared spectrum of an enzyme, lactic dehydrogenase. In this study it was shown that coenzymes such as nicotinamide-adenine dinucleotide reduce exchange rates and partially protect against detergent denaturation as well. These results suggest that the coenzyme acts to shift the equilibrium between conformations in favor of the native (slowly-exchanging) form of the protein.

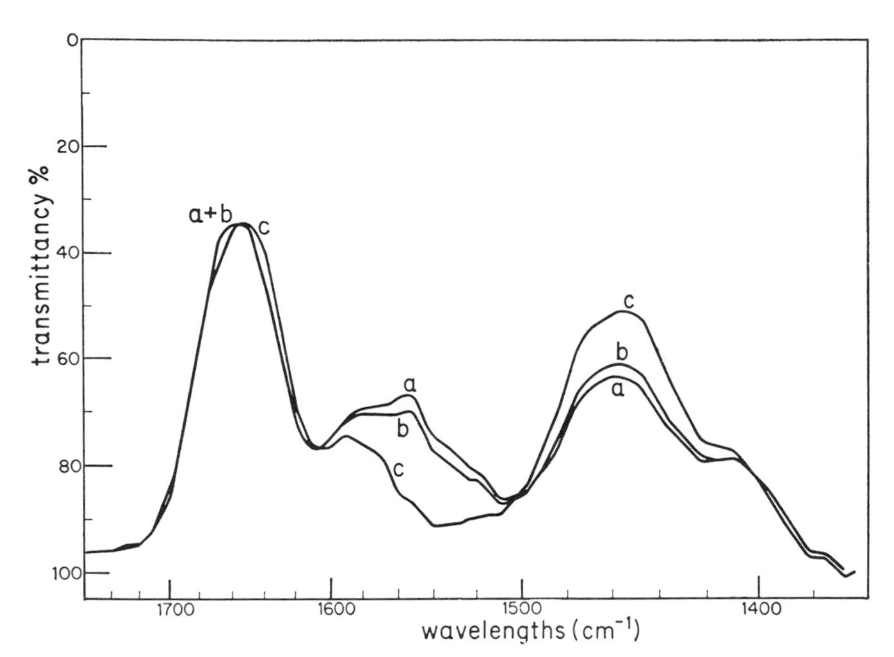

Fig. 11-3 Infrared spectrum of chicken heart lactic dehydrogenase, 1.1×10^{-4} *M*, in D_2O–0.1 *M* sodium phosphate buffer, pH 6.9. Curves: (a) at 42 min of exchange; (b) at 300 min of exchange; (c) at 300 min of exchange in the presence of a detergent, 0.15 *M* sodium dodecyl sulfate. [From G. DiSabato and M. Ottesen, *Biochemistry*, **4**, 422 (1965).]

Another method for studying hydrogen exchange, although limited at the present time to small molecules, is nuclear magnetic resonance. For further information on all aspects of hydrogen exchange in proteins, the various reviews as well as the original literature citations, should be consulted.

REFERENCES

1. A. Hvidt and S. O. Nielsen, "Hydrogen exchange in proteins," *Adv. Protein Chem.*, **21,** 287 (1966).
2. G. DiSabato and M. Ottesen, "Hydrogen exchange," in *Methods in Enzymology*, Vol. 11, C. H. W. Hirs, Ed., Academic Press, New York, 1967, p. 734; M. Ottesen, "Methods for measurement of hydrogen isotope exchange in globular proteins," *Methods Biochem. Anal.*, **20,** 135 (1971).
3. S. W. Englander, "Hydrogen exchange," in *Poly-α-amino Acids*, G D. Fasman, Ed., Marcel Dekker, Inc., New York, 1967.
4. C. Tanford, "Protein denaturation," *Adv. Protein Chem.*, **24,** 1 (1970).
5. M. Praissman and J. A. Rupley, "Comparison of protein structure in the crystal and in solution. II. Tritium–hydrogen exchange of zinc-free and zinc insulin," *Biochemistry*, **7,** 2431 (1968).
6. A. Rosenberg and J. Enberg, "Studies of hydrogen exchange in proteins. II. The reversible thermal unfolding of chymotrypsinogen A as studied by exchange kinetics," *J. Biol. Chem.*, **244,** 6153 (1969).
7. C. K. Woodward and A. Rosenberg, "Studies of hydrogen exchange in proteins. VI. Urea effects on ribonuclease exchange kinetics leading to a general model for hydrogen exchange from folded proteins," *J. Biol. Chem.*, **246,** 4114 (1971).

XII Physical Methods for Special Applications

1 NUCLEAR MAGNETIC RESONANCE AND ELECTRON SPIN RESONANCE[1–6]

The measurement of the absorption of electromagnetic radiation by elementary particles placed in a magnetic field has proved to be an important tool in the study of the nature of matter. Because of the sensitive dependence of the resonance absorption of nuclei and electrons on their electronic environment, the technique permits detailed analysis of chemical structure in the environment of absorbing particles.

The effect of magnetic fields on elementary atomic particles may be illustrated for the simplest case, that of a single electron. We can consider a spinning electron as an elementary magnetic dipole with a north and south pole. This dipole, when placed in an external magnetic field, will become aligned by the interaction of its magnetic moment with the field. Because of quantum restrictions, there are only two allowed values for the spin angular momentum of the electron and thus two allowed energy levels in the field. A given population of identical electrons will be split between the two energy levels. The difference in energy between levels is determined by the magnitude of the applied field H and a constant (g) characteristic of the particular type of particle (e.g., electron, proton):

$$\Delta E = (g)H\mu_B$$

where μ_B is a unit called the Bohr magneton which equals 9.27×10^{-21} erg/gauss. For protons, μ_N (nuclear Bohr magneton) is given by $\mu_B/1830$.

If an electromagnetic field of high frequency (ν) is superimposed perpendicular to the constant field H, transitions of electrons from the lower to the higher energy level will occur if the energy $h\nu$, where h is Planck's constant, equals the difference in energies ΔE. Thus when ν is chosen such that

$$h\nu = gH\mu_B \tag{12-1}$$

absorption of the high frequency radiation occurs. Because this phenomenon represents energy transfer between two oscillators oscillating at the same frequency, it is termed resonance. For free electrons ($g = 2$) and with magnetic fields of a magnitude easily obtainable in the laboratory (e.g., 3000 gauss), the resonance frequency ν falls in the microwave region of the spectrum, about 9×10^9 cycles per second [Hertz (Hz)]. For proton magnetic resonance ν is in the FM band, 60 to 100 MHz (or at 220 MHz in modern high field instruments). In practice the measurement of the magnetic resonance of nuclei or electrons is carried out by holding the frequency of the oscillating field constant and varying H through the range over which absorptions occur.

The components of an NMR apparatus are illustrated in Fig. 12-1. The magnetic field is applied between the poles of an electromagnet and is

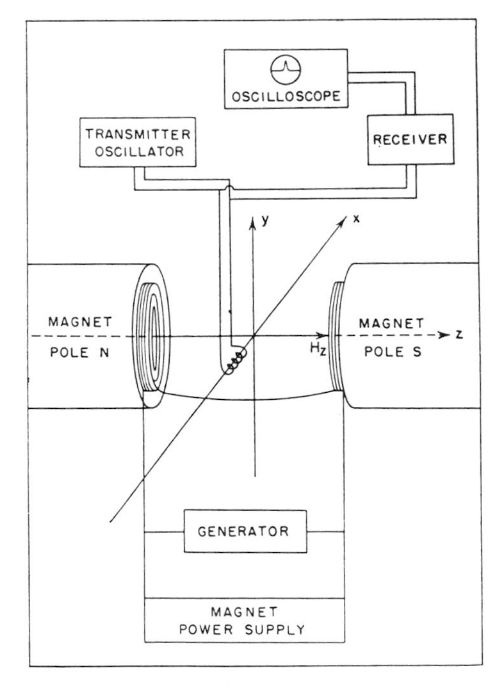

Fig. 12-1 Block diagram of the equipment used in the measurement of magnetic resonance.[3]

controlled by varying the flow of current. The oscillating field is supplied in a coil of wire wrapped around the sample. An oscilloscope records the radio-frequency signals induced in this coil or in a separate coil placed at right angles, indicating when the resonance condition is achieved.

Resonance absorption can be readily observed for electrons in molecules with unpaired electrons [*electron spin resonance* (ESR) or *electron paramagnetic resonance* (EPR)] and for all nuclei which have nonzero spin angular momentum [nuclear magnetic resonance (NMR), nuclear spin resonance]. For biological applications the most important of the nuclei which produce resonance is the proton ^{1}H. Others are ^{2}H (deuterium), ^{13}C, ^{14}N, ^{17}O, and ^{31}P. The common isotopes of carbon and oxygen (^{12}C, ^{16}O) have zero spin and do not exhibit magnetic resonance. The resonance absorption of a given atomic nucleus may be conveniently expressed in terms of the *gyromagnetic ratio* γ, which relates the magnetic moment of the nucleus to its nuclear spin. With this parameter Eq. 12-1 becomes

$$\nu = \gamma H/2\pi$$

The importance of NMR in the study of molecular structure is due to the fact that the field strength at which resonance is observed is highly sensitive to the electronic environment of the nucleus. When identical nuclei lie in different environments and thus are nonequivalent in chemical terms, a *chemical shift* of the resonance bands is observed (i.e., a shift of the resonance peak from the field strength at which it occurs in some reference condition for that nucleus). The shift is determined by the difference between the applied field and the effective field that exists at the nucleus. The effective field may be reduced due to shielding by electrons in the immediate vicinity (*diamagnetic effect*), causing the resonance to occur at higher applied fields. Alternatively, the field at the nucleus of a chemically bonded atom may be increased through induction by the applied field of magnetic moments associated with orbital motion of the electrons, which leads to an unshielding of the nucleus and a shift of the resonance to lower field strength (*paramagnetic shift*). Other factors that influence the observed chemical shift are delocalization of electrons, as in the π-electron system of benzene, or the proximity of charged groups or dipoles that polarize the electron cloud around the nucleus. The latter is thought to account for the chemical shifts in proton resonance observed upon formation of hydrogen bonds. Figure 12-2 illustrates chemical shifts of proton resonance in different atomic groupings of alanine. Studies of chemical shifts in the proton magnetic resonance spectra of amino acids have provided valuable information about the behavior of these molecules in solution. For example, existence of amino acids in the zwitterion form in neutral solution has been directly confirmed. Because of the dependence of chemical shifts on the ionization state, it is possible to

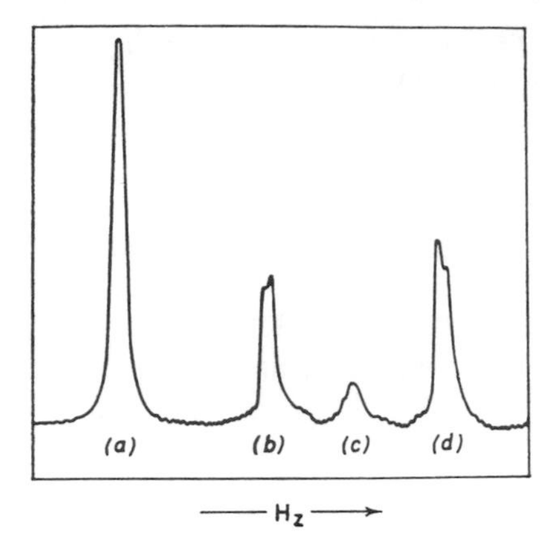

Fig. 12-2 Nuclear magnetic resonance spectrum of alanine in D_2SO_4, illustrating the chemical shift. The four lines show the position of proton resonance in four different environments—solvent (*a*) and three chemical groupings in the alanine molecule: NH_3^+ (*b*), α-CH (*c*), and CH_3 (*d*).[3]

follow titrations by measurement of NMR and to unequivocally identify titrating groups such as the various ionizable groups of amino acid side-chains.

Additional structural information is obtained in NMR spectroscopy as a consequence of spin-spin coupling. This interaction causes the characteristic resonance band of a particular nucleus to be split into a group of bands determined by the number, distance, and symmetry of neighboring nuclei in the molecule. Coupling is most sensitive to the environment produced by near neighbors of the absorbing particle; it becomes negligible when the number of intervening bonds reaches four or five except in cases of electron delocalization. Coupling produces characteristic multiplet groups in place of single bands; for protons, for example, the number of lines in a multiplet is $N + 1$ where N is the number of equivalent neighboring protons. Thus analysis of band splitting may permit differentiation between conformations such as the various possible staggered rotational isomers (*rotamers*) corresponding to energy minima in the rotation about single bonds.

The shape of the absorption signal is used to obtain another characteristic NMR parameter, the relaxation time. It represents the rate of approach to equilibrium in the transfer of particles from the excited (higher energy) level to the ground state. The characteristic time for reorientation with respect to the z axis (see Fig. 12-1) is called the *longitudinal relaxation time*. Another characteristic, the *correlation time*, refers to the time constant of exponential decay of local fields about a nucleus. In some cases these quantities can be used to obtain rates of molecular tumbling in solution and have had application to biological problems. In general, however, relaxation phenomena are complex and difficult to interpret.

A great deal of empirical data have been amassed on chemical shifts, spin coupling constants, and other resonance properties that can be used for

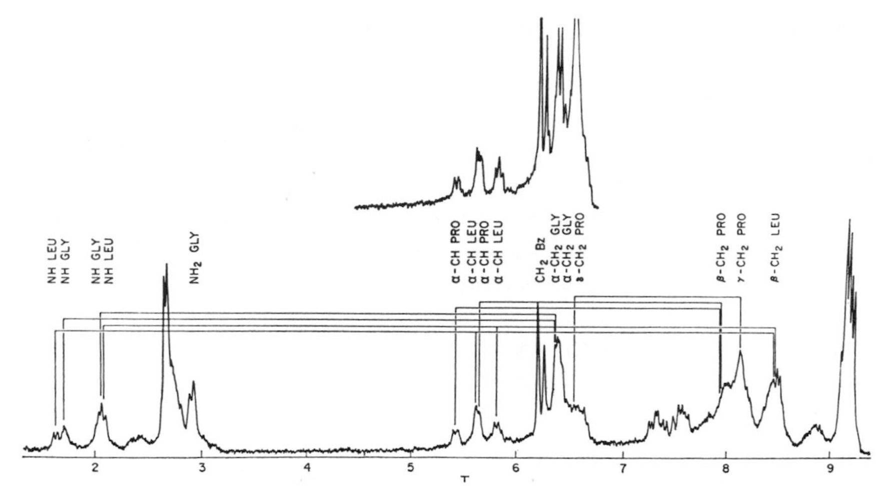

Fig. 12-3 Nuclear magnetic resonance spectrum in the 220 MHz region of (*S*-Bzl)-L-Cys-L-Pro-L-Leu-Gly(NH_2). Resonance lines are identified above the spectrum. [From V. J. Hruby, A. I. Brewster, and J. A. Glasel, *Proc. Nat. Acad. Sci.*, *U.S.*, **68,** 450 (1971).]

structural inferences in molecules where little other structural information is available. This knowledge is now being used in increasing numbers of applications to biological problems. Of the several resonance phenomena, the most often studied in biological molecules has been proton NMR. Figure 12-3 illustrates the 220-MHz proton magnetic resonance spectrum of a tetrapeptide studied as a model compound for the polypeptide hormone oxytocin. Assignment of bands is based upon the spectra of the constituent amino acid residues and on empirical data on the effects of peptide bond formation (primary structure) and three-dimensional conformation on the proton resonances. Systematic studies of chemical shifts in small oligopeptides have been carried out to aid in identification of proton resonances in polypeptides.[7] Conformational studies have been done with a variety of synthetic polypeptides, and it has been found that changes in magnetic resonance spectra provide a sensitive assay for α-helix formation in some cases.

Resolution of the overall proton resonance spectrum in proteins is difficult because of the very large number of protons, almost every one of which is in a slightly different electronic environment. In addition to the overlapping of signals, the proton resonances are broadened by dipole-dipole interactions, and certain regions of the spectrum show a nearly continuous field of signals. A practical difficulty, also, has been the high concentration of sample required for sufficient sensitivity in the measurement. This problem, however, is being increasingly overcome by the development of more sensitive resonance spectrometers and by the use of higher magnetic field strengths. For proteins an important application of magnetic resonance techniques has been in the study of the structural relationships involved in protein–ligand interactions. A particularly advantageous situation is that in which the ligand (usually a small molecule) has one or more resonance bands that are well separated from the bulk of the protein absorption. This may be brought about artificially by selective deuteration of one of the interacting species. For example, deuteration of the imidazole C2 hydrogen of histidine 12 in the *S*-peptide (residues 1–20) of ribonuclease S permitted the identification of its resonance in the spectrum and determination of the p*K* of that residue in the recombined protein. The review by Roberts and Jardetzky[5] is recommended for further information on these interesting studies.

Considerable attention is being given to modifications of ligand spectra produced upon binding as a means of probing the structure and interactions involved at the protein binding site. Changes in chemical shifts of ligand nuclei, band splitting, and differences in line widths between free and bound species may be observed. For example, the splitting and upfield shift of the acetamide methyl proton resonance of 2-acetamide-2-deoxy-D-glucose that occur when this inhibitor is bound to lysozyme provide information on binding strength and magnetic environment at the binding site.[8] In another case, the effect of enzyme-bound manganese on the proton resonance of water in the presence and absence of a coenzyme was used to gain insight into the microenvironment at the metal ion-coenzyme site on the protein.[9] Benzoyl and phthaloyl groups have been tested as proton resonance markers in polypeptides. These may be valuable in the determination of specific side-chain interactions involving aromatic amino acid residues in polypeptides. Halide ions such as chloride (^{35}Cl) or bromide (^{81}Br) have also been suggested as useful NMR probes in proteins.[10] The use of such labels can provide information on the accessibility of a particular region of a macromolecule to solvent and also the freedom of motion of the attached label.

Carbon-13 NMR spectroscopy is now being applied to protein structure problems. Because of the larger range of chemical shifts observed for ^{13}C, better resolution and greater ease in resonance assignment is possible. Protein spectra based on the natural abundance of ^{13}C in the molecules show a major separation of resonances into classes arising from carbonyl groups, unsaturated side chains, α-carbon atoms, and saturated side-chain residues; detailed analysis of oligopeptides is being carried out to aid in band assignment.[11] Interesting results with ESR are being obtained by the introduction of specific markers or "spin labels" into proteins. Paramagnetic nitroxide radicals have been used in studies of bovine serum albumin and hemoglobin.[12] Results for the latter have revealed a small structural change upon oxygenation which had not been resolved by crystal structure studies.

Although the effective application of magnetic resonance methods to proteins is relatively recent, increasing activity in this area may develop. Correlation of results obtained here with those of X-ray diffraction should be especially interesting. As with some of the other physical methods discussed in this chapter, however, its use will be restricted to specialists in the area for some time to come.

2 MÖSSBAUER SPECTROSCOPY[13–15]

The Mössbauer effect refers to the phenomenon of resonant absorption of nuclear gamma rays that occurs in certain stable nuclear isotopes. It was discovered by Rudolph Mössbauer in 1957 during his graduate work at Heidelberg and brought him the Nobel Prize in physics in 1961. Although no Mössbauer isotopes occur in high proportion in biological molecules, some of these isotopes may be introduced by enrichment methods (e.g., ^{57}Fe, ^{129}I), thus permitting application of the technique to the study of proteins. A complete analysis of this phenomenon would require a thorough background in nuclear physics; here we present a brief explanation of the Mössbauer effect, which may be helpful in evaluation of its biological applications.

Nuclei are capable of transitions between certain nuclear energy levels with the absorption or emission of electromagnetic radiation (γ-rays), much as atoms absorb and scatter X-rays due to their electronic structure. The emission of a γ-ray from a nucleus is normally accompanied by recoil of the nucleus (for conservation of momentum), an energy-requiring process. If the energy of the emitted photon is denoted E_γ, then the energy R_E of the recoil is given by

$$R_E = E_\gamma^2/2Mc^2 \tag{12-2}$$

where M is the nuclear mass and c is the velocity of light. Since both energies E_γ and R_E must be derived from the available energy of the transition, E_γ will be less than the transition energy ΔE:

$$E_\gamma = \Delta E - R_E \tag{12-3}$$

Because of this energy loss, the emission does not have sufficient energy to excite another nucleus of the same type, and no further absorption can occur. What Mössbauer discovered was that certain low energy γ-ray emissions take place without recoil because their recoil energy is not great enough to allow the nucleus to move as a free particle. Instead, the momentum of recoil is transferred to the entire lattice (e.g., the entire molecule or the crystal lattice) in which the nucleus is bound. Because of the large mass of the lattice, the recoil energy loss is negligible. Thus the emitted γ-ray has an energy very close to that of the transition, and can be reabsorbed by another atom of the same kind. As a result, when γ-ray absorption is measured as a function of energy of the incident radiation, a very sharp resonance absorption line will be observed at the transition energy of a Mössbauer transition.

The characteristic features of a Mössbauer spectrum are influenced by the local environment of the nucleus producing the Mössbauer effect. In general, the absorption line is broadened into a band of variable line width depending on the environment. The band may be split into multiple bands (*hyperfine* or *quadrupole splitting*), and may be shifted slightly in energy relative to a standard energy source (*isomer shift*). These effects are strongly dependent on temperature and reveal a great deal of precise information about local interactions within the nucleus producing the Mössbauer emission and in the lattice around it.

The measurement will be illustrated for the case of ^{57}Fe, the Mössbauer isotope of most interest biologically because of the natural occurrence of iron in a number of proteins (e.g., myoglobin, hemoglobin, cytochromes, ferredoxin). The radiation source used in studies of ^{57}Fe is its parent nucleus ^{57}Co. Cobalt-57 decays in two steps to the first excited state of ^{57}Fe. This state has a lifetime of 10^{-7} seconds and produces a recoil-free emission of 14.4 keV, which serves as the Mössbauer photon source. The absorption of this radiation by the ^{57}Fe in the sample being studied is then measured at various temperatures, and over a small range of energies on either side of the transition energy. In order to vary the energy of the 14.4 keV source emission, the source is moved with respect to the sample (absorber) at a constant velocity v. This produces a Doppler effect, the energy E of the γ-ray being shifted a small amount δE according to the relation $\delta E = (v/c)E$, where c is the velocity of light. The velocity v is positive when the source moves toward the absorber.

Mössbauer spectra are therefore usually presented as the transmission of the source radiation (counted in a γ-ray spectrometer on the other side of the absorber) plotted against the velocity at which the source (or absorber) is moved. For ^{57}Fe, velocities in the range of a few millimeters per second are suitable to scan the energy range of the Mössbauer spectrum. The resulting profile is illustrated in Fig. 12-4 for methemoglobin. In a Mössbauer spectrum

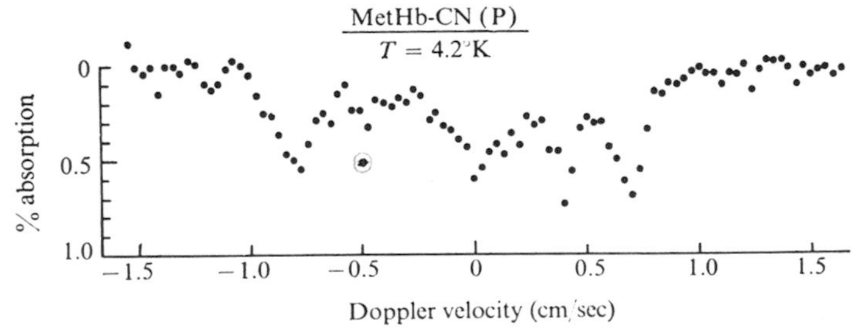

Fig. 12-4 Mössbauer spectrum of human methemoglobin cyanide at 4.2°K.[15]

the energy of the incident radiation is expressed in terms of a Doppler shift velocity instead of wavelength or frequency.

The study of iron-containing proteins by Mössbauer spectroscopy normally requires enrichment of the protein with ^{57}Fe. The method has been applied to the study of a number of heme proteins and other iron-containing proteins, and has proved to be sensitive for distinguishing different classes of cytochromes.[16] Further studies with ^{57}Fe and with other Mössbauer isotopes (e.g., ^{127}I, ^{129}I, and ^{125}Te) in biological systems are being made. For example, the Mössbauer effect of heme–iodide and related complexes has been examined as a model system for the interaction of iodide with hemoproteins such as chloroperoxidase which catalyze halogenation reactions.[17]

3 RELAXATION SPECTROMETRY[18,19,23]

The study of chemical kinetics following perturbation of a system at equilibrium by rapid change of an external parameter (e.g., temperature) was developed in the 1950s by Eigen and his collaborators and brought him the Nobel Prize in Chemistry in 1967. The importance of the method is that very rapid reactions can be investigated; with the temperature jump method, reactions with half-lives as short as 10 μseconds can be followed. Other types of perturbations extend the range of analysis to 0.1 nanosecond (10^{-10} sec). Fast reactions of biological interest include enzyme–substrate interactions and conformational changes in macromolecules. Figure 12-5 shows a comparison of various methods for studying reaction kinetics and the time ranges in which they may be applied.

The experimental quantity of interest is the relaxation time of the reaction. It is determined from the change in a concentration (reactant or product) as a function of time after perturbation of a system initially at equilibrium. For example, for the reaction

$$A + B \underset{k_{-1}}{\overset{k_1}{\rightleftharpoons}} C$$

a perturbation such as a fast temperature change will cause the concentrations of A, B, and C to change to a new set of equilibrium values corresponding to the new conditions. If the concentration of A, for example, is followed as a function of time, a relaxation time τ for the process may be obtained by the use of

$$\Delta(A) = \Delta(A)_0 e^{-t/\tau} \tag{12-4}$$

where $\Delta\ (A)_0$ is the total concentration difference between the equilibrium positions and $\Delta(A)$ is the difference between the concentration of A at time t and its final equilibrium value. For this reaction τ is related to the rate

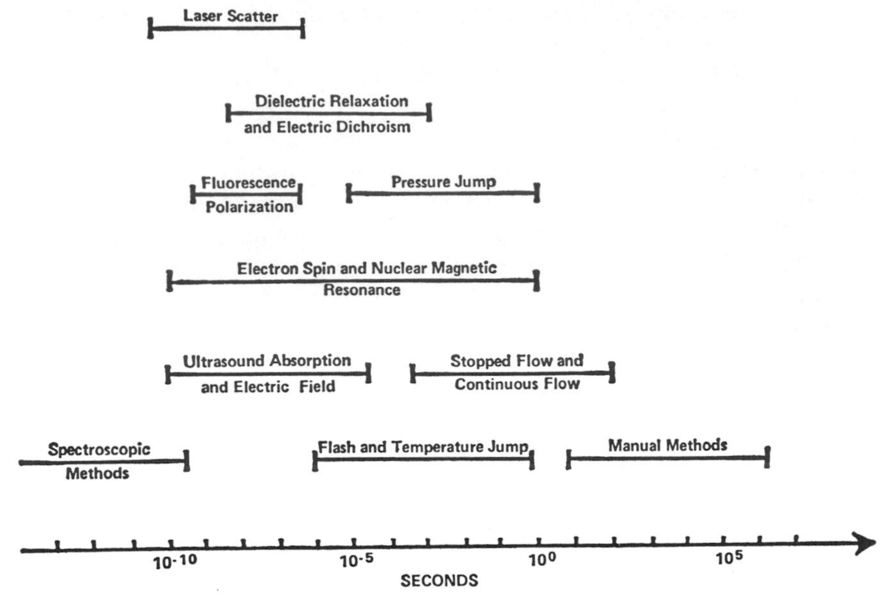

Fig. 12-5 Comparison of various methods of the measurement of fast processes in terms of the accessible time ranges. [From A. N. Schechter, *Science*, **170,** 273 (1970).]

constants of the reaction by

$$1/\tau = k_1[(\bar{A}) + (\bar{B})] + k_{-1} \tag{12-5}$$

where $(\bar{A})$ and $(\bar{B})$ represent equilibrium concentrations of the reactants. By determination of τ at different concentrations, the rate constants of both the forward and back reactions are obtained and thus also the equilibrium constant $K = k_1/k_{-1}$. Analyses of this type have also been made for more complex reactions.[18]

Figure 12-6 shows an oscilloscope tracing obtained for the reaction of ribonuclease with cytidine-3′-phosphate following perturbation of the equilibrium by a temperature jump of 8°. The jump was produced by applying a high voltage pulse across the enzyme and substrate solution; the spectral changes produced upon formation of the enzyme–substrate complex were then detected with a photomultiplier and recorded on the oscilloscope screen. From the relaxation times for the process under various conditions (ranging from 0.1 to 1.0 msec) rate and equilibrium constants were calculated for different models. These results were used to develop a mechanism for the reaction.[20] Rate constant ranges for a variety of reactions of biological interest have been tabulated by Havsteen and are given in Table 12-1. Many reactions are found to approach the calculated limiting rate for diffusion-controlled encounter.

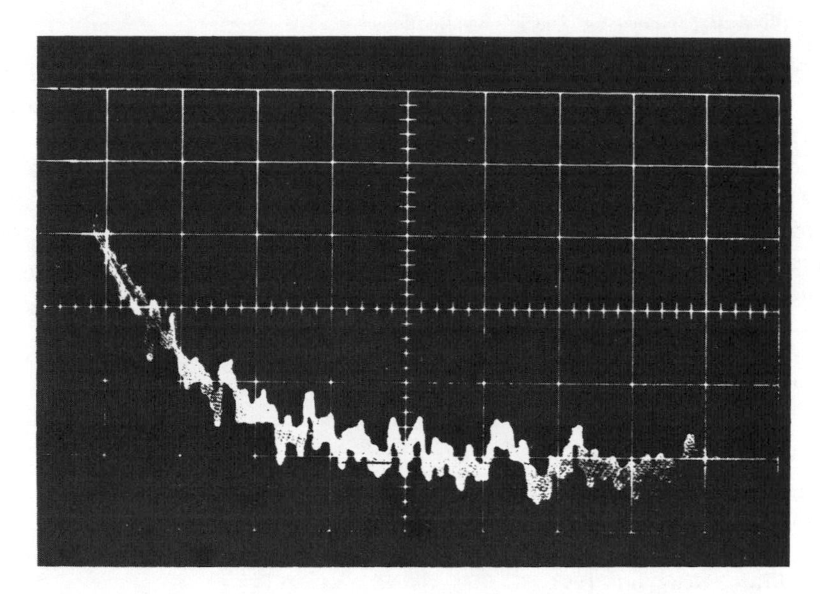

Fig. 12-6 Oscilloscope recording of the absorbance change at 255 nm as a function of time produced by temperature-jump perturbation of the equilibrium in the reaction of ribonuclease with cytidine-3′-phosphate. (Courtesy of R. Cathou and G. G. Hammes.)

Table 12-1 Rate constants for some reactions of biological interest, as determined by relaxation methods[19].

	Bimolecular reactions	
	Combination rate (M^{-1} sec^{-1})	Dissociation rate (sec^{-1})
Complex formation		
Diffusive encounter	10^{10}–10^{11} (calculated)	
Proton transfer (involving H^+ or OH^-)	10^8–10^{11}	10^{-2}–10^7
Proton transfer (involving H-bond chelates)	10^5–10^{10}	1–10^4
Acid/base catalysis	10^8–10^{11}	1–10^7
Hydration and hydrolysis	10^{-2}–10^{12}	10^{-2}–10^7
Charge interactions	10^9–10^{11}	10^{-7}–10^7
Hydrophobic interactions	10^8–10^9	10^2–10^4
Dye binding to proteins	10^6–5×10^8	10^2–10^4
Enzyme–substrate, antibody–hapten interactions	10^7–10^8	10^2–10^4
Metal–ligand interactions	10^{-5}–10^8	10^{-7}–10^5
Electron transfer	10^3–10^8	

First-order rate constants for monomolecular protein isomerization reactions (in either direction) are found in the range of 10^{-2} to 10^7 sec^{-1}.

Other systems that have been studied are the helix-coil transition of poly-L-glutamic acid (average relaxation time in the range of 0.05–10 μsec)[21] and the oxygen uptake of the highly reactive hemoglobin produced by photochemical decomposition of carboxyhemoglobin.[22] The application of the technique to study the kinetics of ligand binding by allosteric enzymes should be valuable in distinguishing between various models of allosteric interaction. For further information the review articles by Hammes and by Havsteen should be consulted.

REFERENCES

NMR and ESR

1. F. A. Bovey, *Nuclear Magnetic Resonance Spectroscopy*, Academic Press, New York, 1969.
2. A. Carrington and A. D. McLachlan, *Introduction to Magnetic Resonance*, Harper and Row, New York, 1967.
3. O. Jardetzky and C. D. Jardetzky, "Introduction to magnetic resonance spectroscopy methods and biochemical applications," *Meth. Biochem. Anal.*, **9,** 235 (1962).
4. J. C. Metcalfe, "Nuclear magnetic resonance spectroscopy," in *Physical Principles and Techniques of Protein Chemistry*, Part B, S. J. Leach, Ed., Academic Press, New York, 1970.
5. G. C. K. Roberts and O. Jardetzky, "Nuclear magnetic resonance spectroscopy of amino acids, peptides, and proteins," *Adv. Protein Chem.*, **24,** 448 (1970).
6. C. C. McDonald and W. D. Phillips, "Proton magnetic resonance spectroscopy of proteins," in *Fine Structure of Proteins and Nucleic Acids* (Biological Macromolecules Series, Vol. 4), G. D. Fasman and S. N. Timasheff, Eds., Marcel Dekker, Inc., New York, 1970, p. 1.
7. A. Nakamura and O. Jardetzky, "Systematic analysis of chemical shifts in the nuclear magnetic resonance spectra of peptide chains. II. Oligoglycines," *Biochemistry*, **7,** 1226 (1968).
8. F. W. Dahlquist and M. A. Raftery, "A nuclear magnetic resonance study of association equilibria and enzyme-bound environments of *N*-acetyl-D-glucosamine anomers and lysozyme," *Biochemistry*, **7,** 3269 (1968).
9. M. C. Scrutton and A. S. Mildvan, "Pyruvate carboxylase. XI. Nuclear magnetic resonance studies of the properties of the bound manganese after interaction of the biotin residues with avidin," *Biochemistry*, **7,** 1490 (1968).
10. T. R. Stengle and J. D. Baldeschwieler, "Halide ions as chemical probes for NMR studies of proteins," *Proc. Nat. Acad. Sci., U.S.*, **55,** 1020 (1966).
11. F. R. N. Gurd, P. J. Lawson, D. W. Cochran, and E. Wenkert, "Carbon 13 nuclear magnetic resonance of peptides in the amino-terminal sequence of sperm whale myoglobin," *J. Biol. Chem.*, **246,** 3725 (1971).
12. T. J. Stone, T. Buckman, P. L. Nordio, and H. M. McConnell, "Spin-labeled biomolecules," *Proc. Nat. Acad. Sci., U.S.*, **54,** 1010 (1965). J. C. A. Boeyens and H. M. McConnell, "Spin-labeled hemoglobin," *Proc. Nat. Acad. Sci., U.S.*, **56,** 22 (1966).

Mössbauer Spectroscopy

13. H. Frauenfelder, *The Mössbauer Effect*, Benjamin, New York, 1963.
14. U. Gonser and R. W. Grant, "Application of the Mössbauer effect to biological systems," in *Mössbauer Effect Methodology*, Vol. 1, I. J. Bruverman, Ed., Plenum Press, New York, 1965.
15. J. E. Maling and M. Weissbluth, "The application of Mössbauer spectroscopy to the study of iron in heme protein," in *Solid State Biophysics*, S. J. Wyard, Ed., McGraw-Hill, New York, 1969.
16. T. H. Moss, A. J. Bearden, R. G. Bartsch, and M. A. Cusanovich, "Mössbauer spectroscopy of bacterial cytochromes," *Biochemistry*, **7,** 1583 (1968).
17. M. Pasternak, P. G. Debrunner, G. DePasquali, L. P. Hager, and L. Yeoman, "Application of ^{129}I Mössbauer effect to biological systems: Studies with heme models," *Proc. Nat. Acad. Sci., U.S.*, **66,** 1142 (1970).

Relaxation Spectrometry

18. G. G. Hammes, "Relaxation spectrometry of biological systems," *Adv. Protein Chem.*, **23,** 1 (1968).
19. B. H. Havsteen, "Perturbation and flow techniques," in *Physical Principles and Techniques of Protein Chemistry*, Part A, S. J. Leach, Ed., Academic Press, New York, 1969.
20. R. E. Cathou and G. G. Hammes, "Relaxation spectra of ribonuclease. I. The interaction of ribonuclease with cytidine-3′-phosphate," *J. Amer. Chem. Soc.*, **86,** 3240 (1964).
21. J. J. Burke, G. G. Hammes, and T. B. Lewis, "Ultrasonic attenuation measurements in poly-L-glutamic acid solutions," *J. Chem. Phys.*, **42,** 3520 (1965).
22. Q. H. Gibson, "The photochemical formation of a quickly reacting form of haemoglobin," *Biochem. J.*, **71,** 293 (1959).
23. A. F. Yapel, Jr., and R. Lumry, "A practical guide to the temperature-jump method for measuring the rate of fast reactions," *Methods Biochem. Anal.*, **20,** 169 (1971).

XIII Electron Microscopy[1–6]

The electron microscope is an extremely valuable tool for the study of materials throughout the biological and physical world. Instrumentation has markedly improved in the past 20 years, and new sample preparation methods have been developed that make electron microscopy increasingly important in the examination of proteins and other macromolecules.

Transmission Electron Microscopy. In an electron microscope used in the normal transmission mode (Fig. 13-1), electrons emitted from a hot tungsten filament (cathode) are accelerated toward an anode at ground potential. A condenser lens with an aperture system to control the angular width of the beam is used to direct the beam onto the specimen. During passage through the specimen, electrons are deflected from their path by electrostatic interactions primarily with the nuclei of the atoms where positive charge is concentrated.

The specimen stage is followed by an objective lens that serves to magnify the image about 100-fold. Contrast is improved by the presence of an aperture, placed after the objective lens, that serves to prevent large numbers of scattered electrons from blurring the image produced by the transmitted electrons. Finally, one or two projector lenses are used to produce an image magnified about 1000 to 200,000× on a photographic plate or fluorescent screen. Lenses in the electron microscope may be of the magnetic (most common) or electrostatic type in which the incoming electrons are focused by interaction with a magnetic field or electrostatic potential distribution along the axis. As in other microscopes, image quality is limited by spherical and chromatic aberrations and by diffraction effects. Other technical difficulties involve variations in the field strength of the lenses, effect of the electron beam on the sample, and scattering from contamination. The latter effect is largely due to a coating of the specimen exposed to the electron beam with products produced by decomposition of residual oil vapors in the vacuum chamber. Contamination is reduced to tolerable levels by trapping such vapors on a liquid nitrogen cooled trap located near the specimen stage.

Resolution and Contrast. Electron microscopes capable of 10 Å resolution have been available since the mid-1940s. Today there are many commercially available microscopes with capability for point-to-point resolution of 3 Å or less on ideal specimens. A brief account of the historical development of electron microscopes is given by Hall.[1] It is important to realize that there may be a large difference between the resolution defined by instrumental capability and that obtained by a microscopist in practice. Even when the instrument is in top operating condition, resolution will be limited by the degree to which the operator can correct astigmatism and achieve focus. In light microscopy, lack of proper focus (defocus) results in a fuzzy image which, although it may be unsatisfactory in resolving details, is unlikely to be misinterpreted. Defocused electron micrographs, on the other hand, are often interpreted erroneously. The edge of a hole in the carbonized supporting film of a sample grid may be used for focusing. The underfocused image (objective current too low) of a hole in a carbon film is characterized by a "halo" appearance of increased intensity just inside the edge, whereas in an overfocused image a dark fringe is seen inside the edge of the hole. Neither effect is detected at focus. If the objective-lens field is asymmetrical, the image will be focused at different levels, with greatest difference in two mutually perpendicular directions. Such an image is termed "astigmatic," and the condition is readily detected by the nonuniform disappearance of the overfocused fringe as the objective-lens current is decreased stepwise toward underfocus and/or by the asymmetric appearance of the phase image (see below). Contour phenomena are produced by Fresnel diffraction which occurs when the electron beam encounters abrupt discontinuities in refractive index. These discontinuities also exist at the edge of certain particles to produce contrast effects that are superimposed on the subject image.

Another phenomenon that complicates the interpretation of electron micrographs is phase contrast (also called phase grain or phase image). This gives rise to an apparent background structure even in regions of constant scattering power containing no projecting edges. The apparent granularity increases with defocusing and at focus becomes comparable in size to the resolution limit of the instrument. Combinations of phase contrast and contour phenomena at images of negatively stained molecules result in spurious contrast, which may be incorrectly interpreted as a subunit structure in a protein molecule. The size of the phase contrast image varies with the degree of defocus (Fig. 13-2) and meaningful detail cannot be interpreted for

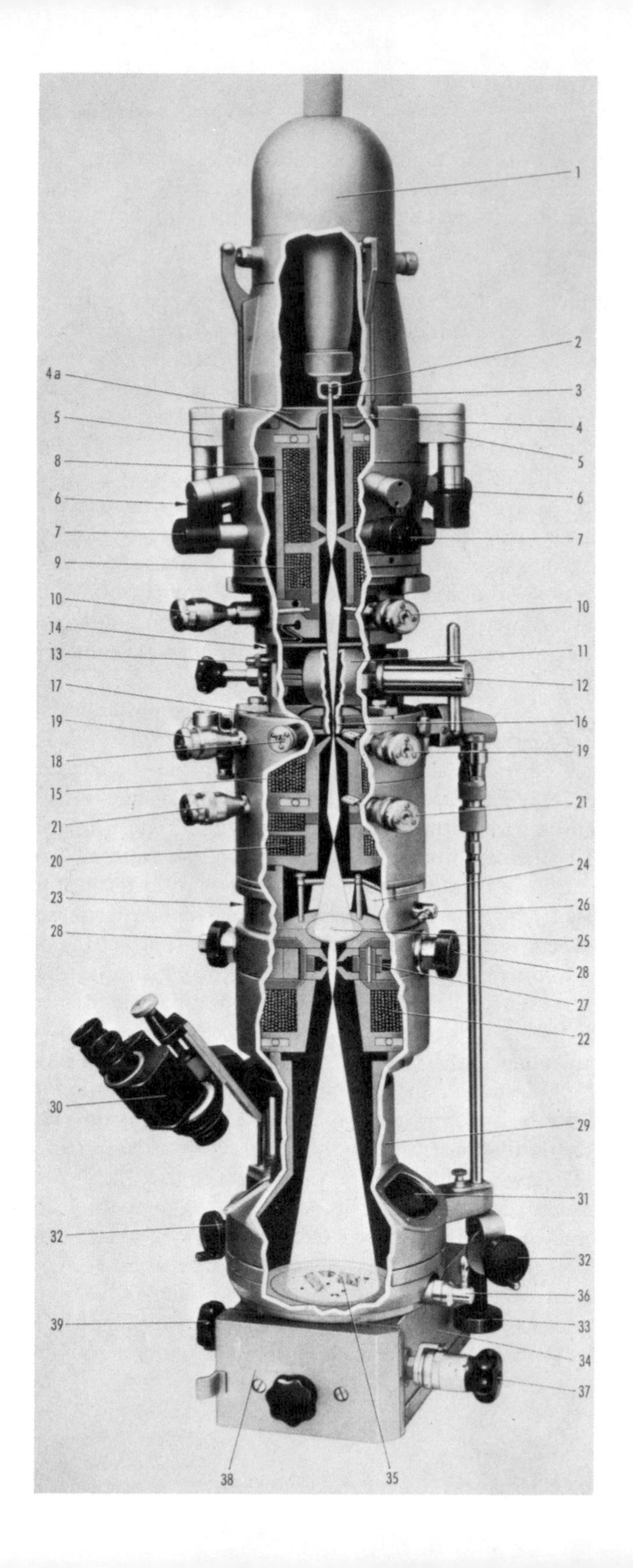

1
2
4a
3
5
4
5
8
6
6
7
7
9
10
10
14
11
13
12
17
19
16
19
18
15
21
21
20
24
23
26
28
25
28
27
22
30
29
31
32
32
36
39
33
34
37
38
35

1	Electron gun
2	Cathode
3	Bias shield (Wehnelt cylinder)
4, 4a	Anode and anode plate
5	Drives for adjustment of electron gun
6, 7	Drives for fine-focus condenser
8, 9	Condensers 1 and 2
10	Condenser aperture controls
11, 12	Specimen air lock and lock handle
13, 14	Ventilation and pre-evacuation valves
15	Objective
16, 17	Specimen cartridge and stage
18	Stereo drive
19	Objective aperture controls
20, 21	Intermediate lens and aperture controls
22	Projector
23	Viewing ports
24	Mirror for intermediate image observation
25, 26	Intermediate image screen and control
27, 28	Pole piece turret system and drives
29	Projector tube
30	Binocular magnifier
31	Observation window
32, 33	Coarse and fine controls for specimen stage
34	Photographic chamber with air lock
35	Final image screen
36	Control of final image screen
37	Counting device for films and plates
38	Door of photographic chamber
39	Air lock drive

Fig. 13-1 Cutaway diagram of the Siemens Elmiscop 1A electron microscope. [Courtesy of Siemens Corp.]

nonperiodic specimens at a resolution less than the separation between such images.

With proper care it is possible to achieve 5 to 10 Å resolution routinely in electron micrography; however, the problem of contrast remains. Although macromolecules can be visualized directly in the electron microscope when placed on an appropriately thin support, it has not generally been possible to visualize a difference in the scattering of electrons passing through different regions of an enzyme, as would be required to "see" subunit detail. It is therefore necessary to enhance contrast artificially to produce sufficient intensity differences between portions of the molecule to permit visualization of the subunit structure. This may be accomplished by shadow casting, negative staining, or positive staining. Contrast enhancement procedures coupled with the generally harsh conditions of specimen preparation (e.g.,

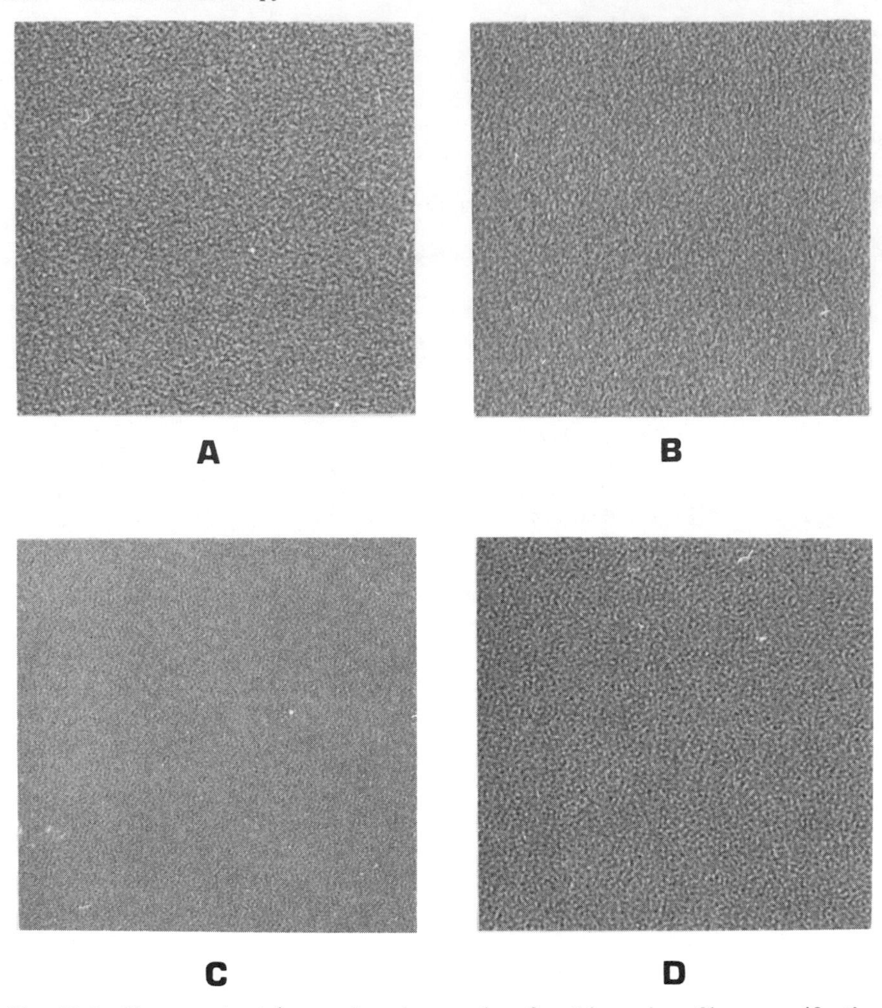

Fig. 13-2 Phase contrast images in micrographs of a thin carbon film, magnification 1,000,000×. A, 0.1 μ underfocus; B, 0.05 μ underfocus; C, nearfocus; D, 0.05 μ overfocus. (Courtesy of E. deHarven.)

drying) and sample thickness all lead to a reduction in the resolution of molecular detail. The practical limit of resolution is thus about 15 Å for protein preparations.

Shadow Casting. The specimen to be shadowed is prepared by spraying or placing a drop of the sample solution on the electron microscope grid (a fine mesh copper screen) that has been previously coated with a support film of collodion, Formvar, or carbon. After evaporation of the solvent, the grid is placed in a vacuum evaporator, and metal atoms evaporated from a hot

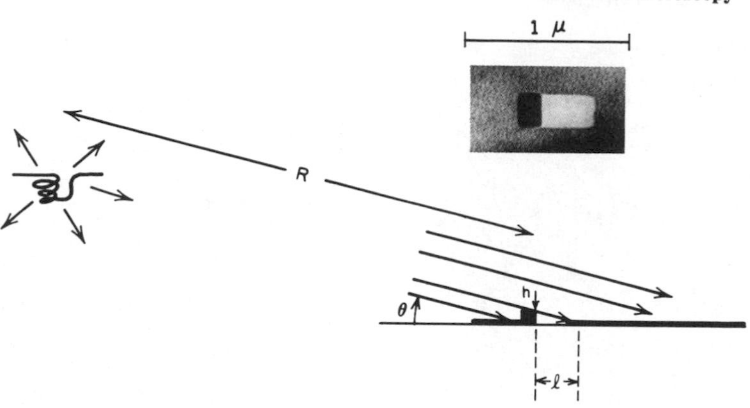

Fig. 13-3 Schematic representation of the geometry used for shadowing specimens. Metal atoms are vaporized *in vacuo* from a hot filament at distance R from the sample and strike the sample at a shallow angle θ. The particle height h can be obtained from the shadow length l by $h = l \tan \theta$. Inset, shadow cast MgO crystal on a collodion film.[1]

filament are caused to impinge on the molecules of the sample at a suitable angle. Metal deposits thus build up on surfaces directly exposed to the filaments, as shown in Fig. 13-3, and are completely absent on the protected side of the molecules. When examined in the electron microscope, the shadowed specimen scatters electrons in regions where metal deposits have formed. The protected areas cause little scattering and the resulting transmitted beam produces a darkening of the photographic plate in these regions. This appears as a lighter area on the screen and as a shadow on the plate. The image is often printed in reverse contrast to produce shadows on the final print that simulate those cast by an object in bright sunlight (Fig. 13-4). The length of a shadow is a function of the height of the surface that blocked the metal emission and of the shadowing angle used. (See Fig. 13-3.)

Detection of protein subunit detail by shadow casting is difficult. One problem is the fact that the substrate background itself has irregularities with dimensions of similar magnitude to the structure that one wishes to resolve. Another complication is the granularity of the shadowing metals, especially those that are most easily evaporated, such as chromium. Reasons for this granularity are not entirely understood, but it may be due to the tendency of atoms to slide after striking the surface until they meet other atoms and form small crystalline areas. Another possibility is that crystallization is induced by the heat generated in the electron beam of the microscope. The problem of uneven substrate films was circumvented by Hall with the introduction of the so-called mica-substrate technique, which utilizes freshly cleaved mica as a support for molecules during shadowing. A backing film of carbon is then

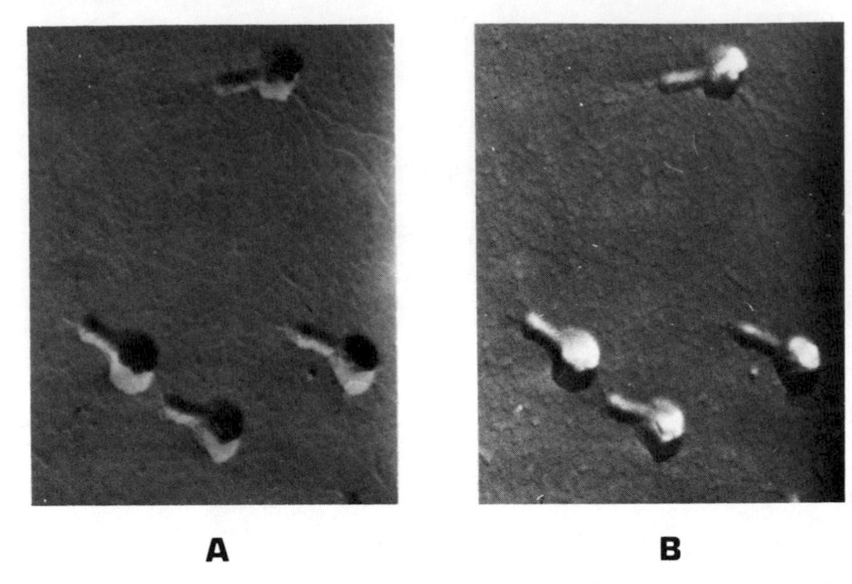

Fig. 13-4 Electron micrographs of bacteriophage T2 shadowed with carbon–platinum pellets. A, Normal print of image as seen on the fluorescent screen. B, Reverse contrast print of A.

evaporated normal to the shadowed surface, and both deposits are simultaneously floated off on water and picked up on grids. The shadow-transfer method has been used extensively to study protein molecules and has been reviewed by Hall.[1]

The classical shadow-casting method has been used extensively to determine the general shape and dimensions of proteins. For example, collagen molecules that are only 13 Å in diameter are clearly seen by the mica-substrate technique. Some other applications of shadowing for contrast enhancement have been the visualization of surface details of protein crystals and of fibrous materials. By reducing the amount of metal deposited (usually platinum), it is sometimes possible to study intramolecular structure in protein molecules. This method was used by Slayter and Lowey to demonstrate the double-headed nature of myosin (cf. Chapter 17).

Positive Staining. Biological macromolecules contain many charged groups that will bind ions of opposite charge. Certain ions containing heavy metals with high electron scattering power are useful as electron stains; the binding of such stains to particular sites on the molecules of the sample is termed positive staining. In order to achieve sufficient contrast, many heavy metals must be located within a given volume element. The method is particularly successful in the case of molecular aggregates in which the molecules are ordered in such a way that strong stain-binding regions adjoin and bands of stain are observed. The results obtained on various collagen aggregates (cf. Chapter 17) are examples of the application of this method for the solution of molecular problems. Positive staining of specimens such as these is analogous to the classical staining procedures of thin sections. Thin-section microscopy also permits localization of protein components by specific enzyme reactions, ferritin-labeled antibody, or by radioautographic methods (e.g., see References 4 and 6).

Negative Staining. The visualization of protein subunit structure by electron microscopy has most frequently and successfully been accomplished by employing some modification of procedures collectively termed negative staining. A number of reviews on the subject have been published (e.g., cited in Reference 4).

The principle of negative staining can be illustrated by one of the simplest preparative procedures. An appropriate concentration of sample is mixed with 1% phosphotungstic acid neutralized with NaOH. About 5 λ of the mixture is applied to a thin carbon film supported on a copper grid, and most of the solution is withdrawn by touching the edge with filter paper to leave a thin liquid film behind. As this film dries, the sodium phosphotungstate solidifies as a glassy, almost structureless, electron dense layer. Surface tension forces at the surface of a protein molecule usually cause preferential buildup of the stain around the protein, and stain enters the crevices between protein subunits (Fig. 13-5). Alternatively in thicker

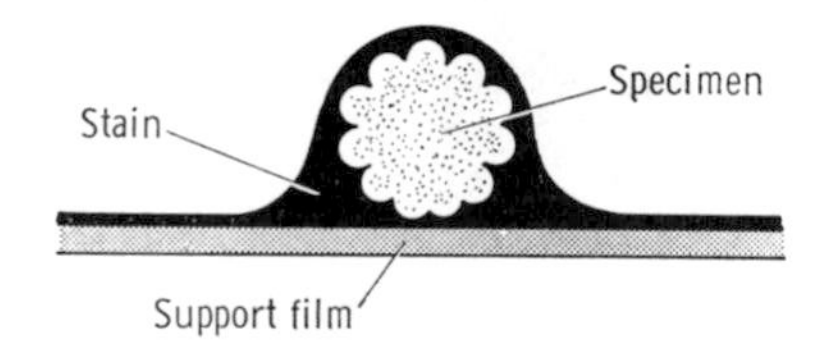

Fig. 13-5 Schematic representation of typical negative stain distribution around a small particle.[4]

areas of stain, the solid protein effectively prohibits the formation of an electron dense layer in the areas of its molecular domain. Scattering of the electron beam in the vicinity of the protein domain is determined by the relative amounts of stain and protein. The solidification of the stain appears to precede the total dehydration of the protein, thereby affording the additional advantage of apparent structural preservation for molecules that are markedly affected by direct drying.

In addition to phosphotungstate, uranyl oxalate at pH 6.8 is an excellent negative stain for use above the isoelectric point of a protein. This stain is of smaller molecular size and may penetrate into narrow crevices to outline subunit detail not seen with phosphotungstate. Uranyl acetate and uranyl formate are the agents most commonly used for negative staining of proteins below their isoelectric point. Some degree of simultaneous positive and negative staining is often observed, particularly when dealing with molecular aggregates.

Negative staining may be used to advantage for the observation of general molecular shape in molecules and molecular aggregates. More recently, it has become a valuable tool for the determination of subunit stoichiometry and the probable symmetry of oligomeric proteins (see review, Reference 5). Determination of structural models from electron micrographs of macromolecules visualized by negative staining must be done with caution. The image seen in two dimensions represents a projection down an axis perpendicular to the support. Thus, as may be seen in Fig. 13-5, contrast differences produced by stain penetration into crevices of both the top and the bottom of the particle are superimposed in the resultant image. Superposition patterns for various models may be used to aid in analysis of the protein images (see discussion by Haschemeyer and Myers in Reference 4). Another problem is caused by the varying orientation of the particles with respect to the electron beam, which in most cases cannot be controlled. Interpretation of the image is easiest for particles oriented with a symmetry axis perpendicular to the support plane.

Elucidation of the stacked hexagon structure of *Escherichia coli* glutamine synthetase[7] provides a good example of the application of negative staining to an oligomeric protein. Figure 13-6 shows the negatively stained

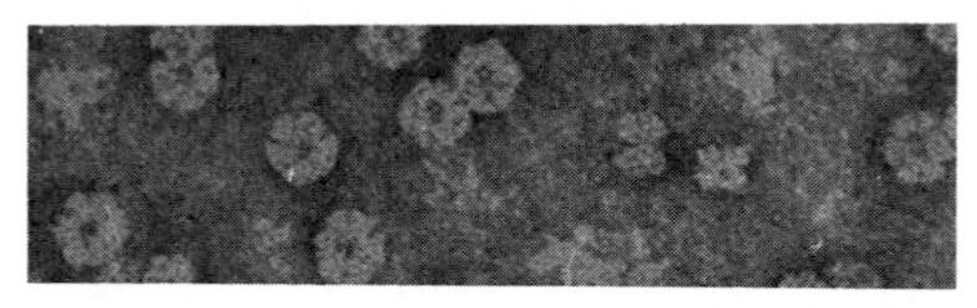

Fig. 13-6 *Escherichia coli* glutamine synthetase negatively contrasted with uranyl acetate. 500,000×.

image of this molecule of 12 subunits and total molecular weight 600,000. The clarity of the hexagonal and tetrad presentations is due to the direct superposition of top and bottom subunits when viewed along a (presumably) six- and twofold axis, respectively.

Dark Field and Phase Contrast Electron Microscopy. Thus far we have discussed only direct transmission electron microscopy in which, at the present time, the limiting factor for imaging molecular detail is one of contrast. However some successful applications of high resolution electron microscopy of small protein molecules also have been reported in which phase contrast and dark field techniques have been used. In dark field electron microscopy the image is formed by focusing the scattered electrons themselves while preventing the transmitted beam from reaching the photographic plate. This results in a remarkable contrast gain, since the number of electrons imaged at a point is roughly proportional to the mass of the specimen from which they originated. The scattered electrons suffer momentum loss to various degrees due to inelastic collisions, thereby producing chromatic aberrations. Consequently, the technique is currently limited to very thin specimens. Ottensmeyer[8] has used ultrathin specimens and support film to obtain high resolution images of nucleic acids and small proteins without staining and with remarkable detail. Figure 13-7 shows complexes of RNA polymerase with DNA detected by this method. Phase contrast electron microscopy is similar in principle to phase microscopy in light optics. A phase shift is produced between parts of the beam that have been transmitted through a phase plate (e.g., a carbon film) and waves that have been deflected by a structure of interest. Interference of these components at the image plane produces an image of the structure that is dark with a bright outline or the converse. Another important development has been the resolution and contrasting of individual heavy atoms with a high resolution scanning

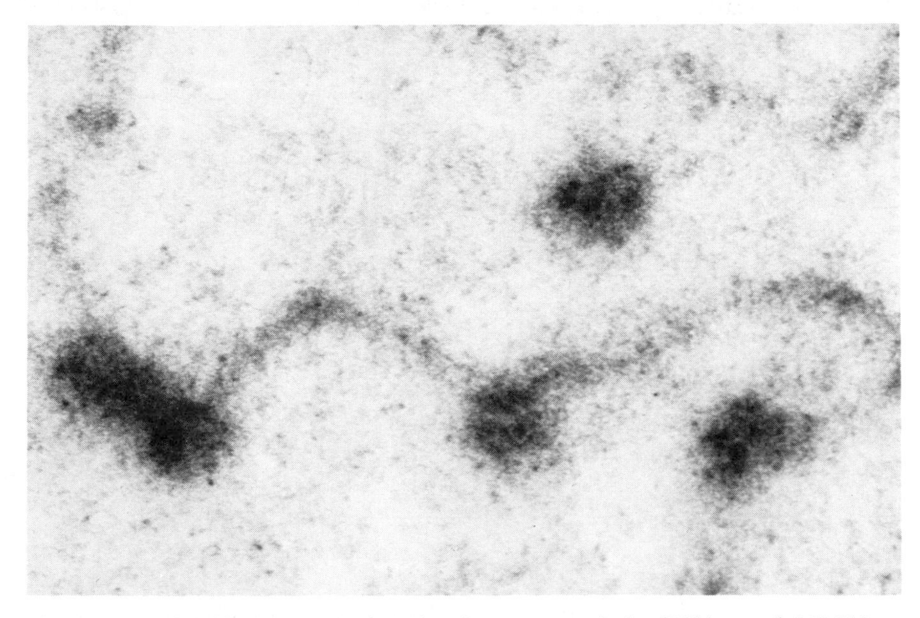

Fig. 13-7 Dark field micrograph of RNA polymerase attached to DNA strand. 940,000×. [From J. Dubochet, M. DuCommum, M. Zollinger, and E. Kellenberger, *J. Ultrastruct. Res.*, **35,** 147 (1971).]

electron microscope. Application of this method to protein structure problems has great potential. The operation of the microscope is reviewed by Crewe.[9]

REFERENCES

1. C. E. Hall, *Introduction to Electron Microscopy*, 2nd ed., McGraw-Hill, New York, 1966.
2. D. H. Kay, Ed., *Techniques for Electron Microscopy*, 2nd ed., Davis, Philadelphia, 1965.
3. B. M. Siegel, Ed., *Modern Developments in Electron Microscopy*, Academic Press, New York, 1964.
4. R. H. Haschemeyer and R. J. Myers, "Negative staining" in *Principles and Techniques of Electron Microscopy, Biological Applications*, Vol. 2, M. H. Hayat, Ed., Van Nostrand-Reinhold, New York, 1972, p. 101.
5. R. H. Haschemeyer, "Electron microscopy of enzymes," *Adv. Enzymol.*, **33,** 71 (1970).
6. F. S. Sjostrand, *Electron Microscopy of Cells and Tissues*," Vol. 1, Academic Press, New York, 1967.
7. R. C. Valentine, B. M. Shapiro, and E. R. Stadtman, "Regulation of glutamine synthetase. XII. Electron microscopy of the enzyme from *Escherichia coli*," *Biochemistry*, **7,** 2143 (1968).
8. F. P. Ottensmeyer, "Macromolecular fine structure by dark field electron microscopy," *Biophys. J.*, **9,** 1144 (1969).
9. A. V. Crewe, "A high resolution scanning electron microscope," Sci Amer. **224,** 26 (1971).

XIV X-Ray Diffraction

X-ray diffraction methods provide a powerful approach to the solution of molecular structure problems, through determination of accurate bond lengths and angles within molecules and of symmetry and other relationships between molecules in the solid (or semisolid) state. Results obtained in recent years have clearly demonstrated the increasing importance of X-ray diffraction techniques in biology. Analyses of the crystal structures of biological compounds have ranged from amino acids with molecular weights around 100 to proteins with molecular weights in the tens of thousands. Cocrystallization methods have been used to obtain information on the interaction between compounds as well as the exact molecular structure of the individual components. A second important area has been the study by fiber diffraction of noncrystalline materials such as fibrous proteins, nucleic acids, and polysaccharides that contain some degree of order. In addition, certain limited structural information can be obtained from the diffraction produced by suspensions or gels of very large particles such as the spherical viruses.

Although these are methods not likely to be used by anyone but specialists in the field, some knowledge of X-ray diffraction theory and terminology should be helpful to most biological scientists. The structural information obtained by X-ray crystallography, of course, is essentially of a static nature, whereas most biological questions about proteins are concerned with dynamic states, occurring in solution or in noncrystalline organized structures in cells. In general, however, experience has indicated that structure provides a valuable guide to the understanding of function. As for other structural methods of study, crystallographic information may permit inferences about mechanisms of action or may provide the needed insight to approach the dynamic processes of the molecule by other methods. A classic example is the experiment of Meselson and Stahl, familiar to all molecular biology students, that showed the mode of replication of DNA, based upon the structure proposed by Watson and Crick. In this case, X-ray diffraction data obtained by Franklin and Wilkins provided the necessary structural clues. The more specific question of the exact relationship of the crystallographic structure of a protein to its possible conformations in solution will be considered in the last section of this chapter.

1 X-RAY ANALYSIS OF CRYSTALLINE SUBSTANCES[1–5]

The techniques of X-ray diffraction have developed rapidly since 1912 when Friedrich and Knipping, at Laue's suggestion, first demonstrated diffraction from copper sulfate and from zinc blende. These experiments confirmed the existence of structural order within crystals. Such order, or at least the presence of some kind of repeating structure in the specimen under examination, is a prerequisite for the use of this powerful method. This requirement is ideally met in the study of single crystals, where order exists in three dimensions. The crystalline organization then can be described in terms of a fundamental building block, the *unit cell*, which by translations in three directions is able to account for the structure of the entire specimen. This array of identical units, arranged along parallel straight lines, makes up the crystal *lattice*. From the diffraction of monochromatic X-rays, one is able to determine the dimensions of the unit cell and the nature of its contents in terms of an electron density function that reveals the exact arrangement in three dimensions of all atoms present. A brief treatment of the theoretical basis for this determination will be presented in the following two sections.

1-1 Laue Equations

X rays are characterized by wavelengths of the same order as interatomic distances in molecular structures. A commonly used source is the 1.542 Å X-ray emission produced by electron bombardment of copper, which is collimated into a narrow parallel beam. Most of the beam passes through the crystal specimen as the undeflected direct beam. However, at various angles to the direct beam, weak diffraction beams appear and can be measured with some suitable detector (e.g., photographic film). What is happening at the atomic level may be thought of as follows: The electromagnetic field of the X-ray beam interacts with electrons of each atom and causes them to oscillate with resultant scattering of the incident radiation. The instantaneous electric field acting on a given atom n in a particular unit cell may be represented in exponential form as:

$$E_n = E_0 \exp\,(2\pi i[\nu t - d_1/\lambda]) \tag{14-1}$$

where E_0 is the amplitude, ν is the frequency, and λ the wavelength of the incident X-irradiation; d_1 represents the distance of atom n from an arbitrary origin plane. The electric field of the scattered radiation at an arbitrary point P, a distance R from the chosen origin O, may now be determined with the aid of the diagram shown in Fig. 14-1. $\overline{R_n}$ denotes a vector from the origin to atom n.

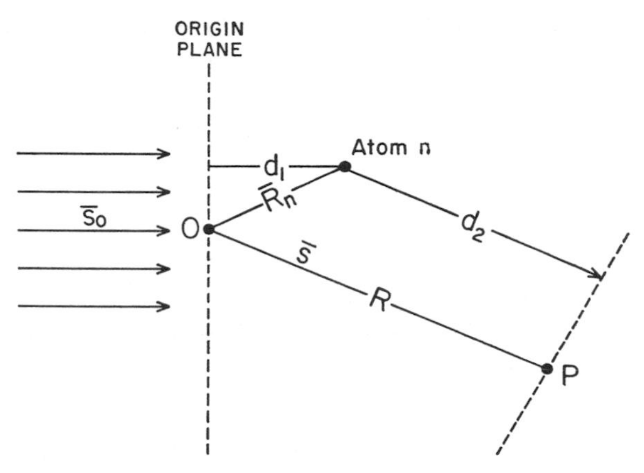

Fig. 14-1 Geometry for the derivation of the equations of X-ray diffraction. Incident X-rays of direction $\mathbf{S}_0$ are scattered by atom n at $\mathbf{R}_n$ from the origin $\mathbf{O}$, and detected at the plane of $\mathbf{P}$.

From electromagnetic theory the scattered electric field at P will be given by

$$E_p = E_0 f_n \left(\frac{e^2}{mc^2R}\right) \exp\,\{2\pi i[\nu t - (d_1 + d_2)/\lambda]\} \tag{14-2}$$

where f_n is the scattering power of atom n (related to the total number of electrons Z in the atom but less than Z because of interference effects among the electrons), e and m are the charge and mass of an electron, respectively, and c is the velocity of light. The exponential term or phase factor now involves a pathlength of $(d_1 + d_2)$ from the origin plane to the observation plane. If the direction of the incident and the scattered radiation is denoted by the unit vectors, $\bar{S}_0$ and $\bar{S}$, respectively, the distance $(d_1 + d_2)$ can be written

$$\begin{aligned} d_1 + d_2 &= \bar{R}_n \cdot \bar{S}_0 + R - \bar{R}_n \cdot \bar{S} \\ &= R - (\bar{S} - \bar{S}_0) \cdot \bar{R}_n \end{aligned} \tag{14-3}$$

In order to obtain the total scattered intensity at P due to the crystal specimen, one must sum the amplitude defined by Eqs. 14-2 and 14-3 over all atoms in the unit cell and over all unit cells in the crystal. The square of this resultant sum of amplitudes is the intensity of the diffracted beam I_p. The dimensions of the crystal are assumed to be small compared with the distance R to the point of observation.

It is convenient to describe the crystal shape in terms of a parallelopiped with sides given by the vectors $N_1\bar{a}_1$, $N_2\bar{a}_2$, and $N_3\bar{a}_3$ parallel to the three edges of the unit cell $\bar{a}_1$, $\bar{a}_2$, and $\bar{a}_3$. N_1, N_2, and N_3 represent the number of times the unit cell (of size of the order of Ångströms) is repeated by translation in each direction in the macroscopic crystal being studied. The resultant of the summations is then found to be:

$$I_p = \frac{I_0 e^4 F^2}{m^2 c^4 R^2} \left[\frac{\sin^2 \frac{\pi}{\lambda} (\bar{S} - \bar{S}_0) \cdot N_1 \bar{a}_1}{\sin^2 \frac{\pi}{\lambda} (\bar{S} - \bar{S}_0) \cdot \bar{a}_1} \right] \times \left[\frac{\sin^2 \frac{\pi}{\lambda} (\bar{S} - \bar{S}_0) \cdot N_2 \bar{a}_2}{\sin^2 \frac{\pi}{\lambda} (\bar{S} - \bar{S}_0) \cdot \bar{a}_2} \right] \left[\frac{\sin^2 \frac{\pi}{\lambda} (\bar{S} - \bar{S}_0) \cdot N_3 \bar{a}_3}{\sin^2 \frac{\pi}{\lambda} (\bar{S} - \bar{S}_0) \cdot \bar{a}_3} \right] \tag{14-4}$$

where I_0 is the intensity of the incident X-rays. The quantity F, called the structure factor, is defined by

$$F = \sum_n f_n \exp \left[\frac{2\pi i}{\lambda} (\bar{S} - \bar{S}_0) \cdot \bar{r}_n \right] \tag{14-5}$$

where f_n is the scattering power of atom n, and $\bar{r}_n$ is a vector from atom n in a given unit cell to the origin in that cell.

With the aid of Eq. 14-4 the student is now in a position to obtain some understanding of the fundamental principles of X-ray crystallography. The three quotients in square brackets of Eq. 14-4 are observed to be of the form $\sin^2 Nx$ divided by $\sin^2 x$. N is a very large integer (either N_1, N_2, or N_3). When this function is plotted against x, it is found to be practically zero everywhere except for high sharp maxima of value N^2 at $x = 0, \pi, 2\pi, \ldots$ (or any value of x given by an integer multiplied by π). Since all three quotients must simultaneously exhibit maxima in order to obtain a nonzero intensity I_p, one obtains directly the three Laue equations that define the directions at which X-ray diffraction maxima will be observed:

$$\begin{aligned} (\bar{S} - \bar{S}_0) \cdot \bar{a}_1 &= \lambda h \\ (\bar{S} - \bar{S}_0) \cdot \bar{a}_2 &= \lambda k \\ (\bar{S} - \bar{S}_0) \cdot \bar{a}_3 &= \lambda l \end{aligned} \tag{14-6}$$

where the symbols h, k, and l (*Miller indices*) have been chosen to represent the integers for the three directions. They are always whole numbers or zero.

These equations illustrate the following basic properties of X-ray diffraction from crystals.

1. Diffraction beams will occur and be detected (on a film, for example) only at discrete points characterized by sets of integers h, k, and l.
2. The directions of diffracted beams (and hence the position of the spots observed on a film) are determined by the size and shape of the unit cell (since the unit cell parameters are included in the Laue equations), but not upon the arrangement of atoms in the unit cell.

In addition, examination of Eqs. 14-4 and 14-5 shows that:

3. The intensity of each diffracted beam, as measured, for example, by the degree of blackening of the film in a given time interval, depends upon the arrangement and kind of atoms in the unit cell because of the quantities $\bar{r}_n$ and f_n that occur in the structure factor F. The size of the crystal and other geometrical factors also influence the intensities.[1]

1-2 Bragg Law and the Reciprocal Lattice

With the foregoing mathematical treatment providing an idea of the nature of X-ray diffraction in terms of electromagnetic theory, we can turn to a more convenient representation in terms of Bragg's Law. Here diffraction is treated as a reflection from planes. The equation of Bragg's Law is as follows:

$$\lambda = 2d \sin \theta \tag{14-7}$$

It states that for a set of planes separated by a distance d, diffraction maxima (reflections) are observed only when θ, the angle the incident beam (or the diffracted beam) makes with the planes, satisfies the equation. λ is the wavelength of the incident X-rays. The quantity $2 \sin \theta$ is simply the absolute value of the difference in directions of the Laue derivation, that is, $|\bar{S} - \bar{S}_0|$. Some practical considerations are indicated by Eq. 14-7. In order to detect reflections from all possible sets of planes in three dimensions, the crystal must be rotated in the incident beam (thus the full range of θ is utilized). Also apparent is the fact that the number of measurable diffraction maxima is ultimately limited by the value of λ.

The concept of crystal planes leads one to a convenient mathematical construct used in crystallography that describes the diffraction maxima produced by the real crystal lattice in terms of a second lattice, the reciprocal lattice. The crystal is represented as consisting of sets of planes, each set described by Miller indices (hkl). Each set is comprised of parallel equidistant planes with spacing d_{hkl} and the following orientation: the first passes through the origin of the unit cell, the second makes intercepts a_1/h, a_2/k, and a_3/l on the three axes, and so on. The reciprocal lattice itself is an array consisting of the terminal points of a set of vectors H_{hkl} where each vector H_{hkl} has a direction given by the normal to the corresponding set of planes (hkl) and a length equal to the reciprocal of the spacing d_{hkl}. The value of this complicated construct lies in the fact that X-ray photographs provide, in fact, a direct representation of this reciprocal lattice. Thus a distance measured on the film can be directly related (through a simple conversion factor based on the geometry of the apparatus) to a spacing within the unit cell. A large intensity observed in a diffraction spot at a particular distance and direction from the origin of the film indicates a high electron density (the presence of many atoms) near the planes in the crystal corresponding to that spot. This provides the basis for the determination of important spacings in large molecules, such as the characteristic 1.5 Å spacing of the α-helix observed in precession photographs of protein crystals or in fiber diagrams of certain synthetic polypeptides.

1-3 Symmetry Considerations

Symmetry is an important factor in X-ray work, not only the translational symmetry of the unit cell in a crystal, but the symmetry in the arrangement of molecules within the unit cell and the symmetry of repeating units in a large biological polymer. In most crystals the unit cell contains more than one molecule (of the same kind) and thus it is useful to define a simpler unit, the *asymmetric unit*, which is the basic structure to be determined in an X-ray analysis. This is usually a single molecule but may be more or less depending on the nature of the crystal. The set of symmetry operations by which these units are related to one another within the unit cell is known as the *space group* of the crystal. Most biological compounds contain asymmetric (or optically active) centers (cf. Chapter 9) and, as a result, are limited in their crystalline forms to space groups that contain no symmetry elements other than rotation axes or screw axes. For example, a common symmetry element, the mirror plane, could not occur in these crystals since it is impossible to orient two identical and optically active molecules so that each forms a mirror image of the other.

A crystal possesses an n-fold rotation axis if it presents exactly the same appearance after a rotation about that axis of $360°/n$. With an n-fold screw axis, identity is achieved after a rotation of $360°/n$ coupled with a translation parallel to the axis. Figure 14-2 illustrates a hexagonal lattice, an arrangement of points in two dimensions that possesses two-, three-, and sixfold rotation axes. In crystals, rotation and screw axes may have n values of 2, 3, 4, or 6 only. However, in other materials where the arrangement of molecules

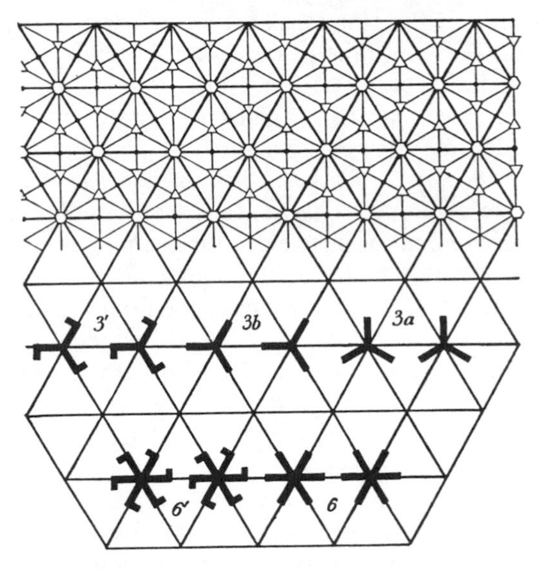

Fig. 14-2 Two-dimensional hexagonal lattice. Five symmetry groups are illustrated that depend on which simple figure (3', 3*a*, 3*b*, 6, 6') is placed at the sixfold axis locations. (From H. Weyl, *Symmetry*, Princeton Univ. Press, Princeton, N.J., 1952.)

does not repeat in all three dimensions, other symmetries may appear. Nonintegral screw axes have been found in a number of fiber structures.

1-4 Crystal Structure Determination

From the preceding theoretical treatment, X-ray diffraction data can be seen to consist of a set of intensities of diffracted beams. In general, the number of independent measurements, although limited by the Laue conditions (since $\sin\theta$ cannot exceed 1), far exceeds the number of parameters to be determined, that is, the coordinates of the atoms in the unit cell. The only thing that prevents immediate and straightforward determination of a structure is the fact that the experimental measurement yields only the amplitude but not the phase of the diffracted beam. Since both must be known to deduce the molecular structure of the diffracting material, the primary concern of the crystallographer is to find ways of solving or circumventing the "phase problem."

An especially valuable tool in crystal structure determinations is the *Patterson function*, $P(x, y, z)$, which may be calculated directly from the observed intensities. In the Patterson function each pair of atoms in the unit cell produces a peak with a magnitude proportional to the product of their scattering factors; the position (x, y, z) of the peak on a Patterson map is determined by the relative positions in the unit cell of this pair of atoms.

The result is a map in three-dimensional space representing all interatomic vectors in the unit cell. Since the peaks produced by different pairs of atoms may be superimposed, interpretation of the map is usually very difficult. However, enough clues may be obtained to devise a trial structure, which may then provide the basis for successive approximations to the final solution by means of *Fourier synthesis*. In this method, phases calculated for a trial structure are combined with the experimental amplitudes and carried through a Fourier summation to yield a three-dimensional electron density map for the unit cell. The trial structure may consist of only a few atoms of the molecule; under favorable circumstances the Fourier map may then reveal the positions of other atoms in the molecule. When the general nature of the structure is known, *refinement* is carried out, usually by the method of *least squares*, in which the atom positions are varied by small increments until the best agreement between observed and calculated amplitudes is obtained. Agreement may be tested by calculation of a *difference Fourier* summation which represents the difference in electron density between that responsible for the observed diffraction intensities and that calculated for the deduced structure. When the differences between observed and calculated amplitudes are small, the difference Fourier will show essentially no resultant electron density in the unit cell (or will reveal light atoms such as hydrogen if they have not been included in the structure calculations).

The student can probably appreciate that such methods will become more and more difficult the more atoms there are in the molecule under investigation, and it is here that crystal structure determinations are greatly aided by the so-called *heavy atom methods*. A heavy atom containing a large number of electrons, such as iodine, mercury, or lead, produces such strong diffraction that its position in the unit cell usually may be determined immediately from the Patterson function. In favorable circumstances the phases calculated for the heavy atom when combined with the experimental amplitudes will yield an electron density map revealing the positions of all atoms except hydrogen. The structure analysis of vitamin B_{12}[6] provides a good example of the heavy atom method in which the use of phases based on the cobalt atom in the molecule provided the starting point for successive approximations leading to complete solution of the structure. Although the presence of a heavy atom has less effect on phases in a really large structure, this method has provided the basis for the solution of protein structures through isomorphous replacement.

2 X-RAY STUDIES OF PROTEIN COMPONENTS

2-1 Amino Acids and Small Peptides[7–10]

An important application of X-ray crystallography to biochemistry has been in the determination of amino acid and small polypeptide structures

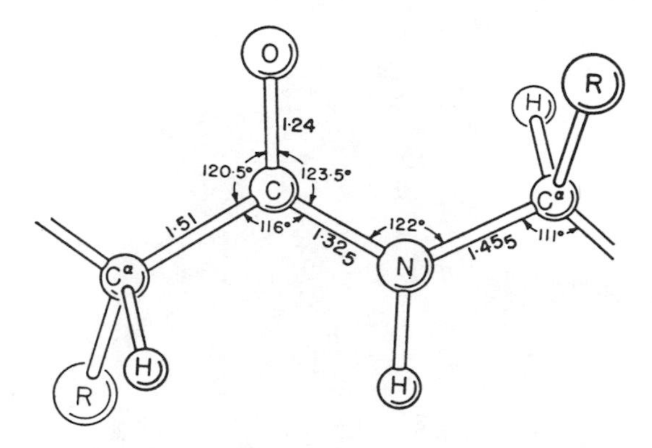

Fig. 14-3 Dimensions of the amide group of peptides (bond distances given in Angstroms).[8]

Extensive study of such compounds by Pauling and Corey and their collaborators led to determination of the fundamental dimensions of the peptide bond. Averaged results based on additional more recent data are shown in Fig. 14-3. The amide group itself, including the C—N bond and the substituent oxygen atom, imino hydrogen and α-carbon atoms, is found to be planar within experimental error. The α-carbon atoms lie in the *trans*-conformation about the peptide linkage. This conformation is sterically favored since as shown in Fig. 14-3, an oxygen atom occurs as third neighbor to the α-carbon at the right rather than the more bulky α-carbon next in the chain. The distinction, however, does not apply in the case of proline (or hydroxyproline); and a polymer of proline has in fact been found in the *cis*-form. The arrangement of two planar peptide groups is illustrated in Fig. 14-4. The structural studies of small molecules also revealed a strong tendency for N—H · · · O hydrogen bonding between neighboring peptide groups. These findings proved of great importance in developing models of polypeptide structure.

Crystal structures have been determined for all of the 20 amino acids commonly found in proteins, in addition to many analogs and a variety of small peptides. An interesting feature is the observation that the $C—NH_3^+$ bond, which appears in the zwitterion form of amino acids and peptides, is significantly longer than the standard C—N single bond length. Information has also been obtained on the structure of the disulfide S—S bond, a bond of special importance in the tertiary structure of some proteins, and on the structures of a variety of metal–peptide complexes.

2-2 The α-Helix and α-Polypeptides

The structure of the α-helix as shown in Fig. 1-2 represents the culmination of attempts to devise a model structure for polypeptides based on the results of

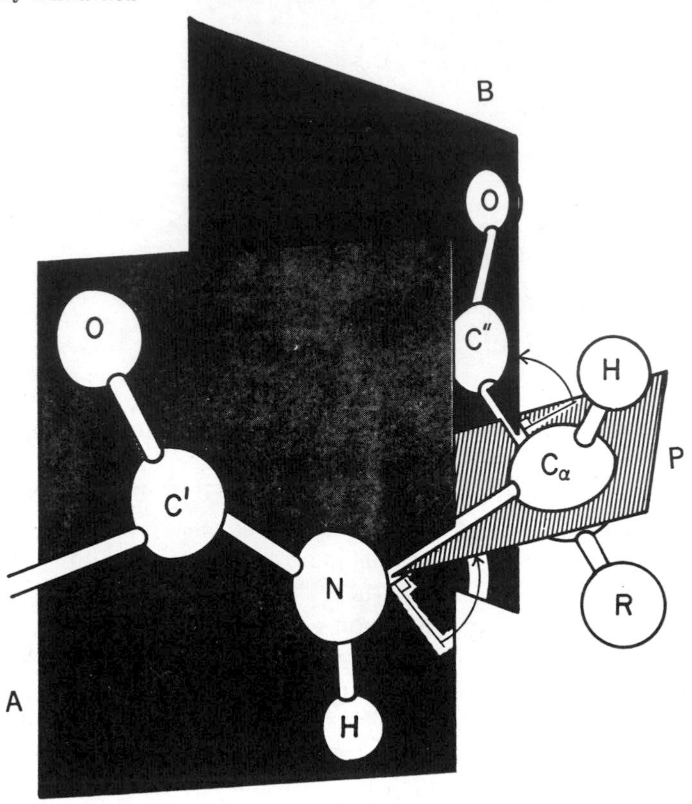

Fig. 14-4 Relationship of planes of successive peptide groups in a polypeptide of L-amino acids. Angles where rotation about single bonds in the backbone may occur are indicated. (From J. A. Schellman and C. Schellman, "The Conformation of Polypeptide Chains in Proteins," in *The Proteins*, Vol. 2, H. Neurath, Ed., Academic Press, New York, 1964.)

amino acid and small peptide crystal studies.[11] Adjacent peptide units are related by a rotation of about 100° about the longitudinal axis and a translation of 1.5 Å along the axis. In a complete turn of the helix (360°) there are 3.6 amino acid residues with a total rise (*pitch*) of 5.4 Å. The helical backbone thus possesses the symmetry of a 3.6-fold screw axis. This arrangement permits the formation of hydrogen bonds (as indicated by dotted lines) between each amide group and the third amide group from it along the peptide chain. These bonds serve to connect the turns of the helix and stabilize the structure. The side chains of the amino acids (denoted by R in Fig. 1-2) project outward and thus do not interfere with the helical structure of the polypeptide backbone. The structure thus accommodates all true α-amino acids (not including proline and hydroxyproline), provided that all residues have the same absolute configuration (all L or all D).

The actual occurrence of the α-helix structure in synthetic polypeptides (the α-polypeptides) has been well established from X-ray diffraction studies of fibers, and for many cases excellent agreement is obtained for observed and calculated parameters of the α-helix. In other cases deviations are found (e.g., in poly-γ-L-methylglutamate), presumably due to influence of the side chains.

2-3 β-Polypeptides and the Pleated Sheet Structures

In addition to the α-helix, two types of so-called pleated sheet models were found to satisfy the requirements of peptide bond geometry and maximization of hydrogen-bonding.[12] One of these, the antiparallel pleated sheet (Fig. 14-5), has alternate chains running in opposite directions. The second model, the parallel pleated sheet, has all chains running in the same direction. In both cases the polypeptide chains are very nearly fully extended with only a slight degree of puckering of the polypeptide backbone required to achieve favorable hydrogen bonding.

It is highly likely that pleated sheets form the basis for the structure of the β-polypeptides and β-keratin. These β-forms, which are readily obtained by appropriate choice of solvent during isolation or by stretching fibers of the α-form, generally show repeat distances of 6.6 to 7.0 Å, consistent with the

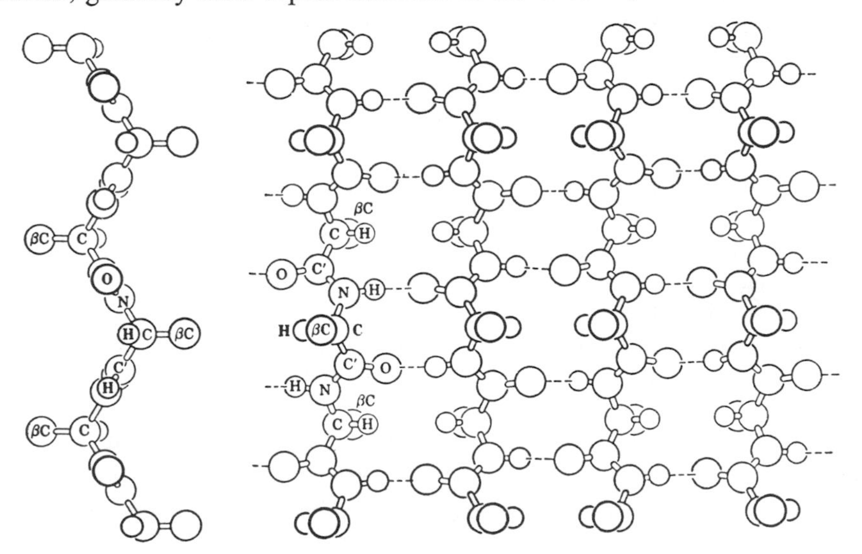

Fig. 14-5 Ball-and-stick model of the β-antiparallel pleated sheet. [From R. E. Marsh, R. B. Corey, and L. Pauling, *Biochim. Biophys. Acta*, **16,** 1 (1955).]

notion of slight puckering from the fully extended configuration (which has a twofold screw axis and a translational repeat of 7.2 Å).

2-4 Other Polypeptides

Fiber diffraction data for poly-L-proline, poly-L-hydroxyproline, and a type of polyglycine obtained under special conditions of isolation have established the existence of repeating secondary structures different from the usual α- and β-forms.[13] These structures are of interest in relation to the structure of collagen, which contains a high percentage of these particular amino acids. Polyglycine II has been described as a somewhat extended helix with a threefold screw axis (i.e., three residues per turn). The translation or rise per residue is 3.1 Å and all amide groups are in the *trans*-configuration. Such a structure was earlier found in poly-L-proline from analysis of partially oriented films and fibers. This form, denoted poly-L-proline II, has the same spatial characteristics as polyglycine II but occurs only as a left-handed helix. The other known structure of this polymer, poly-L-proline I, consists of right-handed helices with a residue translation of 1.8 Å and all peptide bonds *cis*.

2-5 Conformational Analysis of Polypeptide Structures[10]

The relative orientation in space of two linked peptide groups (e.g., those lying in Planes A and B of Fig. 14-4) is conveniently described by two dihedral angles, denoted ϕ and ψ. The angle ϕ represents a clockwise rotation about the N—C_α bond which brings Plane A to Plane P; ψ is the angle of clockwise rotation about C_α—C″ necessary to bring Plane P to Plane B. The measurement is thus relative to the planar or fully extended conformation of the polypeptide chain (corresponding to a distance of 7.2 Å between first and third α-carbon atoms). Additional angles have been defined to account for nonplanarity of the peptide unit and for the orientation of the side chains. These angles provide a useful set of parameters for systematic description of polypeptide geometry.

Although free rotation about the N—C_α and C_α—C″ bonds (Fig. 14-4) is theoretically allowed, stereochemical restrictions are imposed due to energetically unfavorable close contacts between substituent atoms. The distances of closest approach between nonbonded atoms may be described in terms of van der Waals radii or a set of standard allowed contact distances based on crystal structure data (e.g., the normal limiting distance between nonbonded oxygen atoms is 2.7 Å). When these standard values are compared with substituent contact distances in a polypeptide calculated as a function of the two angles ϕ and ψ, one obtains a map revealing the allowed ranges for rotation about the N—C_α and C_α—C″ bonds. In general, known polypeptide structures such as the α-helix, β-structure, and others satisfy the calculated restrictions on ϕ and ψ, and thus the formulation can be taken as a fairly reliable tool for protein model building. A degree of uncertainty remains because of the use of standard contact distances or van der Waals radii; a significant variation in these distances is found in different crystal structures of small molecules as well as in known protein structures. This uncertainty does not affect the distinction between fully allowed structures (no close contacts) and those fully disallowed (multiple contacts closer than the extreme limits which are set 0.1 to 0.2 Å less than normal contact distances), but blurs the boundary between the two.

Values of ϕ and ψ for the amino acid residues in protein crystal structures (e.g., myoglobin and lysozyme) show good agreement with predicted ranges, provided allowance is made for extreme contact limits. Amino acid residues found in the interior of α-helical segments have values within a small area of the total $\phi\psi$ map centered at about $-50°$, $-50°$. The frequency of occurrence of residues of known structures in this area as a function of the identity of the nearest neighbors on either side has been used to make predictions of helix probability based on primary sequence data alone.[14] For example, if residues which occur between alanine and valine residues show a high frequency of ϕ, ψ values which fall in the α-helical domain of the $\phi\psi$ map, one might predict that such a sequence in a protein of unknown conformation would tend to be α-helical, providing adjoining residues were also favorable toward α-helix formation. Such systematic classification of parameters from known protein structures should aid greatly in the problem of prediction of three-dimensional conformations from the primary structures of proteins.

3 PROTEIN CRYSTALLOGRAPHY[15–19a]

Most protein crystals that have been studied in detail have contained a considerable amount (40–60%) of the aqueous solvent from which they were crystallized. The presence of water gives support to the supposition that the structure observed in the crystalline state is, for the most part, the same as that which occurs in solution. Wet crystals give good resolution of fine structure whereas crystals obtained after drying show disorder. Most of the water in the crystal is apparently in a liquid state, although some water is bound so strongly that it is impermeable to salt ions.

Because of the tremendous number of atoms in a protein molecule, the degree of crystallographic resolution that can be achieved is considerably less than that expected for structures of small molecules. Protein crystal analyses usually proceed slowly through a number of stages of increasing resolution. Collection and analysis of the large amounts of data required for solution of the structure have been aided by the development of automatic data-collecting apparatus for measurement of intensities and of high-speed computers for the calculations. This has led to increasingly higher refinement in protein structures and the determination, for a number of structures, of the positions of almost all atoms of the protein molecule (not including hydrogen atoms). In some cases a great deal of information may be obtained before exact atomic positions are determined, as illustrated during the course of analysis of myoglobin and hemoglobin.

3-1 Isomorphous Replacement and Protein Structures[19,19a]

Protein crystal analyses have depended to a great extent upon the method of isomorphous replacement. This is based upon the comparison of the X-ray diffraction observed for the unmodified protein crystal with that obtained from crystalline derivatives in which heavy atoms have been introduced at one or more specific sites on the protein molecules. In order to be useful, the derivatives must be essentially identical to the original protein and must assume the same crystal form, that is, they must be *isomorphous*. Calculations based on differences in intensities between the protein and its derivatives are then made to obtain the desired phase information. Even when satisfactory derivatives can be obtained, however, the amount of work required for the analysis is formidable.

3-2 Solution of the Myoglobin Structure[15,16]

Sperm whale myoglobin was the first protein whose structure was solved by X-ray crystallography, a project extending over about 20 years in the laboratory of J. C. Kendrew. The steps by which the myoglobin structure was solved will illustrate the general approach to protein crystals. The globin portion of the molecule was known to consist of a single polypeptide chain of 152 amino acid residues. The heme prosthetic group is a substituted tetrapyrrolic ring (ferrous protoporphyrin). An iron atom is linked through four of its coordination sites to a nitrogen of each of the four pyrrolic rings (Fig. 14-6). The fifth valence site interacts with a histidine residue of the globin chain. In the ferrous state, the sixth valence site may be occupied reversibly by molecular oxygen; when not occupied by oxygen, the site is apparently empty. Oxidation to the ferric state (metmyoglobin) prevents combination with oxygen, and water occupies the sixth valence site.[20]

Although myoglobin could be readily crystallized from salt solutions, preparation of heavy-atom derivatives was complicated by the absence of

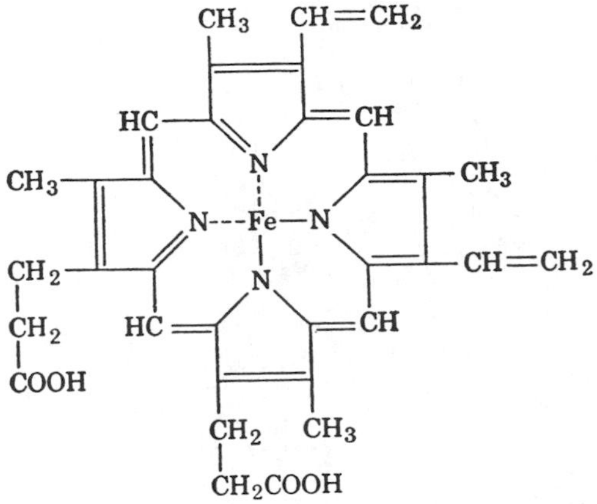

CH_3 $CH{=}CH_2$
HC— =CH
CH_3 N CH_3
N---Fe—N
CH_2 N $CH{=}CH_2$
CH_2 HC= =CH
COOH
CH_2 CH_3
CH_2COOH

Fig. 14-6 Ferrous protoporphyrin (heme).

free sulfhydryl groups where mercury might be easily introduced. Experimental difficulties were encountered in attempts to attach heavy atom ligands to the heme group. Successful introduction of heavy atoms was finally achieved by crystallizing the protein in the presence of ions such as mercuri-iodide (HgI_4^{2-}) which was found to bind to one of the two methionine residues of myoglobin. With the data from this derivative and from others prepared with *p*-chloromercuribenzene sulfonate ($Cl—Hg—C_6H_4—SO_3H$), aurichloride ($AuCl_4^-$), and "mercury diamine," a complex prepared by dissolving HgO in hot concentrated $(NH_4)_2SO_4$ solution, phases were determined for the 400 reflections closest to the center of the diffraction pattern and a 6-Å resolution electron density map, revealing the essential features of the tertiary structure, was obtained.

In the next stage of the myoglobin study, 10,000 reflections for the protein and for each of four derivatives were measured to obtain an electron density map at 2 Å resolution. Some amino acids could now be identified, and a pitch of 5.4 Å was determined for the helical segments of the polypeptide chain, in exact agreement with the Pauling-Corey α-helix. The question of the sense of the helix could also be settled: all distinct lengths of α-helix in myoglobin were found to be right-handed. The orientation of the heme group and the coordination of the iron with a histidine side-chain could be clearly visualized. Figure 14-7 shows a diagram of the structure in which the position of each α-carbon atom of the polypeptide chain is indicated by a large dot. The regions of smooth helix (α-helical segments) can be readily distinguished from the short nonhelical connecting sequences which are shown with zigzag lines. The structure has been refined further by using the remaining measurable data out to a spacing of 1.4 Å or a total of 17,000 measurable reflections. This has permitted determination of the positions of

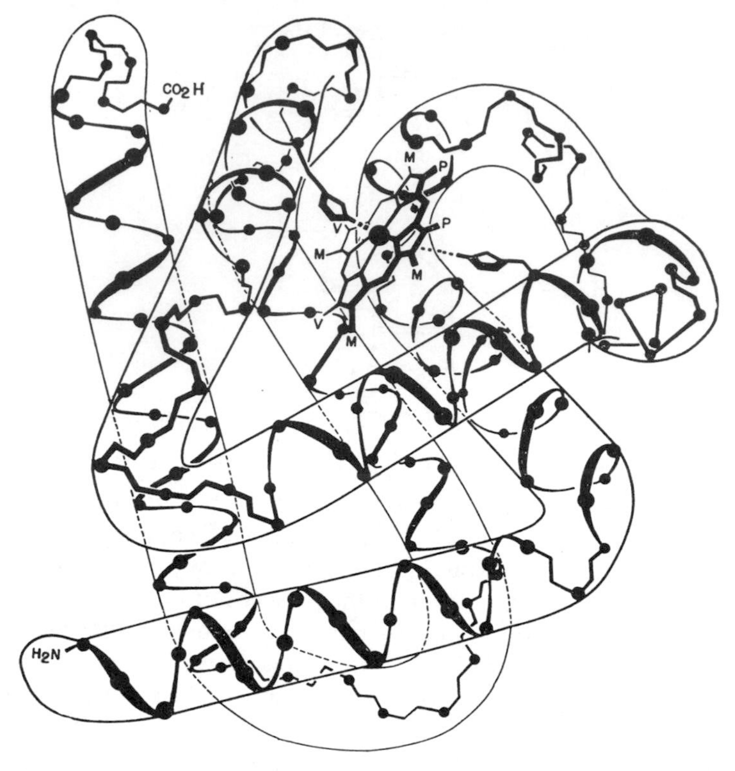

Fig. 14-7 α-Carbon diagram of myoglobin molecule obtained from 2 Å analysis. Large dots represent α-carbon positions. Side groups of the heme are identified: M = methyl, P = propionic acid, V = vinyl. [From J. C. Kendrew, H. C. Watson, B. E. Strandberg, R. E. Dickerson, D. C. Phillips, and V. C. Shore, *Nature*, **190,** 663 (1961); reproduced from Dickerson.[15]]

almost all of the 1260 nonhydrogen atoms in the molecule. The structure based on this refinement is shown in Fig. 1-3.

3-3 The Structure of Hemoglobin[21]

Hemoglobin, the protein responsible for the transport of oxygen and carbon dioxide in the blood, contain four polypeptide chains (each with its bound heme group) of two types, designated α and β in normal adult mammalian hemoglobin. The total molecular weight is 64,500. Studies on the structure of hemoglobin were first concentrated on the oxygenated form *oxyhemoglobin* from horse and the reduced form *deoxyhemoglobin* from normal humans. Solution of the structure was aided by the successful preparation of mercury derivatives and by comparison with the myoglobin structure. Investigation is being extended to other species and to some of the many human hemoglobin variants. For example, the structure of human oxyhemoglobin H containing four β chains is found to resemble reduced hemoglobin and to be unaffected by deoxygenation.

The structure of horse oxyhemoglobin has been determined at 2.8 Å resolution. Each of the four polypeptide chains assumes a configuration very close to that of sperm whale myoglobin. These four subunits are arranged in a tetrahedral manner forming a compact spheroidal molecule of dimensions 64 Å × 55 Å × 50 Å, illustrated in Fig. 14-8. Deoxyhemoglobin (at 5.5 Å

Fig. 14-8 Model of the structure of horse oxyhemoglobin. [Courtesy of M. F. Perutz, from *Nature* **185,** 416 (1960).]

resolution) has a very similar structure; within experimental error the tertiary structures of the component α- and β-chains are essentially identical in the two forms. However, a clear difference in quaternary structure is found, which has been interpreted in terms of rotations of the subunits (9.4° for the α-chains and 7.4° for the β-chains) about separate axes. Subunit contacts are quite different as a result of these apparent rotations. What role the structural modification plays in influencing oxygen affinity in hemoglobin is yet to be clarified.

The sequences of a large number of normal and abnormal hemoglobins from various species have been compared by Perutz, Kendrew, and Watson in an attempt to correlate sequence with structure.[22] Sequence determinations have shown that only nine of the more than 140 sites in the globin chain are essentially invariant, i.e., occupied by the same amino acid residue in almost every species examined. These nine include the critical histidine residue linked to the heme group, although even this residue is replaced in certain hemoglobin variants. Yet, it is likely that the tertiary structure of the molecule is essentially the same in all species. Model building indicates that all known hemoglobin sequences can be accommodated by the observed structure. A number of general features of the structure have also been identified.[21,22] The internal part of the molecule is almost entirely apolar with residues closely packed to achieve a maximum number of van der Waals contacts between atoms. Apolar amino acids capable of hydrogen bonding that occur in the interior are found to be oriented favorably for hydrogen bond formation with suitable acceptors. The surface of the molecule is "studded" with polar residues, generally interacting with H_2O molecules rather than with each other. Substitution of nonpolar for polar residues is found in some hemoglobins with apparently no disrupting influence on the structure; reverse substitutions are also observed among the residues at the surface. Proline residues occur only in nonhelical regions; a histidine, glutamine, or aspartic acid residue is found at the ends of helical sections, suggesting that the presence of one of these residues may be a necessary (but not necessarily sufficient) condition for helix termination. Apparently a desirable condition for formation of helical segments on the surface of the molecule is the occurrence of apolar residues at intervals averaging 3.6 residues (the number of residues per turn in the α-helix), the other residues being primarily polar ones. Thus a helix can form that is apolar on one side for interactions with the apolar core of the molecule and polar on the side facing the solvent.

Several other features of the myoglobin and hemoglobin structures are of particular interest. Most of the heme group lies within an apolar crevice, and the surrounding amino acid residues (including a high proportion of leucine, valine, and phenylalanine) afford plentiful apolar contact with the pyrrole rings. Charge interactions with the propionic acid side chains of the heme (cf. Fig. 14-6) are provided by polar and basic amino acids (histidine in the α-chain, and serine and lysine in the β-chain of hemoglobin). Hemoglobin contains a central cavity of about 25 Å in length and 5 to 10 Å in diameter. There are about 14 polar residues in the cavity and their charges probably cancel to give zero net charge. The nature of solvent exchange in this region is not known. One tyrosine and one histidine residue (normally uncharged) are also found here. The hemoglobin crystal undergoes a lattice transformation when transferred from pH 7 to a pH below 5.8. This has been shown to involve a relative translation of adjacent molecules by about 10 Å and is due to the titration of a single histidine that forms a salt bridge to an aspartic acid residue in a neighboring molecule. This observation illustrates how a simple primary event can produce a dramatic conformational alteration as a secondary effect.

3-4 The Structure of Lysozyme[23,24]

The first successful structure determination of an enzyme was that of lysozyme, a small basic protein (molecular weight 14,600) that occurs in egg white and in human mucosal secretions. Lysozyme causes the lysis of bacteria such as *Micrococcus lysodeikticus* by catalyzing the hydrolysis of a cell-wall polysaccharide, a β-(1-4)-linked alternating copolymer of *N*-acetylglucosamine and *N*-acetylmuramic acid. The linkage between the two types of amino sugars is broken at the position shown in Fig. 14-9. Lysozyme also hydrolyzes homopolymers of *N*-acetylglucosamine (*chitin*).

Hen egg-white lysozyme consists of a single polypeptide chain of 129 amino acid residues. Its primary structure is shown in Fig. 14-10. The protein is readily crystallized from 1 *M* NaCl solution at pH 4.7 to form tetragonal crystals containing 33% liquid of crystallization by weight. The structure has been solved to the level of atomic resolution (2 Å) with the aid of isomorphous crystals of three different heavy atom derivatives.[23] X-ray studies

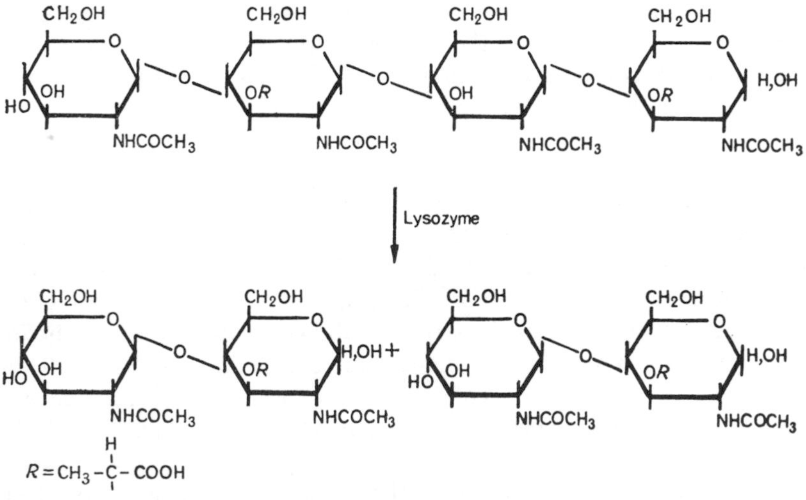

Fig- 14-9 Action of lysozyme on a cell-wall tetrasaccharide [From N. Sharon, *Proc. Roy. Soc., Ser. B*, **167,** 402 (1967).]

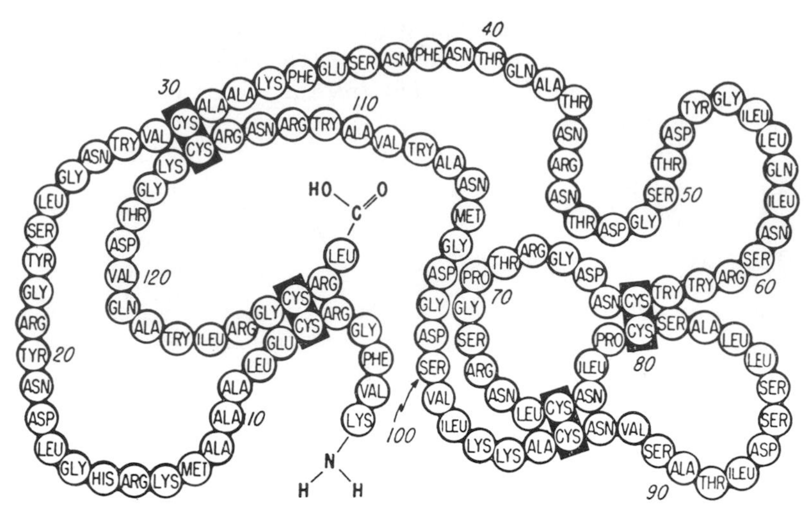

Fig. 14-10 Primary structure of hen egg-white lysozyme. [From R. E. Canfield and A. K. Liu, *J. Biol. Chem.*, **240,** 1997 (1965).]

have also been done on derivatives formed by diffusion of small molecules that resemble the natural substrate into the lysozyme crystal. These have included the monosaccharides *N*-acetylglucosamine and N-acetylmuramic acid and certain di- and trisaccharides containing amino sugars, all of which act as inhibitors of the enzyme. By diffusing substances into preformed enzyme crystals, it was possible to obtain complexes isomorphous to the original enzyme crystal. These could be rapidly solved by combining the new intensities of the reflections with the phases determined for the enzyme alone. These studies have constituted the first direct examination of the interactions of substratelike molecules at the active site of an enzyme.

A schematic representation of the three-dimensional structure of lysozyme is shown in Fig. 14-11. It is immediately apparent that the conformation is of a more complex nature than that of myoglobin. At least six helical segments are found; in some of these the α-helix geometry is markedly distorted in favor of hydrogen bond pairing between every fourth residue, apparently to allow greater stabilization by apolar interaction of amino acid side chains. Another region of the structure shows pleated sheet-type hydrogen bonding between antiparallel segments of the chain (Fig. 14-12). These segments are joined by an extended loop and interact by apolar contacts with another segment of the polypeptide chain that is wrapped around the pleated sheet portion.

The inside and outside domains of the molecule (with respect to solvent accessibility) are more difficult to define for lysozyme than for the heme proteins but, in general, regions that are shielded from solvent are apolar;

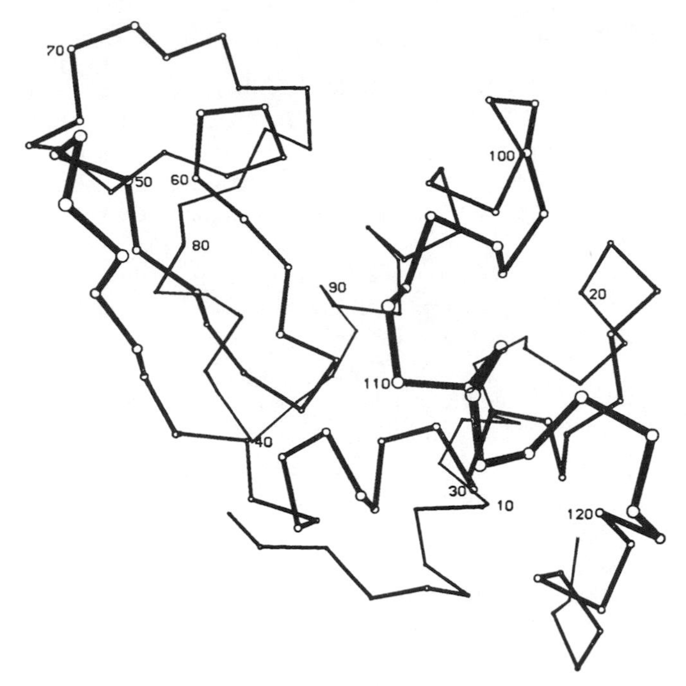

Fig. 14-11 α-Carbon diagram of the hen egg-white lysozyme structure based on crystallographic analysis at 2 Å resolution. Numbers refer to positions in the sequence beginning from the N-terminal at the lower left. The substrate cleft runs from top to bottom in the center of the structure. (Courtesy of D. C. Phillips.)

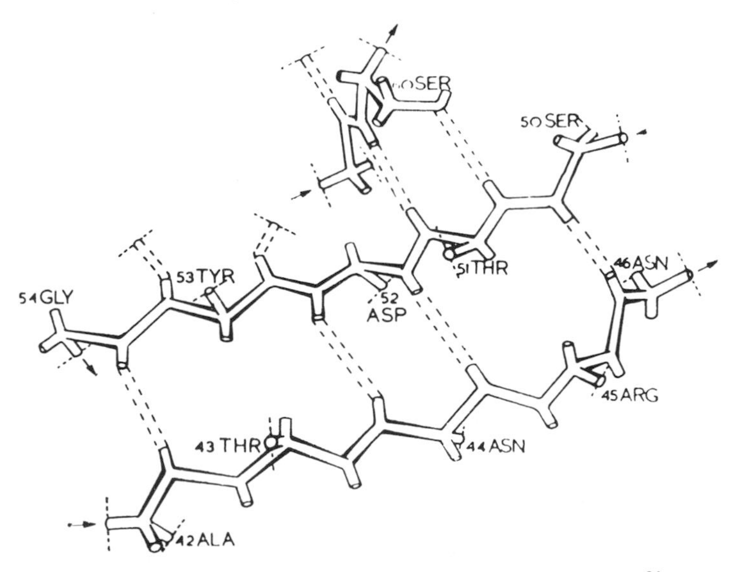

Fig. 14-12 Region of antiparallel pleated sheet in the lysozyme structure.[24]

all of the ionizable groups and all polar residues except one serine and one glutamine are external. On the other hand, a group of hydrophobic residues including four tryptophans is found on the surface. It is of interest that accessibility to solvent by a number of tryptophan residues had been demonstrated earlier by difference spectroscopy in conjunction with other physical measurements.[25]

The lysozyme molecule consists of two fairly distinct regions divided by a rather deep cleft. The specific binding site for substrate is located in the cleft area, thus providing multiple contact possibilities much like those for the heme group in hemoglobin. The binding of two inhibitors, *N*-acetylglucosamine and tri-*N*-acetylchitotriose, has been studied at 2 Å resolution. The latter forms a stable complex in which the reducing group of the trisaccharide points downward in the cleft. Although the structure is probably not that of a productive enzyme–substrate complex, the results nonetheless offer clues about possible types of interactions.

The structural findings for lysozyme and its complexes have been extended by model building in an attempt to understand the nature of the catalytic site.[24] Figure 14-13 illustrates schematically the approximate geometry for the binding of a hexasaccharide molecule in the cleft area of the enzyme. It has been proposed that two carboxyl groups close to the inhibitor binding site function as acid and base to effect hydrolysis, possibly via a carbonium ion intermediate. The structural results also suggest the possibility that the susceptible glycosidic bond is put under strain when the substrate is

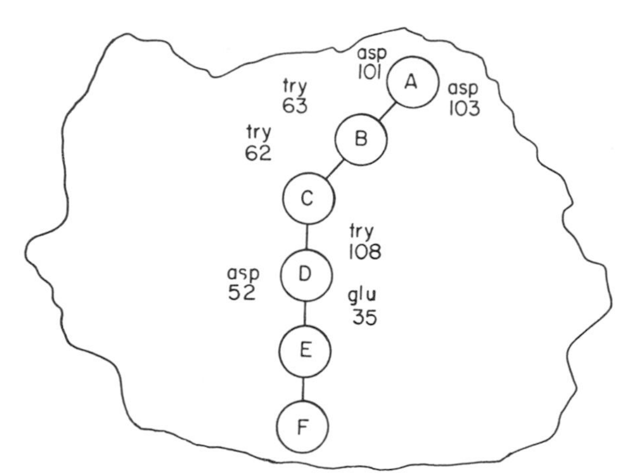

Fig. 14-13 Schematic illustration of the orientation of a bound hexasaccharide ligand with respect to the lysozyme molecule. The saccharide units are indicated by the letters A through F from the nonreducing end of the molecule; important amino acid residues within the active site are shown. (From Rupley and Gates,[26] based on the data of Blake et al.[23])

forced into the relatively rigid enzymatic mold. Discussion of the chemistry of the reaction and various mechanistic pathways has been presented by Rupley and Gates[26] and Raftery and Rand-Meir.[27] Nuclear magnetic resonance spectroscopy and fluorescence studies on enzyme–substrate interactions in lysozyme have also contributed to understanding of the process (cf. Chapters 9 and 12). Measurement of partial specific volume changes has been used to assess conformational alteration of the enzyme during the reaction with substrate.[27a]

3-5 General Features of Protein Crystallographic Structures[28,29]

X-ray studies of proteins have proved to be surprisingly fruitful in spite of the difficulties faced with such large molecular structures. High resolution structures have been determined for α-chymotrypsin (molecular weight 25,000), ribonuclease (13,400), carbonic anhydrase (30,000) carboxypeptidase A (34,600), and papain (22,000) among others. A large number of proteins from a variety of species of organisms are under study.[18] Analysis of many protein structures, however, has been complicated by the absence of any repeating conformation in the polypeptide backbone that might be revealed at low resolution (e.g., the α-helical segments of myoglobin). Thus it is more commonly necessary to reach the level of atomic resolution (2 Å) before the structure becomes comprehensible. Nonetheless, with improved apparatus and greater experience in preparation of heavy-atom derivatives, protein structures are being solved in much shorter times, and larger molecules are being attempted. Results for a number of enzymes are reviewed in two volumes of *The Enzymes*,[30] and a symposium on the subject has been published.[31]

Some tentative generalizations about protein structure can be made on the basis of available structures. As indicated for the hemoglobin structure, apolar residues are always found to be concentrated in the interior of the molecule, lending support to the theory that apolar interactions play a major role in protein folding. Ionizable and other polar side chains are mostly found on the surface of the molecule; others are found in clefts that form active sites for the binding of ligands. Polypeptide backbone conformations generally satisfy the steric restrictions upon free rotation discussed earlier (Section 2-5), although notable exceptions occur in lysozyme. Secondary structure elements found in the protein structures include the α-helix, the antiparallel β-pleated sheet (cf. Fig. 14-5), and the 3_{10} helix (ten residues in three turns of the helix). The most common is the α-helix, although its content varies widely (5% in α-chymotrypsin, 75% in myoglobin and hemoglobin). A very high degree of hydrogen bonding is found within the structures (based on the geometry of possible hydrogen bond donors and acceptors). Figure 14-14 shows a schematic diagram of the probable hydrogen bonds existing

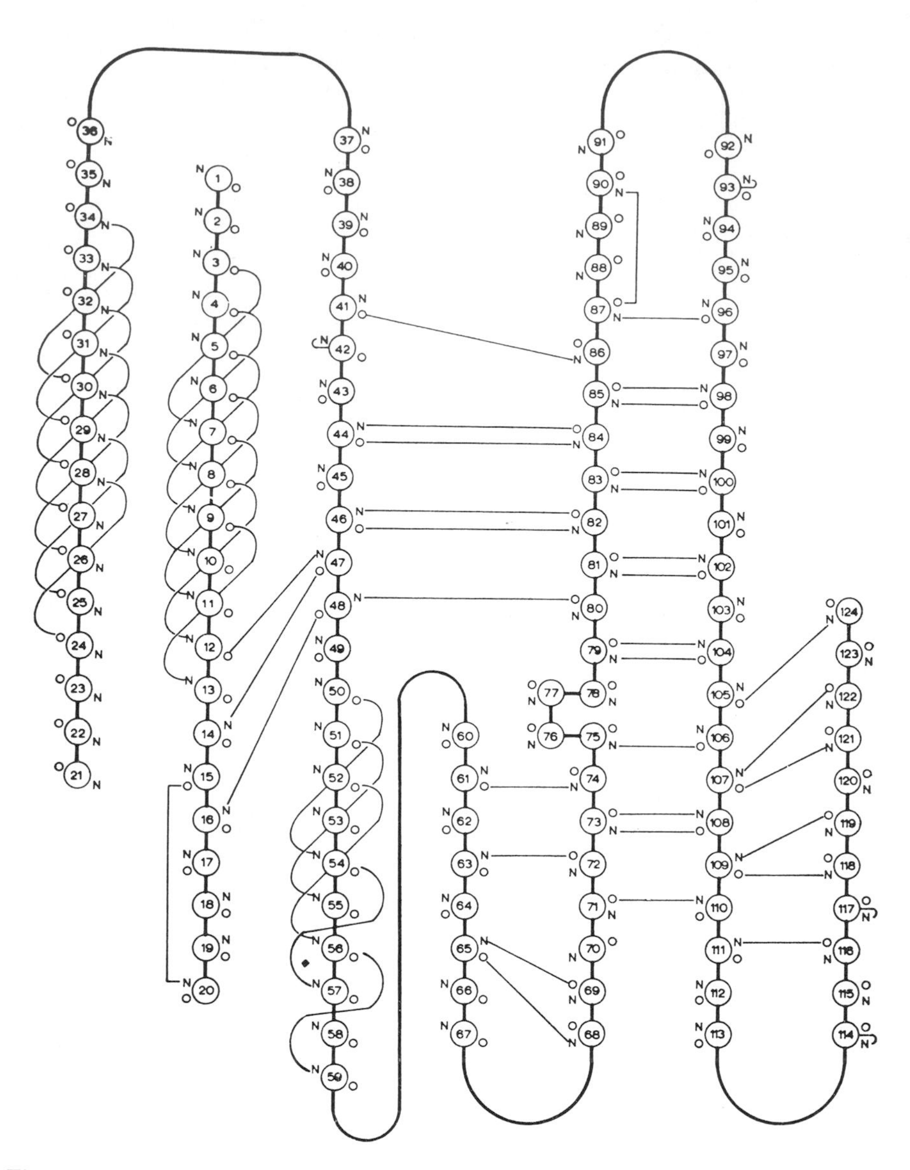

Fig. 14-14 Illustration of the probable hydrogen bonding by atoms of the polypeptide backbone in ribonuclease-S. Two helical regions involving 11 residues each occur toward the *N*-terminus in the sequence, one falling within the *S*-peptide (residues 1–20). An additional small helical segment (α-helix and 3_{10} helix) occurs in the middle of the sequence. The bulk of the structure from residue 41 to the *C*-terminus appears as a somewhat irregular antiparallel β-pleated sheet. [From H. W. Wyckoff, D. Tsernoglou, A. W. Hanson, J. R. Knox, B. Lee, and F. M. Richards, *J. Biol. Chem.*, **245,** 305 (1970).]

between atoms of the main chain in ribonuclease-S. Almost all peptide groups shown that are not involved in interpeptide hydrogen bonds belong to polar residues at the surface of the molecule, and are thus available for hydrogen-bonding with water. This apparent maximization of hydrogen bonds is in agreement with thermodynamic prediction (cf. Chapter 6).

Another important consideration is the question of the relationship or protein structure in the crystalline state to that existing in solution *in vitro* of *in vivo*. The demonstration of catalytic activity or of ligand binding (e.g., substrates and inhibitors) by crystalline enzymes indicates that, in general, conformations found in the crystalline state actually do represent the active molecules. A detailed review of the relevant data in this area has been made by Rupley.[32] Crystal structure results also have confirmed many predictions made about protein conformations based on physical properties observed in solution. Thus, with the exception of some surface details, solution and solid state conformations of proteins may be regarded as essentially identical. The only aspect of the problem still in doubt is the question of probable conformational motility in the active site of enzymes and of the possible role of generalized "breathing" of the structure (cf. Chapter 11) in protein function. Whether some proteins may possess a true "microheterogeneity" *in vivo* is yet to be cleared up. For those proteins that can be crystallized, however, theoretical analyses and experimental studies in solution, together with the X-ray structure, may eventually permit predictions to be made on the range of possible conformational variation.

REFERENCES

1. M. J. Buerger, *Crystal-Structure Analysis*, Wiley, New York, 1960.
2. R. W. James, *Optical Principles of the Diffraction of X-rays*, Cornell University Press, Ithaca, N.Y., 1958.
3. S. C. Nyburg, *X-ray Analysis of Organic Structures*, Academic Press, New York, 1961.
4. G. H. Stout and L. Jensen, *X-ray Structure Determination, A Practical Guide*, Macmillan, New York, 1968.
5. W. H. Zachariasen, *Theory of X-ray Diffraction of Crystals*, Dover, New York, 1967.
6. D. C. Hodgkin, M. J. Kamper, M. Mackay, J. Pickworth, K. N. Trueblood, and J. G. White, "Structure of vitamin B_{12}," *Nature*, **178,** 64 (1956).
7. L. Pauling, *The Nature of the Chemical Bond*, Cornell University Press, Ithaca, N.Y., 1960.
8. R. E. Marsh and J. Donohue, "Crystal structure studies of amino acids and peptides," *Adv. Protein Chem.*, **22,** 235 (1967).
9. H. C. Freeman, "Crystal structures of metal-peptide complexes," *Adv. Protein Chem.*, **22,** 257 (1967).

10. G. N. Ramachandran and V. Sasisekharan, "Conformation of polypeptides and proteins," *Adv. Protein Chem.*, **23,** 283 (1968).
11. L. Pauling, R. B. Corey, and H. R. Branson, "The structure of proteins: Two hydrogen-bonded helical configurations in the polypeptide chain," *Proc. Nat. Acad. Sci., U.S.*, **37,** 205 (1951).
12. L. Pauling and R. B. Corey, "Configurations of polypeptide chains with favored orientations around single bonds: Two new pleated sheets," *Proc. Nat. Acad. Sci., U.S.*, **37,** 729 (1951).
13. W. F. Harrington, R. Josephs, and D. M. Segal, "Physical chemical studies on proteins and polypeptides," *Ann. Rev. Biochem.*, **35,** 599 (1966).
14. T. T. Wu and E. A. Kabat, "An attempt to locate the non-helical and permissively helical sequences of proteins: application to the variable regions of immunoglobulin light and heavy chains," *Proc. Nat. Acad. Sci., U.S.*, **68,** 1501 (1971).
15. R. E. Dickerson, "X-ray analysis and protein structure," in *The Proteins*, 2nd ed., Vol. 2, H. Neurath, Ed., Academic Press, New York, 1964.
16. J. C. Kendrew, "Side-chain interactions in myoglobin," *Brookhaven Symp. Biol.*, **15,** 216 (1962).
17. R. D. B. Fraser and T. P. MacRae, "X-ray methods," in *Physical Principles and Techniques of Protein Chemistry*, Part A, S. J. Leach, Ed., Academic Press, New York, 1969.
18. D. Eisenberg, "X-ray crystallography and enzyme structure," in *The Enzymes*, Vol. 1, P. D. Boyer, Ed., Academic Press, New York, 1970.
19. C. C. F. Blake, "The preparation of isomorphous derivatives," *Adv. Protein Chem.*, **23,** 59 (1968).
19a. G. N. Ramachandran and R. Srinivasan, "*Fourier Methods in Crystallography*, Wiley-Interscience, New York, 1970.
20. B. Chance, R. W. Estabrook, and T. Yonetani, Eds., *Hemes and Hemoproteins*, Academic Press, New York, 1966.
21. M. F. Perutz, "The hemoglobin molecule," *Proc. Roy. Soc., Ser. B*, **173,** 113 (1969).
22. M. F. Perutz, J. C. Kendrew, and H. C. Watson, "Structure and function of haemoglobin. II. Some relations between polypeptide chain configuration and amino acid sequence," *J. Mol. Biol.*, **13,** 669 (1965).
23. C. C. F. Blake, G. A. Mair, A. C. T. North, D. C. Phillips, and V. R. Sarma, "On the conformation of the hen egg-white lysozyme molecule," *Proc. Roy. Soc., Ser. B*, **167,** 365 (1967).
24. D. C. Phillips, "The hen egg-white lysozyme molecule," *Proc. Nat. Acad. Sci., U.S.*, **57,** 484 (1967).
25. A. Kurono and K. Hamaguchi, "Structure of muramidase (lysozyme). VII. Effect of alcohols and related compounds on the stability of muramidase," *J. Biochem.* (*Tokyo*), **56,** 432 (1964).
26. J. A. Rupley and V. Gates, "Studies on the enzymic activity of lysozyme. II. The hydrolysis and transfer reactions of *N*-acetylglucosamine oligosaccharides," *Proc. Nat. Acad. Sci., U.S.*, **57,** 496 (1967).
27. M. A. Raftery and T. Rand-Meir, "On distinguishing between possible mechanistic pathways during lysozyme-catalyzed cleavage of glycosidic bonds," *Biochemistry*, **7,** 3281 (1968).

27a. W. M. Neville and H. Eyring, "Hydrostatic pressure and ionic strength effects on the kinetics of lysozyme," *Proc. Nat. Acad. Sci. U.S.* **69**, 2417 (1972).

28. L. Stryer, "Implications of X-ray crystallographic studies of protein structure," *Ann. Rev. Biochem.*, **37**, 25 (1968).

29. M. F. Perutz, "X-ray analysis, structure and function of enzymes," *Eur. J. Biochem.*, **8**, 455 (1969).

30. P. D. Boyer, Ed., *The Enzymes*, 3rd ed., Vols. 3 and 4, Academic Press, New York, 1971.

31. *Structure and Function of Proteins at the Three-Dimensional Level*, Cold Spring Harbor Symp. Quant. Biol. **36** (1971).

32. J. A. Rupley, "The comparison of protein structure in the crystal and in solution," in *Structure and Stability of Biological Macromolecules*, S. Timasheff and G. Fasman, Eds., Marcel Dekker, Inc., New York, 1969.

XV Dynamics of Protein Conformations

In thinking about protein structure it is important to remember that, because of the high molecular weight and the large variety of amino acid components generally occurring in nonrepeating sequences, a protein molecule is capable of assuming a great variety of three-dimensional conformations. The form (or forms) that exists in cells is the result of an enormous number of interactions, primarily noncovalent, that occur between amino acid residues and with components of the cellular environment. Even though these forces apparently impart only a marginal level of stability to the molecule, proteins are nevertheless able to assume specific compact structures that are stable over a range of conditions *in vivo* and *in vitro*. However, protein structure should not be regarded as static and fixed. In fact, in regard to biological function, it is helpful to think of the structure of a protein as a dynamic state, with the capacity to respond in certain determined ways to the needs of the organism for expression of function by that protein, for regulation of its activity, or for its destruction, when required.

Cooperative forces in proteins exist at various levels of organization: between adjacent residues (as in a helical segment), between regions (as the interaction among helical regions in hemoglobin), between subunits in multichain structures, between molecules (as in fibrils or protein crystals), and between molecular superstructures. These systems are thus capable not only of conformational changes within one level but also of coupled conformational changes resulting from interactions between regions, subunits, and molecules. At each level there may also be possibilities for interactions with small molecules, which may then be expressed at other levels. All of these factors must be considered when one approaches problems of conformational dynamics in macromolecules.

It may appear from this that the complexity of the system far exceeds our capacity for meaningful investigation. However, although a protein may be capable of assuming many different conformations and of participating in many types of interactions, those of biological significance are probably few in number. Thus the subject of conformational dynamics is amenable to experimental studies in which results are compared to a reproducible standard species, usually the biologically functional molecule (assayed *in vitro*). The discussion has been divided into two sections. The first deals with protein denaturation, the types of conditions that may cause loss of function and structure either inadvertently or in a controlled study of conformational alterations. The second section considers evidence on the ways in which protein conformational changes are related to biological properties.

1 PROTEIN DENATURATION[1–5]

The term denaturation applies to any process, not involving the rupture of peptide bonds, that causes a change in the three-dimensional structure of a protein from that which exists in the "native" form. Some authors include disulfide bond rupture or chemical modification of certain groups on the protein under the classification of denaturation provided that these alterations are accompanied by changes in the overall three-dimensional structure as well. Others prefer to limit the definition of denaturation to processes involving noncovalent interactions only. We shall adopt the former, more broad definition here. The term "native" protein refers to the molecule in the exact three-dimensional structure in which it occurs when it is actively carrying out its biological function *in vivo*. The assumption is usually made that a protein isolated *in vitro* possessing a high level of biological activity has a structure sufficiently similar to the *in vivo* protein to term it "native" as well. The latter is pragmatically the most useful definition since a test of true native character is usually impossible.

The degree to which the three-dimensional structure of a protein may differ from the native state may vary from a change in a single noncovalent bond or side-chain orientation to the case where almost no atom exists in the same spatial relationship to others except for the constraints of the primary structure. Denaturation that results in the loss of biological function includes the full range of possibilities; it may involve only a small conformational change that is practically undetectable by available techniques. Denaturation involving loss of structure detectable by physical or chemical techniques may also occur to varying degrees. Modest conformational changes may take place without necessarily affecting biological activity, or partial activity loss may be noted. A qualifying term is often used to describe the degree of denaturation of a particular preparation, for example, slightly denatured.

Most biological scientists will have reason to be concerned about denaturation of proteins in one way or another. Many biological studies depend upon assays of proteins *in vitro*, and preparations are required that can be obtained reproducibly and are as close to the native form as possible. Factors that may cause denaturation, particularly of the more subtle types, must be identified and taken into account. Certain types of denaturation are made obvious by a loss of solubility of the protein in the usual aqueous buffers. However, in most situations a number of different approaches will be needed to detect and identify structural alterations. When a convenient assay for biological activity is available, the purification of the protein to the same constant specific activity by several methods often forms a pragmatic approach to define native character. Specific activity measurements may then be used on the purified protein as one method of following subsequent denaturation processes. Changes in chemical properties may be examined, such as the reactivity of certain groups, or physical properties (optical rotatory dispersion, ultraviolet spectrum, hydrodynamic properties, charge properties, etc.) may be followed. Immunological methods are also frequently employed to assess denaturation. No single one of these properties is ordinarily sufficient to establish the type and extent of the change in structure or to prove that two preparations of a given protein have the same three-dimensional structure in solution. For purposes of characterization and comparison, a number of properties dependent on structure should be examined.

The other major use of the denaturation process is as a means of studying the forces responsible for the tertiary structure of proteins and the role of different types of interactions in determining active sites and other properties of the molecule. Synthetic polymers of certain amino acids or amino acid derivatives and selected mixed polymers have been used extensively as models for the study of protein denaturation. Interpretation of results is simplified because of the repeating primary structure of these molecules. For example, the process of denaturation of an α-helix to a random coil can best be observed in certain synthetic polypeptides that possess no other types of structure. Although proteins may contain α-helical segments, their overall structures are much more complicated and in any denaturation process, changes in conformation other than loss of helix are likely to occur. Synthetic polyamino acids have been useful for the empirical calibration of physical properties, for example, the correlation of optical rotation or ultraviolet absorption with secondary structure. Comprehensive reviews on the study of polyamino acids as models for protein structure are available.[1,2]

The protein chemist is interested in denaturation for a variety of reasons. Perhaps most commonly, the principles and techniques used to measure these processes are applied to establish conditions whereby protein denaturation may be avoided so that chemical, physical and functional studies may be made on the native protein with a minimum of time-dependent structural alteration occurring. In other cases it is desirable to dissociate proteins and reduce their constituent polypeptide chains to the random coil conformation or other denatured state before carrying out certain types of studies. Such experiments might involve the separation of nonidentical polypeptide chains, the separation of charge isomers (chains which differ, for example, in net charge due to modification of residue side chains), chemical reaction of certain residues, or molecular weight determinations for the chains. For example, denaturation with a detergent such as sodium dodecyl sulfate (SDS) produces an SDS–protein complex that can be studied to estimate the molecular weight of the protein chains (cf. Chapter 10).

Many proteins can be induced to undergo reversible structural alterations. Minor perturbations are used to promote dissociation of protein oligomers and are useful in the study of binding sites, the thermodynamics of association, and in the determination of subunit stoichiometry (cf. Chapters 10 and 16). Total denaturation to yield the constituent random coil polypeptide chains can be completely reversed in a number of cases provided that the perturbing influence is removed in a suitable fashion. Such renaturation may restore all or part of the original properties of the native protein. Reversible denaturation of this type has been instrumental in development of the concept that native conformations form spontaneously *in vivo*. Since denaturation and renaturation processes in proteins involve changes in an enormous number of interactions, one may expect the kinetics to be extremely complex. The problem is not quite so difficult, however, because the various individual noncovalent bonds do not act independently. Rather, the evidence suggests that cooperative action of particular groups of "bonds" or contacts are involved in stabilizing various segments of the structure or even the total conformation. Thus conformational transitions are found to pass through a few intermediate states or take place by an all-or-none type of mechanism between two states with no intermediates occurring in substantial concentration. These systems provide thermodynamic data (and, to a lesser extent, kinetic data) for the process of protein folding. For two-state processes involving the native conformation and the random coil, it is possible to obtain equilibrium constants in a straightforward manner. For example, if the optical rotation of the native and random coil conformations is known, measurement of this parameter will give the weight fractions of the two forms in the equilibrium mixtures. Although these data are obtained in the presence of perturbing influences (e.g., denaturing solvents, high temperature), extrapolation may be made to simulated physiological conditions to obtain the desired thermodynamic parameters. These studies tell us much about protein stability and can be compared with calculations based on theoretical models of protein interactions.

Here we first discuss a few examples of the effect of denaturing agents on protein conformations. The mechanism of action is not clear in all cases, but studies with model compounds as well as with proteins permit some generalizations to be made. A more extensive treatment of the subject may be found in the reviews by Tanford,[3,4] where the interested student will also find citations to the original literature. Finally, we discuss briefly some experimental data for proteins that undergo reversible (or partially reversible) denaturation and the implications of these data.

1-1 Temperature-Induced Conformational Changes[3,4]

Proteins in solution can be denatured by raising the temperature to a sufficiently high value (usually 50–60° will produce some effects) for a given period. In most cases it is difficult to reverse the reaction, primarily because of aggregation and precipitation of the protein, although there are cases of temperature-induced denaturation where biological activity and conformation of the molecules have been restored upon cooling. In some proteins denaturation at high temperature promotes disulfide bond rupture or disulfide interchange (particularly at alkaline pH), but usually only noncovalent interactions are affected by heat. Coagulation also occurs in proteins containing no cysteine or cystine. Myoglobin, for example, undergoes reversible heat denaturation at low pH, but coagulation that is not reversible occurs in the thermal denaturation above pH 6 (reviewed by Tanford[3]). The conformation of the molecules that participate in coagulation is not known, but it seems likely that the resultant aggregate is characterized by high potential energy barriers against reversal and is therefore a "frozen-in" or metastable structure thermodynamically. Reversal then would not be observed because the reaction rate is too slow.

Reversible temperature-dependent denaturation studies on proteins are particularly important since they provide thermodynamic information. These processes, however, do not usually involve transition to the completely unfolded random coil. Although the products are quite disorganized, they retain some regions of compact structure. Investigation of the temperature dependence of reversible denaturation induced by other denaturing agents (urea, pH) has shown the process to be characterized by unusually high values of ΔC_p, thus the thermodynamic parameters are highly temperature dependent. The results of these studies, as mentioned in Chapter 6, indicate that many proteins have a temperature of maximum stability. Although it is difficult to interpret the data directly in terms of model compound studies, one may reasonably expect heat denaturation to be associated with the weakening of both hydrogen bonding and apolar interactions. The nature of the residual structure in thermally denatured products and the interactions leading to coagulation (other than that involving disulfide bonds) are not known.

Several examples of thermally induced conformational changes are discussed elsewhere. In the data presented for lysozyme (Fig. 9-4), two conformational changes are revealed by ultraviolet absorbance measurements, whereas only one is evident from optical rotation. Temperature-induced denaturation has been used to compare properties of intact collagen with some enzymically produced fragments (Chapter 17). The heat denaturation of ribonuclease provides an interesting example of a reversible thermal transition. Figure 15-1 shows the effect of pH on this process as followed by

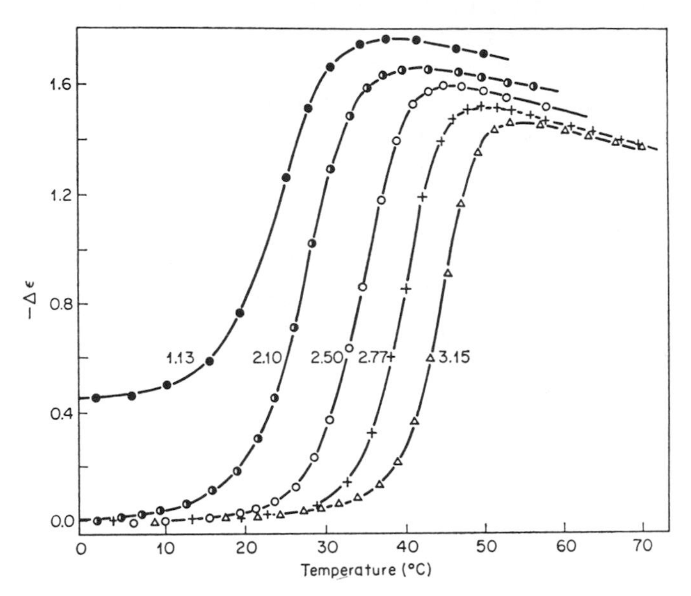

Fig. 15-1 The reversible thermal denaturation of ribonuclease at various low values of pH (pH = 1.13 to 3.15), as followed by absorbance changes relative to the native protein. Conversion to the native protein is achieved at all but the lowest pH value studied. [From Tanford,[3] based on data of J. F. Brandts and L. Hunt, *J. Amer. Chem. Soc.* **89,** 4826 (1967).]

difference spectroscopy. These studies were confined to the low pH range where disulfide bonds remain intact, and thus renaturation is more readily achieved. In this study curves of the same shape were obtained regardless of whether the process was followed by changes in ultraviolet absorption, intrinsic viscosity, or optical rotation. The heat-denatured state under these conditions is probably an example of one that is not a true random coil, but retains regions of ordered structure.[3]

The finding of temperature maxima for stability in proteins (*in vitro*) is of physiological interest, since the values found for some proteins fall within the range experienced by most living organisms (5–40°). Similar observations are made with regard to stability of quaternary structures. These equilibria involve the same types of noncovalent interactions acting within limited domains (intersubunit bonding). The decreased association of the coat protein of tobacco mosaic virus with temperature decrease is a well-known example of this phenomenon.[6] A number of enzymes have been found to be inactivated as a result of cold exposure; in most cases they can be reactivated by warming if sufficient time is allowed. For example, pyruvate carboxylase shows almost no activity when assayed immediately after being held for 1 hour at 0°, but most of the activity is recovered if the sample is rewarmed for 30 minutes before assay. Cold inactivation in this case appears to involve dissociation of the active tetrameric form of the enzyme into four subunits or protomers.[7] This type of behavior in proteins of multiple subunits may prove to be more common than previously imagined.

The effect of temperature on protein stability is of particular interest in relation to physiological processes that are naturally subject to varying temperature (as in organisms that lack body temperature regulation). For example, a great number of studies have been carried out on the effect of temperature on enzyme activity or other properties *in vitro*. Although these studies are important for evaluating protein stability in those environments (usually highly aqueous), caution must be exercised in extrapolating these results to natural environments. Denaturation of purified collagen *in vitro*, for example (see Fig. 15-2), occurs at temperatures that are low compared to

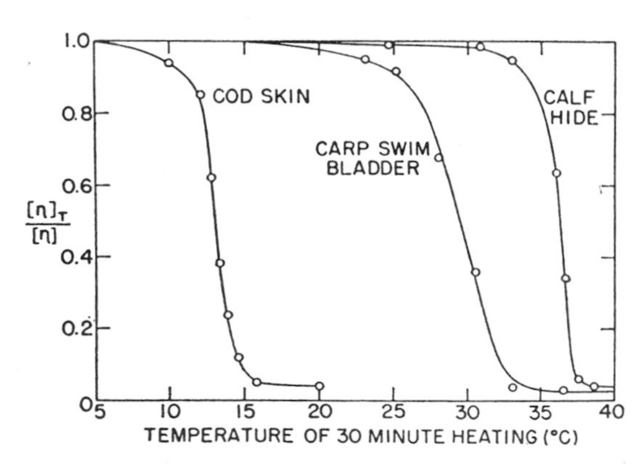

Fig. 15-2 Thermal denaturation of various collagen preparations to form gelatin. The process is followed by the decrease in viscosity of the collagen solution following 30 min heating at the indicated temperature, relative to that of the native molecules. (From P. Doty and T. Nishihara, in *Recent Advances in Gelatin and Glue Research*, G. Stainsby, ed., Pergamon Press, Oxford, 1958.)

body temperatures of the source organisms, although a rough phylogenetic correlation is found. In general, *in vitro* studies of denaturation indicate that the solution environment has a very large effect on the results (cf. Section 1-5 of this chapter). More reliable analysis of phylogenetic variation in thermal stability of proteins might be achieved by kinetic measurements under varying temperature conditions *in vivo* (where possible) or in tissue culture.

1-2 pH-Induced Conformational Changes[1,3]

Since proteins are polyions, there are obviously electrostatic interactions leading to attraction or repulsion between various parts of the polypeptide chain. Net repulsion, which may lead to loss of stability, will occur at pH's below the isoelectric point due to excess positive charge on the molecule and at pH's above the isoelectric point due to excess negative charge. The magnitude of the force is dependent upon the dielectric constant of the medium and the extent of shielding by the counterion atmosphere and the solvent. The occurrence of electrostatic repulsions does not mean that optimal charge stability necessarily is found at the isoelectric point; rather, stability will depend more upon the distribution of charges on the surface of the molecule. As might be expected, the products of pH denaturation vary from extremely minor conformational changes to cases where the denatured form has a nearly random coil conformation (or an expanded random coil at low ionic strength due to charge repulsions). Intermediate cases that contain some ordered and some random coil segments are not uncommon.

The role of electrostatic interactions in the conformational change from α-helix to random coil is demonstrated in the pH dependence of this transition of synthetic polypeptides bearing charged side chains. That of poly-L-lysine is illustrated in Fig. 15-3. Another example is poly-L-glutamic acid. These polypeptides are able to form the helical conformation only under conditions where the side chains are uncharged (pH about 3 for polyglutamic acid and about 11 for polylysine). Proteins differ widely in their susceptibility to denaturation in acidic or alkaline solutions. Lysozyme and ribonuclease, for example, are fairly stable to acid conditions, and at moderate ionic strengths can be titrated to pH 2 without conformational change. Most proteins, however, are stable only in a more narrow range of pH (about 4 to 10); extensive disorganization of structure is found outside this range. In many cases acid or alkaline denaturation has been shown to be associated with the instability of buried groups that develops when the pH goes beyond their normal p*K* by several units. For example, the conformational change of hemoglobin at low pH is correlated with exposure of histidyl residues that are buried in the uncharged form in the native molecule. A similar effect is found in myoglobin and carbonic anhydrase.

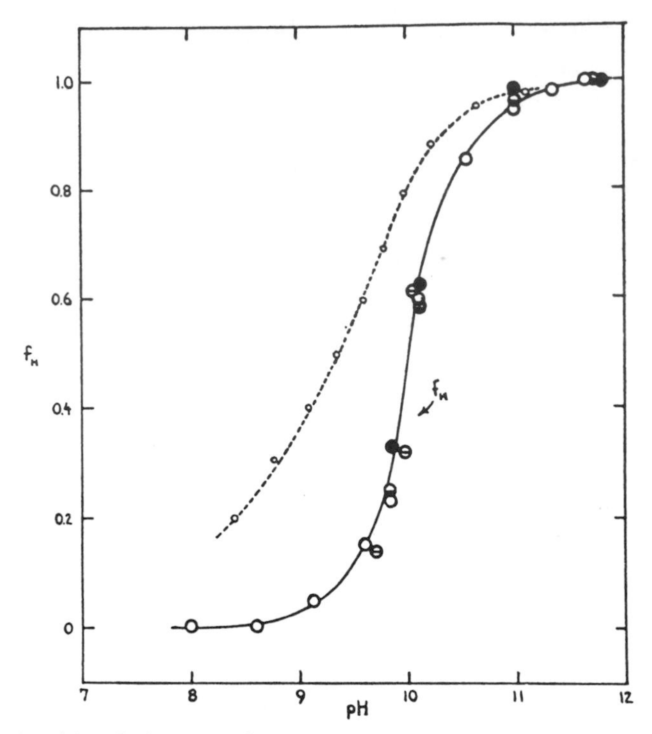

Fig. 15-3 Partial helical content f_H of poly-L-lysine in water as determined by measurements of optical rotatory dispersion as a function of pH. The broken line indicates degree of dissociation. The helical form is stable at high pH where the lysine side-chains have lost their charge. (From J. Applequist and P. Doty, in *Polyamino Acids, Polypeptides and Proteins*, M. A. Stahmann, Ed., Univ. of Wisconsin Press, Madison, Wis., 1962.)

We have expressed the denaturing effect of pH changes on native proteins in terms of repulsive forces of nearby charges. In the random coil these repulsions are presumably reduced due to a larger distance between charges and the greater accessibility of other ions in the solution to the charged groups. More generally, pH-induced shifts in equilibria are treated thermodynamically in terms of differences in binding of protons by the different conformations. This same approach is valuable in the consideration of conformational changes in proteins that may be induced by very small changes in pH, such as in the allosteric effect of hydrogen ions on hemoglobin conformation (cf. Section 2).

1-3 Conformational Changes Induced by Organic Solvents[8]

Nonaqueous solvents for proteins may in general be classified into two broad categories: strongly protic and weakly protic. Singer defines a weakly

Table 15-1 Properties of some nonaqueous solvents used in protein studies.[a]

Solvent	Dielectric constant[b]	B.p. (°C)	M.p. (°C)	Density[b]	Refractive index (n_D)[b]	Viscosity (cP)[b]
Strongly protic acids						
Hydrofluoric	83.6^{0}	19.5	−83	0.9918^{4}	—	0.240^{6}
Formic	58.5^{16}	100.7	8.2	1.2133^{25}	1.3694^{25}	1.966^{25}
Phenol	9.78^{60}	181.8	40.9	1.0576^{41}	1.5418^{41}	4.076^{45}
Dichloroacetic	8.2^{20}	194	9.7	1.5585^{25}	—	—
Acetic	6.15^{20}	117.7	16.6	1.0437^{25}	1.3700^{25}	1.040^{30}
Bases						
Hydrazine	51.7^{25}	113.5	2	1.014^{15}	—	—
Ammonia	22^{-33}	−33.4	−77.7	0.65^{-10}	—	—
Ethylenediamine	14.2^{20}	116.2	11.0	0.891^{25}	1.4513^{30}	1.725^{25}
Pyridine	12.3^{25}	115.6	−41.8	0.9878^{15}	1.5067^{25}	0.829^{30}
Weakly protic alcohols						
Glycerol	42.5^{25}	290.0	18.2	1.2613^{20}	1.4735^{25}	945^{25}
Ethylene glycol	37.7^{25}	197.8	−12.6	1.1171^{15}	1.4331^{15}	26.09^{15}
Methanol	32.63^{25}	64.51	−97.5	0.7961^{15}	1.3266^{25}	0.545^{25}
Ethanol	24.30^{25}	78.32	−114.5	0.7936^{15}	1.3594^{25}	1.078^{25}
Amides						
N-Methylacetamide	178.9^{30}	204	29.7	0.9503^{30}	—	3.885^{30}
Formamide	109.5^{25}	210.5	2	1.1292^{25}	1.4468^{25}	3.302^{25}
N,*N*-Dimethylacetamide	37.8^{25}	165	—	0.9366^{25}	1.4358^{25}	9.610
Miscellaneous						
Dimethylsulfoxide	45	189	18.5	1.100^{20}	1.4787^{21}	1.100^{27}
Dioxane	2.21^{25}	101.3	11.80	1.0269^{25}	1.4202^{25}	1.439^{15}

[a] Based on a table given by Singer.[8]
[b] Superscripts indicate the temperature in °C at which the data apply.

protic solvent as one which at 1 *M* concentration in water gives a pH between 6 and 8.[8] Strongly protic solvents are those whose 1 *M* solutions have a pH below 6 (formic acid, dichloroacetic acid) or above 8 (ammonia). A list of some nonaqueous solvents that have been used in the study of proteins is given in Table 15-1. These studies are important because of the fact that the natural environment of most proteins is not a simple aqueous salt solution but is one containing many organic species such as lipids and carbohydrates. An understanding of the effect of nonaqueous solvents on the conformation of proteins may be approached by considering the expected effects on electrostatic forces, hydrogen bonding, and apolar bonding.

Electrostatic Interactions. Organic solvents influence electrostatic interactions in a protein through their different dielectric constants (compared to water) as well as their effects on counterion atmosphere and on the binding of solvent or other solutes to the macromolecules. In general, decreasing polarity of the solvent and decreasing dielectric constant tends to increase electrostatic repulsive forces and make the protein molecule more highly swollen and unfolded.

Hydrogen Bonding. The reaction for the formation of an interpeptide hydrogen bond in water has been given (cf. Chapter 6). A similar equation can be written for the formation of a protein–protein hydrogen bond in an organic solvent:

$$\text{N—H}\cdots\text{Sol} + \text{Sol}\cdots\text{O═C} \rightleftharpoons \text{N—H}\cdots\text{O═C} + \text{Sol}\cdots\text{Sol}$$

where Sol is a solvent capable of accepting or donating a proton for hydrogen bonding. Protein–protein hydrogen bonds are favored in solvents that cannot themselves form strong hydrogen bonds with the protein. This was clearly shown for the model compound *N*-methylacetamide discussed earlier (cf. Chapter 6), where the strength of its hydrogen bonding was increased in solvents of low hydrogen bonding capacity (chloroform) compared with water. Figure 15-4 shows the specific optical rotation of poly-L-methionine in mixtures of chloroform and trifluoroacetic acid. This polymer, which exists as an α-helix in chloroform, is converted to a random coil by the addition of trifluoroacetic acid, a substance that strongly solvates the random coil form of the polymer. Similar studies have been made with a number of polypeptides and proteins.

Apolar Bonding. As previously discussed (Chapter 6), apolar bonding in a protein apparently results from the clustering of organic side chains of the protein molecule in order to avoid contact with water. When these apolar groups are exposed to the solvent, a highly ordered array of water molecules is formed around them and the process is energetically unfavorable (ΔG positive) due to a negative entropy change. Water is remarkable in the magnitude of this phenomenon; even strongly protic solvents such as

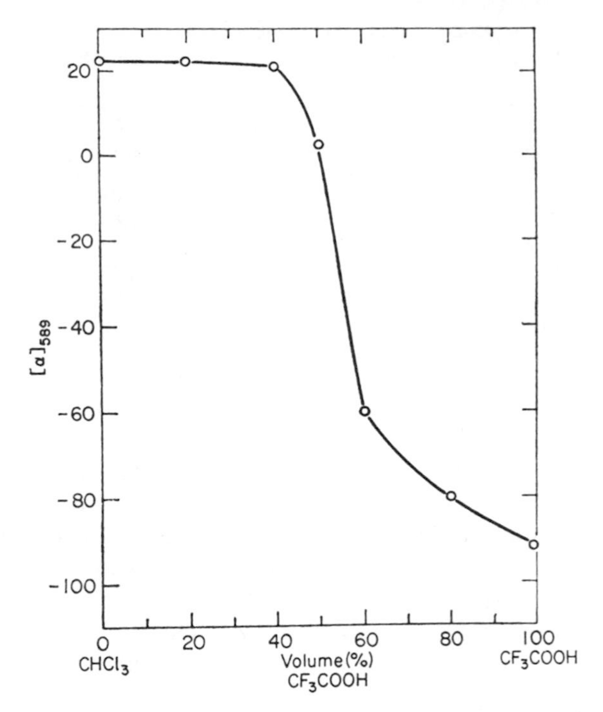

Fig. 15-4 Helix–random coil transition of poly-L-methionine produced by solvent variation and followed by measurements of specific optical rotation at 589 nm. [From G. E. Perlmann and E. Katchalski, *J. Amer. Chem. Soc.*, **84,** 452 (1962).]

anhydrous formic acid do not show such a large effect. Unitary free energy changes are given in Table 15-2 for the transfer of benzene to various solvents. As expected, a significant amount of energy is required when benzene is transferred into water, in which it is sparingly soluble; ΔG decreases as the solvent becomes less and less polar. These data indicate that apolar interactions stabilizing a protein will be weakened in the presence of solvents with less polarity than water.

In comparisons of protein conformations in organic solvents and in water, it is frequently possible to minimize the effect of electrostatic repulsions by control of pH, ionic strength, and dielectric constant. Under such circumstances the collapse of the protein structure induced by nonpolar organic solvents is most likely due to their effect on apolar bonding. There is, however, the additional factor that protein–protein hydrogen bonds are comparatively favored in these solvents. Extensive conformational studies with synthetic polypeptides have been made in pure and mixed organic solvents. The α-helical form of these polymers in pure solvents (e.g., poly-γ-benzyl-L-glutamate in α-chloronaphthalene and poly-L-glutamate in *N*-methylacetamide) is unexpectedly stable and shows no major conformational change

Table 15-2 Unitary free energy changes for the transfer of benzene to water and other solvents of decreasing polarity.[8]

Solvent	Temperature (°C)	ΔG_u (kcal/mole)
Water	18	4.07
Ethylene glycol	25	1.83
Formic acid	25	1.45
Propylene glycol	25	0.99
Methanol	35	1.23
Ethanol	45	0.96
Isopropanol	45	0.88
Acetonitrile	45	0.67
CCl_4	40	0.08

upon heating to over 100°C. As yet no satisfying explanation of the forces responsible for stability of the helix in these solvents is available, particularly in the case of the solvent *N*-methylacetamide which itself contains peptidelike hydrogen-bonding donor and acceptor groups. Another unusual effect observed in several systems is illustrated by the random coil to α-helix transition of poly-γ-benzyl-L-glutamate in a mixed organic solvent (Fig. 15-5). An inverse temperature transition is observed, that is, the random

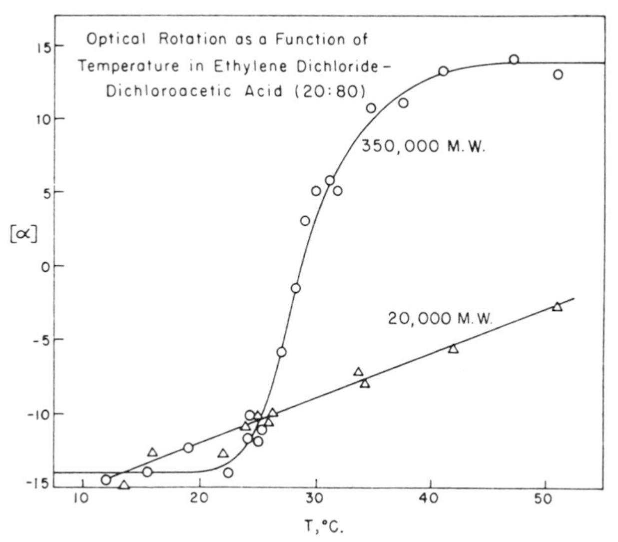

Fig. 15-5 A thermal random coil–helix transition for poly-γ-benzyl-L-glutamate in ethylene dichloride–dichloroacetic acid. Both large (molecular weight = 350,000) and small (molecular weight = 20,000) polymers exhibit the transition, although the latter is spread out over a large temperature range. [From P. Doty and J. T. Yang, *J. Amer. Chem. Soc.*, **78,** 498 (1956).]

coil is the stable form at low temperatures whereas the helix is favored at higher temperatures. A possible explanation is that the coiled form is solvated by a large number of dichloroacetic acid molecules, thereby reducing entropy.

Further theoretical and empirical knowledge is needed before the complex noncovalent interactions that occur within proteins and polyamino acids dissolved in organic or mixed organic solvents can be fully described and quantitatively evaluated. Such studies are of considerable importance because similar interactions may serve an important physiological role in conjugated proteins and in structural complexes such as membranes.

1-4 Denaturation by Urea or Guanidine Hydrochloride[3–5]

Urea and guanidine hydrochloride have probably been used more extensively as effectors of protein denaturation than any other reagents. At high concentrations of these substances (e.g., 8 *M* urea or 5 *M* guanidine-HCl) many proteins adopt a highly unfolded conformation in solution. Proteins of multiple subunits are likely to be separated into their constituent polypeptide chains. Other proteins aggregate upon denaturation in urea or guanidine-HCl; this is frequently due to the formation of disulfide bridges between sulfhydryl groups made accessible by the unfolding of the polypeptide chains. Such reaction may be inhibited by adding an excess of thiol reagents such as mercaptoethanol or dithiothreitol, by reduction followed by alkylation, or by oxidation to the *S*-sulfo derivative. Not all cases of aggregation in these systems, however, can be accounted for by disulfide interchange. In some urea-denatured proteins, noncovalent interchain bonds are able to form which are of sufficient stability to cause aggregation or even precipitation. It is also known that 8 *M* urea, for example, does not destroy all noncovalent bonds in proteins; further structural changes, as measured by optical rotation, are often observed upon heating.

The mechanism of action of urea, guanidine, and other similar denaturing agents such as formamide, dimethylformamide, and diethylformamide on proteins has been the subject of much investigation and considerable controversy. The structure of these compounds, notably urea and guanidine, suggests that they might act as both proton donors and acceptors in the formation of hydrogen bonds with the protein. For many years it was believed that these substances could form stronger H-bonds with the protein than could water, and thus they denatured the protein through rupture of intramolecular hydrogen bonds. However, the almost complete absence of solute–solute hydrogen bonding by *N*-methylacetamide in water or by urea in water[9] argues against a powerful H-bonding capacity in these compounds. It would appear that water would be as effective a denaturant of protein hydrogen bonds as are small amides. On the other hand, alkyl substitution on the nitrogen atoms of urea or guanidine reduces their interaction with the amide groups of model compounds.[10] Thus the presence of hydrogen atoms on the nitrogen atom, with hydrogen-bonding capability, appears to play an important role in the denaturing action of these substances.

A large part of the action of urea, guanidine hydrochloride, and similar compounds is now thought to involve a hydrophobic mechanism that favors exposure to the solvent of nonpolar groups in the interior of the protein molecule. In general, aqueous solutions containing a high concentration of these compounds act as better solvents for nonpolar substances than does water alone. Studies on the transfer of hydrocarbons used as models for the amino acid side chains from water to 7 *M* urea or 4.9 *M* guanidine hydrochloride revealed a favorable ΔG for the process.[11] Although the transfer requires energy ($\Delta H > 0$) at room temperature, it is accompanied by a positive entropy change that overrides the unfavorable enthalpy change. Table 15-3 summarizes the thermodynamic results in this study.

Table 15-3 Values of thermodynamic functions for the transfer of various hydrocarbon compounds from water to the solvents 7 *M* urea and 4.9 *M* guanidinium chloride commonly used for protein denaturation. Values are given on a mole fraction scale (unitary functions); ΔG_u refers to 25°C.[11]

	$H_2O \rightarrow$ 7 *M* urea			$H_2O \rightarrow$ 4.9 *M* GuCl		
Compound	ΔH° (kcal)	ΔS_u° (eu)	ΔG_u° (kcal)	ΔH° (kcal)	ΔS_u° (eu)	ΔG_u° (kcal)
Methane	1.3	4.1	+0.07	1.6	5.9	+0.05
Ethane	1.9	6.5	−0.04	1.9	6.7	−0.06
Propane	1.7	6.3	−0.13	1.6	6.0	−0.20
Butane	1.9	7.2	−0.25	1.8	6.6	−0.35
Neopentane	1.7	6.6	−0.29	1.1	5.5	−0.45
Toluene	1.5	6.6	−0.48	0.9	4.9	−0.61

At the present time it is still uncertain what the exact mechanism of action of these denaturing agents is; however, they have proved to be quite useful in the study of protein structure, particularly in the elucidation of the number and size of protein subunits in multichain proteins. Reliable molecular weights can now be determined in these multicomponent systems by hydrodynamic methods or by light scattering. However, care must be taken to determine the completeness of the denaturation process. Even at the maximum possible concentrations (8–10 *M*), urea solutions often produce only partial denaturation, and other conditions must be changed in order to completely disorganize the protein molecule. On the other hand, as far as we are aware, all nonconjugated and noncrosslinked proteins that have been sufficiently studied do attain a random coil conformation in solutions containing high concentrations of guanidine-HCl. Even more potent than this reagent (i.e., producing a similar degree of denaturation at lower concentration of the denaturant) are several related salts. These include guanidylguanidinium salts and salts of guanidine in which the negative ion is itself a promoter of denaturation, such as guanidine thiocyanate (see the following section).

Figure 15-6 illustrates a denaturation curve for β-lactoglobulin, a protein that is completely denatured by 8 *M* urea under the conditions used. An

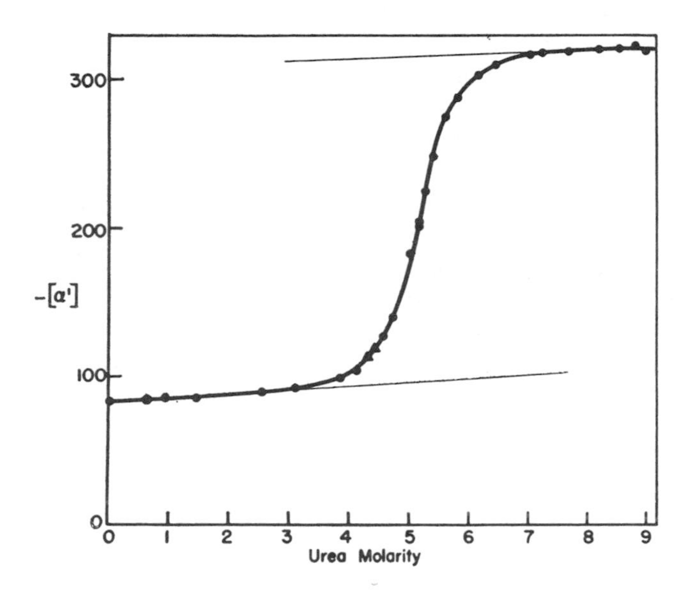

Fig. 15-6 Denaturation of β-lactoglobulin by increasing concentrations of urea, followed by measurement of optical rotation at 365 nm. Temperature 25°C, pH 2.77, ionic strength 0.15. The plateau of the reduced specific optical rotation at −310° corresponds to the completely unfolded form of the protein. [From N. C. Pace and C. Tanford, *Biochemistry*, **7**, 198 (1968).]

interesting thermodynamic analysis has been made using the temperature and urea concentration dependence of the denaturation process; this will be discussed in more detail later in this chapter. Urea and guanidine-HCl denaturations have also been useful in obtaining information about the interior of protein molecules of unknown structure. The determination of numbers of "buried" tyrosine residues has been illustrated (Fig. 9-3) as has the estimation of the number of exchangeable hydrogen atoms (Fig. 11-1). Titration studies in the presence and absence of urea also give information on accessibility of certain side chains in the protein molecule.

1-5 Salt Effects on Conformational Stability[12]

The effect of salts on protein structure is of particular interest to biological scientists because of the common application of "salting-out" methods in protein purifications (cf. Chapter 3). At low concentrations the ions of neutral salts (those that do not cause a pH change in aqueous solution) can be expected to shield the charged groups of proteins and thus have an effect on electrostatic interactions between those groups. Actually, physiological salt concentrations and typical ionic strengths used *in vitro* (0.05–0.20), even as low as 0.01, are usually sufficient to minimize these interactions. As a result, electrostatic forces are considered to play no significant role in determining protein conformation.[4] Yet it is an experimental fact that certain salts have a large effect on protein structure and association equilibria. These effects are specific to particular cations and anions and prove to have a variety of applications. The subject has been reviewed in depth by von Hippel and Schleich.[12]

A convenient method to test for the stabilizing or destabilizing action of added components is to observe their effect on a conformational transition induced by another environmental perturbant. The temperature melting curve for collagen as measured by optical rotation is shown for several concentrations of calcium chloride in Fig. 15-7. The temperature T_m at the midpoint of the transition markedly decreases with increasing $CaCl_2$ concentration, identifying this salt as one that promotes conformational destabilization. In a similar study on ribonuclease a variety of effects were found for different salts. As shown in Fig. 15-8, some salts (e.g., ammonium sulfate) acted to stabilize the conformation as exemplified by an increased T_m, others (NaCl, KCl) had virtually no effect, while most salts tested had a destabilizing effect. Both cation and anion contribute to these effects, generally in an additive fashion. A major problem in interpreting these results is the fact that the native and unfolded forms whose transition is being studied are not the same in the various salt solutions. This is indicated by a general shift in observed specific rotation with increasing salt concentrations. For example, in high salt the unfolded conformation of ribonuclease is considered to be less expanded than a random coil. The folded or "native" form at high salt concentrations (e.g., 2 *M*) also may not be identical with the native conformation that exists at 0.1 *M*.

As experimental data have accumulated on the effects of salts on various processes, it has become clear that the relative effectiveness of ions in stabilizing native proteins against conformational changes is the same as that noted in related processes, such as the stabilization of nucleic acids and the salting-out of proteins. Thus stabilizing action follows the classical Hofmeister series that ranks the effectiveness of ions in salting-out proteins. Some of these results are summarized in Fig. 15-9 where it is noted that increased

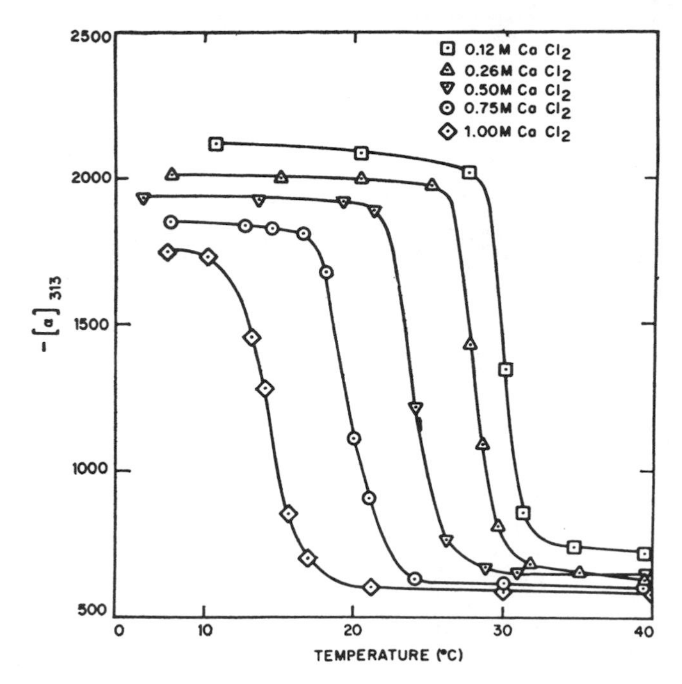

Fig. 15-7 Effect of calcium chloride (a conformation-destabilizing salt) on the thermal denaturation of ichthyocol (carp swim-bladder) collagen at pH 7, followed by optical rotation measurements at 313 nm. [From P. H. von Hippel and K.-Y. Wong, *Biochemistry*, **2,** 1387 (1963).]

conformational stability is associated with loss of solubility (salting-out) whereas decreased stability is associated with greater solubility of the protein (often with denaturation). Although these empirical relations for neutral salts are well established, the mechanism of salt interaction with protein conformations is still obscure. There are two general categories of interaction that have been considered, one being a direct ionic interaction with the protein (the polar groups of the polypeptide backbone that are internalized in the native structure), the second involving indirect effects through changes in solvent structure. The latter would include effects of the ions on interactions between water molecules and apolar side chains of the protein. The problem of mechanism of salt action is, of course, related to the general problem of protein folding, which has still not been adequately explained in terms of small molecule interactions (cf. Chapter 6).

The empirical application of salt perturbation in protein systems has been effectively demonstrated in a variety of studies. Subunit dissociation has been carried out with the use of $CaCl_2$ or KSCN. Under suitable conditions these salts bring about the disruption of quaternary structure of proteins without damage to the somewhat more stable tertiary structure of

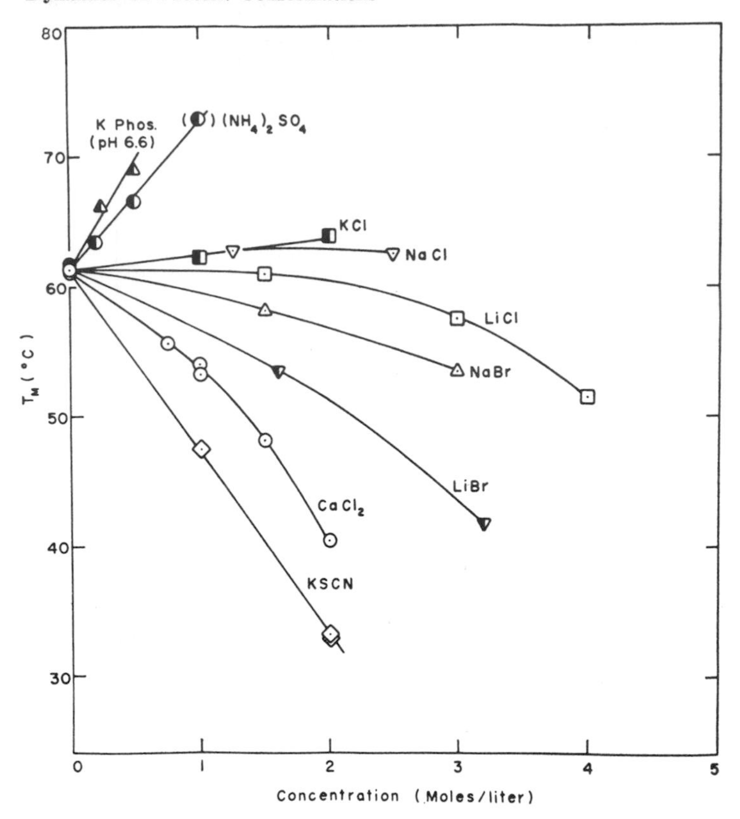

Fig. 15-8 Effect of various salts on the melting temperature T_m for the thermal denaturation of ribonuclease at pH 7.0. All solutions contained 5 mg/ml ribonuclease, 0.15 *M* KCl, and 0.013 *M* sodium cacodylate. [From P. H. von Hippel and K.-Y. Wong, *J. Biol. Chem.*, **240,** 3909 (1965).]

the polypeptide chains. Concentrated salt solutions have proved to be effective in isolation of proteins from complexes with other substances maintained by noncovalent interactions. Salting-out methods have permitted the fractionation and concentration of proteins, often without any significant loss of biological activity. A variety of examples of these methods is discussed and referenced by von Hippel and Schleich.[12]

1-6 Effects of Detergents, Aromatic Rings, and Other Compounds

Long-chain fatty acids such as lauric acid or the corresponding detergents such as sodium dodecylsulfate (SDS) react readily with proteins, frequently

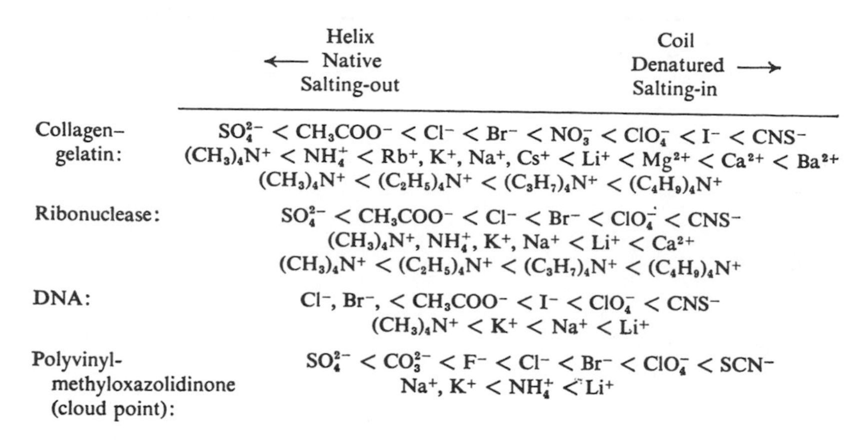

	← Helix, Native, Salting-out	Coil, Denatured, Salting-in →
Collagen–gelatin:	$SO_4^{2-} < CH_3COO^- < Cl^- < Br^- < NO_3^- < ClO_4^- < I^- < CNS^-$	
	$(CH_3)_4N^+ < NH_4^+ < Rb^+, K^+, Na^+, Cs^+ < Li^+ < Mg^{2+} < Ca^{2+} < Ba^{2+}$	
	$(CH_3)_4N^+ < (C_2H_5)_4N^+ < (C_3H_7)_4N^+ < (C_4H_9)_4N^+$	
Ribonuclease:	$SO_4^{2-} < CH_3COO^- < Cl^- < Br^- < ClO_4^- < CNS^-$	
	$(CH_3)_4N^+, NH_4^+, K^+, Na^+ < Li^+ < Ca^{2+}$	
	$(CH_3)_4N^+ < (C_2H_5)_4N^+ < (C_3H_7)_4N^+ < (C_4H_9)_4N^+$	
DNA:	$Cl^-, Br^-, < CH_3COO^- < I^- < ClO_4^- < CNS^-$	
	$(CH_3)_4N^+ < K^+ < Na^+ < Li^+$	
Polyvinyl-methyloxazolidinone (cloud point):	$SO_4^{2-} < CO_3^{2-} < F^- < Cl^- < Br^- < ClO_4^- < SCN^-$	
	$Na^+, K^+ < NH_4^+ < Li^+$	

Fig. 15-9 A ranking of the relative effectiveness of various ions in promoting stable macromolecular conformations (helix, native conformation, salting-out) and in destabilization (random coil or denatured state, salting-in).[12]

causing dissociation into subunits and denaturation of the individual polypeptide chains. Unlike such denaturing agents as urea and guanidine hydrochloride, they are effective at very low concentrations and show a high degree of strong binding to the protein.

The disruption of proteins with SDS has had extensive application and is particularly useful as a prelude to the estimation of molecular weights and subunit stoichiometry by gel electrophoresis (cf. Chapter 10). The binding of SDS to several proteins has been studied by Reynolds and Tanford.[13] Their results indicate that the degree of binding is remarkably similar on a weight for weight basis for a variety of proteins. Binding is a function of the free SDS concentration in solution. This is not necessarily the same as the total concentration since SDS solutions form micelles to varying extents depending on total concentration, pH, ionic strength, and temperature. Typical results for SDS binding to proteins are shown in Fig. 15-10. Two plateau regions are observed, corresponding to 0.4 and 1.4 g SDS bound per gram of protein.

Physical studies of SDS–protein complexes (for noncrosslinked proteins at high binding (about 1.4 g/g) suggest a model of a thin rod of constant diameter and length proportional to the length of the chain. The dimensions are consistent with a helixlike conformation with SDS molecules intercalated along the length of the helix. At very low levels of SDS binding, marked conformation changes do not occur generally; indeed, several proteins have been reported to be stabilized against denaturation by other perturbants as a result of the presence of SDS at low concentration. It seems likely that the early stages of binding involve apolar interactions of the aliphatic side chain

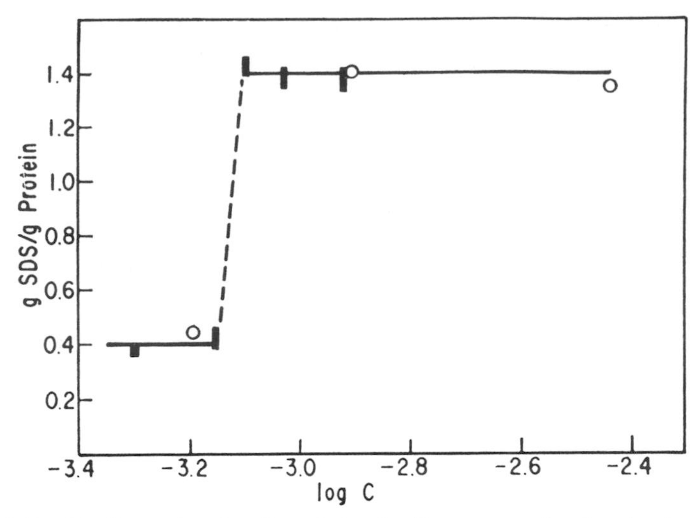

Fig. 15-10 Binding ratio of sodium dodecyl sulfate to proteins for a variety of different proteins, as a function of the logarithm of the equilibrium concentration of SDS in the monomer form.[13]

of SDS with interior domains of the protein, leaving the charged group at the surface to interact with solvent. Although apolar interactions probably play an important role in the stabilization of the limit SDS–protein complex as well, the interactions are of a more complicated nature since they almost certainly entail strong binding all along the polypeptide chain, regardless of the polarity of the side chains. The extensive binding of detergent produces a strong solubilization effect since the highly charged surface of the complex tends to repel other polypeptide chains and prevent aggregation. Positively-charged detergents (e.g., cetyl trimethyl ammonium bromide) and neutral detergents (Triton X-100: isooctoylphenoxypolyethoxy ethanol) have also been used to disrupt proteins or to solubilize them for removal from membranes and other structural components. The neutral detergents are useful for mild disruption of viruses as well.

Strong interaction with proteins is also shown by a class of compounds that contains ring structures. At high concentrations, steroid hormones and their analogs and other polycyclic compounds such as anthracene and phenanthrene cause protein denaturation, possibly through binding to apolar regions of the protein molecule to produce steric interference with normal apolar stabilization. At lower concentrations, some of these compounds may act more specifically, influencing association–dissociation equilibria of protein subunits[14] or affecting substrate specificity of enzymes. Although no proof of physiological significance has yet been established in these studies, the possibility exists that steroids and other hormones with apolar regions (e.g., thyroxine) may function biologically by causing small conformational changes in one or more particular proteins.

Many other small molecular weight compounds, some of which are found in living systems, are capable of strong interactions with proteins *in vitro*. The extent to which such effects have physiological significance is not known in all cases; however, information is accumulating to suggest that conformational changes in response to such interactions may provide a general mechanism for biological regulation. Studies in this area will be discussed in the last part of this chapter. Another important class of protein associations includes those with nucleic acids, lipids and carbohydrates in which complexes are formed involving noncovalent interactions and possible conformational alterations. A systematic study of the precise nature of these interactions as well as their physiological role is still needed.

1-7 Renaturation[3]

The term renaturation refers to the return of denatured portions of a macromolecule to the original three-dimensional geometry present before denaturation. Three general types of renaturation may be distinguished. In complete renaturation, total restoration of the native form is achieved even after denaturation to a random coil conformation. In partial renaturation, species are obtained that contain major portions of the original conformation but that retain regions of altered structure. Another type of partial renaturation is also possible in which a mixture of products is obtained, some essentially identical to the native protein and others with markedly different properties. This situation results when there are multiple pathways for folding, leading to products that are frozen into metastable conformations. Because of large activation energy barriers, these products are converted to the native form slowly or not at all. An example of such a state is the inactive form of ribonuclease obtained by the formation of random disulfide bridges in the presence of urea (cf. Chapter 6). This form is converted only very slowly by thiol reagents to the native conformation, although rapid recovery can be achieved by addition of an enzyme that catalyzes disulfide interchange. In most cases, however, frozen-in inactive states of proteins will be primarily stabilized by large numbers of incorrect noncovalent contacts. No enzyme is known that catalyzes exchange of noncovalent bonds to correct this situation. Presumably, inactive states of this type occur to a minimal degree *in vivo* because the environmental conditions required for correct tertiary structure formation are more adequately met.

Exactly what happens at the site of synthesis of polypeptide chains *in vivo* is important in this respect. It must be considered highly likely that some kind of folding begins before synthesis of the polypeptide chain is completed so that the cell never has to cope with the problem of forming native proteins from chains that are entirely in an unfolded or random coil conformation. Polypeptide chains folded in this way then may not necessarily represent the state of lowest free energy for the molecule as a whole, but could fall in the category of "frozen-in" state. It would then be understandable why it has not been possible to renature most proteins *in vitro* from the random coil conformation. On the other hand, the final conformation of several known proteins is not really consistent with an order of folding in the direction of synthesis (*N*-terminal to *C*-terminal), at least not if this is an irreversible process.[15] Even if folding from the *N*-terminus were initiated during synthesis, there is evidence to indicate that conformational rearrangement to attain a lower free energy state might still occur after release from the ribosome (e.g., see discussion by Teipel and Koshland[16] and by Ikai and Tanford[17]).

In spite of the expected difficulties in the renaturation of proteins, a good deal of attention has been given to the problem. The earliest and most thoroughly studied case has been ribonuclease.[18] Denaturation of the protein was carried out in 8 *M* urea and β-mercaptoethanol under mildly alkaline conditions. The four disulfide bridges were reduced by this treatment and the physical behavior of the protein was that of a random coil (e.g., $[\eta] = 14$ cc/g). Upon dialysis of the preparation to remove the urea, and in the presence of oxygen to permit reformation of disulfide bonds, nearly complete recovery of enzymatic activity was achieved. Peptide mapping showed that the pairing of cysteines in disulfide linkage was identical to that in the original material; crystals of the renatured protein gave the same X-ray diffraction pattern as native ribonuclease.

Renaturation experiments have also been conducted on chemically or enzymatically altered RNAse. This type of approach permits a systematic study of the relationships among sequence, conformation, and thermodynamic parameters. Some chemical modifications that were used are listed in Table 15-4. As expected the enzymatic activity of RNAse is decreased by some substitutions (possibly reflecting conformational changes or an altered charge distribution that changes the probability for effective collisions with the charged substrate); however, the degree to which the modified enzyme may be renatured (in most cases) is quite amazing. The reacting groups are surface located (as obtained from X-ray studies and as indicated by ease of reaction), and apparently play little role in determining conformation. Introducing apolar groups gave similar results (except for the very bulky DNS group).

A fragment containing the first 20 amino acid residues (from the *N*-terminal) can be removed from pancreatic ribonuclease at low pH following hydrolysis of a single peptide bond by the enzyme subtilisin. The remaining protein (*S*-protein), which is inactive, can be denatured to an unfolded,

Table 15-4 Effect of chemical modification of ribonuclease on enzymatic activity and on the capacity for renaturation by reoxidation after reduction of the modified protein.[a]

Preparation	Group covered	Number of groups covered (average)	Charge substitution	Activity of protein (% of native)	Activity after reoxidation (% of initial)
Succinyl ribonuclease	$—NH_2$	6–7 of 11	− for +	15	62
Methylated ribonuclease	—COOH	7–8 of 11	0 for −	18	48
Methylated succinyl ribonuclease	$—NH_2$ —COOH	5–6 of 11 7–8 of 11	− for + 0 for −	2	73
DNS ribonuclease[b]	$—NH_2$	2 of 11	0 for +	100	15
Phthalyl ribonuclease	$—NH_2$	3–4 of 11	− for +	24	75
Butyryl ribonuclease	$—NH_2$	5–6 of 11	0 for +	44	77
Caproyl ribonuclease	$—NH_2$	2–3 of 11	0 for +	96	70

[a] From C. J. Epstein, R. F. Goldberger, and C. B. Anfinson, *Cold Spring Harbor Symp. Quant. Biol.*, **28,** 446 (1963).
[b] DNS = 5-dimethylamino-1-naphthalene sulfonyl.

reduced state and then renatured with re-formation of the correct disulfide linkages. This finding demonstrates a capacity, in ribonuclease at least, for proper folding of a part of the molecule without the obligatory presence of the entire polypeptide chain. The *S*-protein is readily converted to active enzyme by adding back the cleaved fragment (*S*-peptide) at neutral pH. Except for the missing peptide bond, the resulting protein (ribonuclease *S*′) is essentially identical in structure to the native protein (cf. Chapter 14). Studies of the thermodynamics of recombination of the *S*-protein and *S*-peptide indicate a strong temperature dependence of all parameters including ΔC_p.[19] Use of the known three-dimensional structure of ribonuclease-*S* permits some predictions to be made on probable hydrogen and hydrophobic bonding. However exact interpretation of the energetics of the process has not yet been possible.

Reconstitution of multiple chain proteins after tertiary structure disruption can be illustrated by the examples of aldolase and *Escherichia coli* alkaline phosphatase. Aldolase has a molecular weight of about 160,000 and contains four polypeptide chains of similar size. It contains no disulfide bonds. After dissociation into separated random-coil-like chains in a denaturing solvent (e.g., 4 *M* urea or pH about 2), it was possible to renature the protein by dialysis against a suitable buffer and recover as much as 70% of the original activity.[20] At higher protein concentrations (5 mg/ml), the conditions of renaturation were critical. For example, a rapid change of pH from 2 to 5.3 produced inactive aggregates (frozen-in states). At low protein concentrations such aggregates did not form and rapid reconstitution occurred. During reconstitution a species of renatured protein appeared that differed somewhat from the original native material, as indicated by its instability at 37°. This result is one of many that suggests, in larger proteins at least, regions of the protein (in this case, the active site) may reform correctly even though other regions may not be in the fully native conformation. Such species may, in some cases, reflect intermediates in a kinetic process that will eventually produce correctly folded molecules.

Another protein that has proved amenable to this type of study is alkaline phosphatase.[21] The native protein as isolated from *E. coli* is a dimer of molecular weight 86,000, consisting of two identical subunits. Denaturation in 6 *M* guanidine hydrochloride or 0.01 *M* HCl produces unfolded polypeptide chains that become refolded when the solvent is changed to a neutral buffer. The folded monomers are able to reassociate spontaneously to form a dimer; in the presence of Zn^{2+}, four Zn^{2+} are bound per molecule, with the restoration of enzymatic activity. It would thus appear that in this case, as in the earlier examples, it is possible to find *in vitro* conditions that mimic cellular conditions sufficiently to permit formation of an active protein from the separate structureless polypeptide chains. All of these results support the view that polypeptide primary structure alone determines a protein's conformation, provided that required environmental conditions are satisfied.

Many additional examples of protein renaturation have been reported. In studies of the kinetics of refolding of oligomeric proteins from the random coil conformation, it is found that refolding of major structural elements, as evidenced by optical rotation and fluorescence, may be complete in 1 minute, whereas restoration of biological activity occurs more slowly.[16] The latter was found to be influenced by a variety of environmental factors such as the presence of substrates and cofactors, ionic strength, and protein concentration. Thus the final recovery of enzymatic activity is variable. Further discussion of these results and references to related studies in which the formation of conformational isomers is demonstrated are given by Teipel and Koshland.[16]

Renaturation of quaternary structure after dissociation by mild perturbing influences that cause little tertiary structure alteration is often more successful than renaturation from random coils. Such reversible denaturation is observed, for example, in the cold inactivation of enzymes, which we have called attention to earlier. The dissociation of carbamyl phosphate synthetase from *E. coli* provides another interesting example.[22] Potassium thiocyanate has been used to dissociate the enzyme into two subunits of unequal size, which may be separated. After removal of the denaturant, the large subunit is found to contain full activity for catalyzing carbamyl phosphate synthesis provided ammonia is supplied as the amide donor. The reaction is also subject to allosteric control by ornithine and inosine monophosphate, as in the native enzyme. However, the large subunit alone is unable to utilize the normal amide donor glutamine in the synthesis of carbamyl phosphate. The small subunit is found to possess the binding site for glutamine and shows a low level of glutaminase activity. Recombination of subunits restores the native quaternary structure and all enzymatic properties. This system provides an interesting model for the evolutionary development of quaternary structure in proteins and of control mechanisms.

Reversible denaturation systems, particularly those that fit a simple two-state process (native structure ⇌ random coil), are especially important in evaluating various models for folding and for the strength of noncovalent interactions. These studies have been thoroughly reviewed by Tanford,[3] and that reference should be consulted for detailed consideration of experimental approaches and data interpretation. An interesting example of this type of study is the temperature dependence of denaturation of β-lactoglobulin at various concentrations of urea (Fig. 15-11). Here, denaturation is assessed from optical rotation measurements by the increase in $-[\alpha']$. A minimum in the curve at about 35°C suggests that this is the temperature of maximum stability for the protein. Extrapolation to zero urea concentration permits the inference that the native protein has maximum stability near 35° as well.

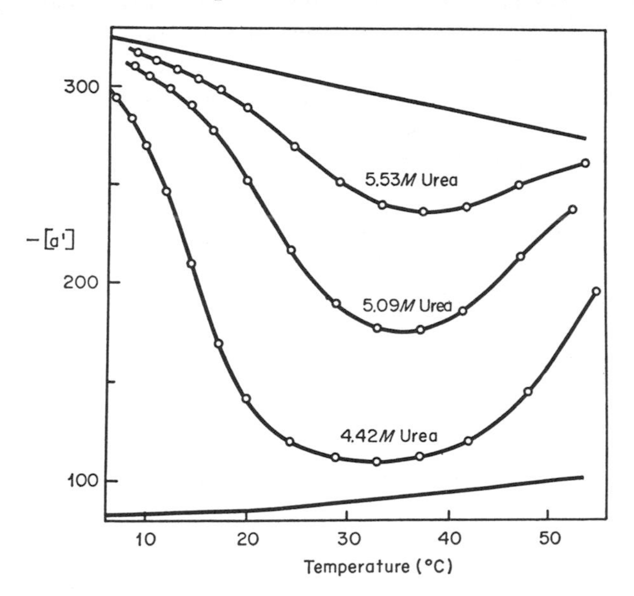

Fig. 15-11 Effect of temperature on the stability of β-lactoglobulin at various concentrations of urea, as measured by optical rotation at 365 nm. The lines at the top and bottom represent the rotations found for completely denatured and completely native protein, respectively. [From N. C. Pace and C. Tanford, *Biochemistry*, **7,** 198 (1968).]

Measurement of the equilibrium constant in the absence of urea is not possible at this temperature because a negligible proportion of random-coil form is present. The results also show an anomalously large ΔC_p of denaturation [about 2.1 kcal/(deg)(mole)] which is nearly constant over the temperature range of 15 to 55°C. These data have implications in regard to the role of apolar bonding in the native protein (cf. Chapter 6). A similar approach is possible in the study of subunit interactions. Here, also, the equilibrium is usually so far in favor of the oligomer under physiological conditions that it is necessary to induce dissociation by environmental perturbants in order to measure K. Such results have been reviewed by Klotz and associates[23] (see also Chapter 16).

2 PROTEIN–LIGAND INTERACTIONS; ALLOSTERY

From our consideration of protein denaturation processes, it is clear that proteins have a large number of conformational possibilities available to them. It may be expected that some of this conformational flexibility has evolved as the structural basis for advantageous functional properties. For example, comparison of oxygenation curves of hemoglobin and myoglobin reveals that the former protein, with its multiple subunits, is better adapted for reversible oxygen transport than a protein like myoglobin would be (compared to myoglobin, hemoglobin has less affinity for oxygen: it thus gives up O_2 more readily at low O_2 levels, as in most body tissues, but binds strongly at high O_2 levels, as in the lungs).

The sigmoid shape of the oxygen dissociation curve of hemoglobin compared with the simple hyperbolic curve of myoglobin (Fig. 15-12) is

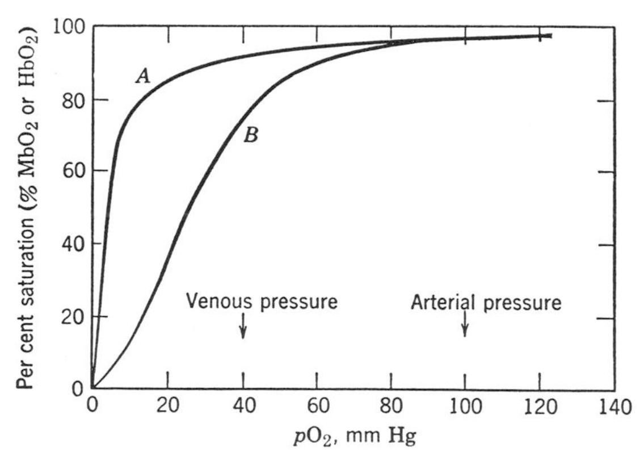

Fig. 15-12 Oxygen dissociation curves of myoglobin (A) and hemoglobin (B) as a function of partial pressure of oxygen. (From J. S. Fruton and S. Simmonds, *General Biochemistry*, Wiley, New York, 1961.)

consistent with the existence of some kind of linkage among the four oxygen-accepting sites which causes the equilibrium constant for oxygenation at any one site to be influenced by oxygenation of other sites. Direct steric interaction between heme groups is excluded by the finding of wide separations between hemes in the hemoglobin structure (the shortest distance between iron atoms is 25 Å). Instead, the crystallographic results indicate that a change in quaternary structure involving rotation of globin subunits is associated with oxygenation (cf. Chapter 14).

Other types of studies also led to the realization that conformational changes in proteins were implicated in functional behavior, particularly among enzymes capable of complicated regulatory controls. Classical work on phosphorylase *b* from muscle had indicated that enzyme activity may be influenced by interaction with small molecules at sites other than the active site. This property of proteins was also demonstrated by studies of feedback inhibition, where the product of a sequence of reactions acted to inhibit activity of an enzyme functioning early in the pathway and thus turn off the entire pathway. Consideration of these findings led Monod, Changeux, and Jacob to propose a theory of allosteric interactions.[24] By this, they meant that the interaction or binding of a small molecule (ligand) at one site on a protein molecule could induce a transition or shift in equilibrium toward a different conformational state of the protein, with resultant modification of function at another site (e.g., the active site) on the molecule. For example, in feedback inhibition, when an excess of a metabolic product has accumulated, its binding to an enzyme functioning early in the pathway causes a conformational change in that enzyme and loss of activity. The process is reversed when the concentration of the product falls to a low level. The regions where interactions occur, either with regulator molecules or substrates, may be quite distant. Thus the effect of a particular ligand on the activity of another site is an indirect one, mediated by the allosteric transition from one allowed conformation of the protein to another.

The idea in its most general form, that two or more areas within a molecule are subject to a "linkage" or interdependence with respect to conformational changes, provides a useful approach to interpretation of numerous phenomena observed in protein systems. In the model of Monod and his associates,[24,25] two classes of allosteric effects were distinguished. Homotropic effects are those in which only one type of ligand is involved (i.e., where the binding of one molecule of this ligand influences the binding of another molecule of the same ligand at the comparable site on another subunit). Heterotropic effects are those of the regulator type, where binding of a regulator ligand at one site affects the binding of a different ligand (e.g., the substrate) at another site. In general, allosteric proteins are expected to have multiple subunits, although this requirement is absolute only in the case of homotropic effects (where the subunits are identical or very nearly identical). It was also proposed that the subunits of an allosteric protein should be related to each other by simple symmetry operations (cf. Chapter 16), these symmetry elements being conserved in the allosteric transition. The term *protomer* was applied to the molecular entity or subunit subject to the symmetry operations (comparable to the asymmetric unit in crystallography). For heterotropic effects there may be different subunits for the binding of regulator and substrate molecules; one polypeptide chain (or other subunit) of each type may be contained in the protomer of the allosteric protein molecule.

2-1 Oxygenation of Hemoglobin

The allosteric system that has received the most detailed analysis is that of the oxygenation of hemoglobin.[26] As noted above, structural differences between oxy- and deoxyhemoglobin have been described that are consistent with the notion of an allosteric transition; homotropic effects also are indicated by the sigmoidal oxygen saturation curve (Fig. 15-12). These results are interpreted by postulating that the molecule exists in two forms, one having all subunits in one conformation, the other having all subunits in another conformation. With the assumption that one form binds oxygen more strongly than the other does, it is possible to obtain a fit of the oxygen saturation data to the theoretical equation of the model.[25]

Another physiological function of hemoglobin is the transport of carbon dioxide. When oxygen is released from oxyhemoglobin at pH 7.4, protons are taken up (nearly 1 mole per mole of oxygen released), and combination with carbon dioxide occurs through a carbamylation reaction. Because of the relationship between proton-affinity and the oxygen equilibrium, hemoglobin acts as a buffering agent in the blood (neutralizing carbonic acid). This mechanism is responsible for the common observation that oxygen affinity of hemoglobin is reduced at lower pH (*Bohr effect*). The effect of protons on the oxygen binding can be regarded as a heterotropic allosteric effect. Both this and the homotropic effect of oxygen are lacking in hemoglobin H, a form consisting of four β chains; thus the presence of multiple subunits is not alone sufficient for the generation of allosteric effects. Wyman suggests that interactions within the $\alpha\beta$ pair of chains play the major part in the allosteric behavior of normal mammalian hemoglobin.[26] His analysis indicates that all of the oxygenation data of hemoglobin, both homotropic and heterotropic effects, can be adequately accounted for by allosteric transition between conformational states, provided that true equilibrium is assumed to exist. The complete tetramer, however, appears to be required for the effect. Cooperative effects are not found in dimers produced by the symmetrical dissociation of hemoglobin ($\alpha_2\beta_2 \rightleftharpoons 2\alpha\beta$) in high salt or in modified $\alpha\beta$ dimers produced by removal of *C*-terminal arginine of the α-chains. It has been suggested that constraints imposed by salt bridges involving terminal residues in the tetrameric form are critical for cooperation in deoxyhemoglobin.[27] The reader is referred to these references and to several reviews[28] for a thorough discussion of the linked functions of hemoglobin.

2-2 Aspartate Transcarbamylase

Allosteric behavior has been demonstrated in a number of cases, primarily with enzymes that are subject to feedback control. A particularly well documented example, which will serve to illustrate some pertinent points, is aspartate transcarbamylase (ATCase). ATCase catalyzes the first step of a metabolic pathway that leads eventually to pyrimidine biosynthesis. The enzyme is inhibited by a product of the pathway, cytidine triphosphate (CTP), and is activated by adenosine triphosphate (ATP).[29] Structural studies have shown that the enzyme (molecular weight = 310,000) is composed of two distinct types of subunits that dissociate upon treatment of the enzyme with a mercurial reagent, *p*-mercuribenzoate.[30] The dissociated mixture still has catalytic activity but is not subject to inhibition by CTP. Upon isolation of the subunits, one type (the catalytic subunit) was found to possess the enzymatic activity of the original molecule, the other type (the regulatory subunit) was able to bind CTP. Renaturation of the original molecule was possible; under appropriate conditions the subunits recombine in good yield to restore the original properties of the enzyme, including inhibition of activity by CTP. By the use of bromocytidine triphosphate (with strong ultraviolet absorption at 298 nm) in place of the usual inhibitor, Gerhart and Schachman were able to make effective use of the ultraviolet optical system of the analytical ultracentrifuge to assay binding of inhibitor to the subunits.[30]

Further studies of the ATCase system have shown that allosteric effects of both the homotropic and heterotropic types occur. Each of the two kinds of subunits obtained by mercurial cleavage contain multiple identical polypeptide chains and multiple interaction sites. The catalytic subunit (molecular weight about 96,000) consists of three polypeptide chains; the regulatory subunit (molecular weight about 34,000) contains two chains. The complete molecule consists of two catalytic and three regulatory subunits for a total of six chains of each type.[31] Clearly, this is a molecule of impressive structure, and it is not surprising to find it capable of many complex interactions. As with other regulatory proteins, it exhibits a sigmoidal dependence of reaction velocity on substrate concentration, suggestive of cooperative effects between multiple substrate sites.[29]

Direct evidence has been obtained for an allosteric conformational change in the enzyme which is induced by substrate and antagonized by CTP. The ATCase data have been successfully interpreted in terms of the allosteric theory in which coordinated transitions occur between two conformational states.[32] The results of this analysis suggest that only one of the two conformational states accessible to the enzyme is able to bind the substrate; the other conformational state has negligible substrate binding capacity but binds CTP about twofold more strongly than the first state does. The activator (ATP) apparently acts to stabilize the catalytically-active conformation.[33] Thus the level of enzyme activity depends upon the equilibrium between the two conformational states, which, in turn, depends upon substrate, activator, and inhibitor concentrations.

2-3 Alternative Models

The analysis of hemoglobin binding and other systems in terms of allosteric transitions has been further extended theoretically.[34] It is found that data from ligand saturation curves are not sufficient to distinguish between a number of alternative models involving different geometries and different types of subunit interaction. For example, the transition between conformations may be "concerted" (all subunits shifting simultaneously to the new conformation) or "sequential" (a stepwise transition involving a series of intermediates). This is no sharp demarcation between the two types, but the former may be associated with transitions that principally involve quaternary structure changes with little modification of subunit tertiary structure. More extensive tertiary structure changes are likely to produce sequential transitions.

The theoretical treatment of protein–ligand interactions in terms of multiple equilibria is the same as that already discussed in relation to hydrogen ion equilibria (cf. Chapter 10). The general equation expressing the number of moles of ligand bound per mole of protein ($\bar{\nu}$) in terms of the equilibrium constants for consecutive associations (Eq. 10-5) has been reduced to fit a number of specific allosteric models.[35] For example, in the case treated by Monod, Wyman, and Changeux,[25] the allosteric protein is a tetramer of identical subunits which can exist in two possible conformations A and B. The tetrameric forms are interconverted by a concerted transition, that is, $A_4 \rightleftharpoons B_4$. If both conformations are capable of binding the ligand (as described by association constants K_A and K_B), then an expression for $\bar{\nu}$ as a function of ligand concentration is obtained in terms of three unknowns, K_A and K_B for the ligand associations and K_W, the equilibrium constant of the conformational transition. The results for this and other current models are given by Magar and Steiner.[35] Although it is possible to obtain a unique set of parameters to fit certain models for allosteric interaction, the analyses ordinarily do not rule out other possible models, even when equilibrium situations are the only ones considered. Flexibility in the conformation of the ligand introduces additional complications.[36]

Kinetic models provide another means of analysis of regulatory properties of proteins. It has been pointed out that it is not necessary to invoke subunit interactions in an enzyme in order to explain a sigmoid relationship between reaction velocity and substrate concentration. Alternative models have been proposed in which active sites are independent, but there exists more than one reaction pathway leading to substrate binding at the sites.[37] A more extensive analysis of kinetic variation has been developed in terms of a concept of hysteresis in enzyme properties.[38] According to this theory, a delay in response by a protein to a rapid change in ligand concentration could bring about a time-dependent buffering of metabolites that might play an important role in the control of pathways utilizing certain protein ligands in common.

REFERENCES

Protein Denaturation

1. E. Katchalski, M. Sela, H. I. Silman, and A. Berger, "Polyamino acids as protein models," in *The Proteins*, Vol. 2, H. Neurath, Ed., Academic Press, New York, 1964.
2. G. D. Fasman, Ed., *Poly-α-amino Acids* (Biological Macromolecules Series, Vol. 1), Marcel Dekker, Inc., New York, 1967.
3. C. Tanford, "Protein denaturation, Parts A and B," *Adv. Protein Chem.*, **23,** 121 (1968).
4. C. Tanford, "Protein denaturation, Part C," *Adv. Protein Chem.*, **24,** 1 (1970).
5. S. N. Timasheff and G. D. Fasman, Eds., *Structure and Stability of Biological Macromolecules* (Biological Macromolecules Series, Vol. 2), Marcel Dekker, Inc., New York, 1969.
6. K. Banerjee and M. A. Lauffer, "Polymerization-depolymerization of tobacco mosaic virus protein. VI. Osmotic pressure studies of early stages of polymerization," *Biochemistry*, **5,** 1957 (1966).
7. J. J. Irias, M. R. Olmsted, and M. F. Utter, "Pyruvate carboxylase. Reversible inactivation by cold," *Biochemistry*, **8,** 5136 (1969).
8. S. J. Singer, "The properties of proteins in nonaqueous solvents," *Adv. Protein Chem.*, **17,** 1 (1962).
9. I. M. Klotz and J. S. Franzen, "Hydrogen bonds between model peptide groups in solution," *J. Amer. Chem. Soc.*, **84,** 3461 (1962).
10. D. R. Robinson and W. P. Jencks, "The effect of compounds of the urea-guanidinium class on the activity coefficient of acetyltetraglycine ethyl ester and related compounds," *J. Amer. Chem. Soc.*, **87,** 2462 (1965).

11. D. B. Wetlaufer, S. K. Malik, L. Stoller, and R. L. Coffin, "Nonpolar group participation in the denaturation of proteins by urea and guanidinium salts. Model compound studies," *J. Amer. Chem. Soc.*, **86,** 508 (1964).

12. P. H. von Hippel and T. Schleich, "The effects of neutral salts on the structure and conformational stability of macromolecules in solution," in *Structure and Stability of Biological Macromolecules*, S. N. Timasheff and G. D. Fasman, Eds., Marcel Dekker, Inc., New York, 1969.

13. J. A. Reynolds and C. Tanford, "Binding of dodecyl sulfate to proteins at high binding ratios. Possible implications for the state of proteins in biological membranes," *Proc. Nat. Acad. Sci., U. S.*, **66,** 1002 (1970).

14. E. Reisler, J. Pouyet, and H. Eisenberg, "Molecular weights, association, and frictional resistance of bovine liver glutamic dehydrogenase at low concentrations. Equilibrium and velocity sedimentation, light-scattering studies, and settling experiments with macroscopic models of the enzyme oligomer," *Biochemistry*, **9,** 3095 (1970).

15. M. F. Perutz, "X-ray analysis, structure and function of enzymes," *Eur. J. Biochem.*, **8,** 455 (1969).

16. J. W. Teipel and D. E. Koshland, Jr., "Kinetic aspects of conformational changes in proteins. I. Rate of regain of enzyme activity from denatured proteins; II. Structural changes in renaturation of denatured proteins," *Biochemistry*, **10,** 792, 798 (1971).

17. A. Ikai and C. Tanford, "Kinetic evidence for incorrectly folded intermediate states in the refolding of denatured proteins," *Nature*, **230,** 100 (1971).

18. F. H. White, "Regeneration of native secondary and tertiary structures by air oxidation of reduced ribonuclease," *J. Biol. Chem.*, **236,** 1353 (1961).

19. R. P. Hearn, F. M. Richards, J. M. Sturtevant, and G. D. Watt, "Thermodynamics of the binding of *S*-peptide to *S*-protein to form ribonuclease *S'*," *Biochemistry*, **10,** 806 (1971).

20. E. Stellwagen and H. K. Schachman, "The dissociation and reconstitution of aldolase," *Biochemistry*, **1,** 1056 (1962).

21. J. A. Reynolds and M. J. Schlesinger, "Conformational states of the subunit of *Escherichia coli* alkaline phosphatase," *Biochemistry*, **6,** 3552 (1967).

22. P. P. Trotta, M. E. Burt, R. H. Haschemeyer, and A. Meister, "Reversible dissociation of carbamyl phosphate synthetase into a regulated synthesis subunit and a subunit required for glutamine utilization," *Proc. Nat. Acad. Sci., U.S.*, **68,** 2599 (1971).

23. I. M. Klotz, N. R. Langerman, and D. W. Darnall, "Quaternary structure of proteins," *Ann. Rev. Biochem.*, **39,** 25 (1970).

Protein–Ligand Interactions; Allostery

24. J. Monod, J. P. Changeux, and F. Jacob, "Allosteric proteins and cellular control systems," *J. Mol. Biol.*, **6,** 306 (1963).

25. J. Monod, J. Wyman, and J. P. Changeux, "On the nature of allosteric transitions: A plausible model," *J. Mol. Biol.*, **12,** 88 (1965).

26. J. Wyman, "Allosteric linkage," *J. Amer. Chem. Soc.*, **89,** 2202 (1967).

27. J. A. Hewitt, J. V. Kilmartin, L. F. Ten Eyck, and M. F. Perutz, "Noncooperativity

of the $\alpha\beta$ dimer in the reaction of hemoglobin with oxygen," *Proc. Nat. Acad. Sci., U.S.*, **69,** 203 (1972).

28. J. Wyman, Jr., "Linked functions and reciprocal effects in hemoglobin: A second look," *Adv. Protein Chem.*, **19,** 223 (1964). A. Rossi Fanelli, E. Antonini, and A. Caputo, "Hemoglobin and myoglobin," *Adv. Protein Chem.*, **19,** 73 (1964). E. Antonini and M. Brunori, *Hemoglobin and Myoglobin in Their Reactions with Ligands*, North-Holland, Amsterdam (Elsevier, New York), 1971.
29. J. C. Gerhart and A. B. Pardee, "The effect of the feedback inhibitor, CTP, on subunit interactions in aspartate transcarbamylase," *Cold Spring Harbor Symp. Quant. Biol.*, **28,** 491 (1963).
30. J. C. Gerhart and H. K. Schachman, "Distinct subunits for the regulation and catalytic activity of aspartate transcarbamylase," *Biochemistry*, **4,** 1054 (1965).
31. K. Weber, "New structural model of *E. coli* aspartate transcarbamylase and the amino-acid sequence of the regulatory polypeptide chain," *Nature*, **218,** 1116 (1968).
32. J. P. Changeux and M. M. Rubin, "Allosteric interactions in aspartate transcarbamylase. III. Interpretation of experimental data in terms of the model of Monod, Wyman and Changeux," *Biochemistry*, **7,** 553 (1968).
33. G. Markus, D. L. McClintock, and J. B. Bussel, "Conformation changes in aspartate transcarbamylase. III. A functional model for allosteric behavior," *J. Biol. Chem.*, **246,** 762 (1971).
34. D. E. Koshland, Jr., G. Nemethy, and D. Filmer, "Comparison of experimental binding data and theoretical models in proteins containing subunits," *Biochemistry*, 5, 365 (1966).
35. M. E. Magar and R. F. Steiner, "Equivalence of certain models in protein ligand equilibria and the possibility of distinguishing between them," *J. Theor. Biol.*, **32,** 495 (1971).
36. N. Laiken and G. Nemethy, "A new model for the binding of flexible ligands to proteins," *Biochemistry*, **10,** 2101 (1971).
37. J. R. Sweeny and J. R. Fisher, "An alternative to allosterism and cooperativity in the interpretation of enzyme kinetic data," *Biochemistry*, **7,** 561 (1968).
38. C. Frieden, "Kinetic aspects of regulation of metabolic processes," *J. Biol. Chem.* **245,** 5788 (1970).

XVI Symmetry in Protein Structures[1–5]

It has been pointed out that the presence of multiple subunits with certain symmetry relationships and the principle of symmetry conservation are important features in the theory of allosteric transitions. In addition, it is now recognized that most proteins, in fact, contain more than one polypeptide chain per molecule, and that many of these proteins can be dissociated under mild conditions into subunits that retain most of the tertiary structure of the original molecule. Multiple polypeptide chains in proteins are normally held together through noncovalent interactions as discussed in Chapter 6. Covalent linkages between chains such as disulfide bridges occur in some cases. Procedures that may be used in determination of subunit stoichiometry in proteins containing multiple chains, whether all of one type or of more than one type, and of the stoichiometry of substrates, allosteric effectors, and cofactors were presented earlier. In this chapter we consider the possible arrangement of subunits within an oligomeric protein.

The unambiguous determination of quaternary structure is possible only by crystallographic methods. The cost involved in such an analysis, as well as experimental difficulties in some cases, limits its general application, and other approaches must be explored. If subunits in a protein could assume any arbitrary geometrical relationship to one another, the problem would be insurmountable with presently available methods. However, a number of elegant arguments based on evolutionary and structural grounds have been presented to support the view that only certain symmetrical arrangements of subunits can exist in protein quaternary structures (e.g., see References 1 and 2, and the discussion by Monod in Reference 3). Available crystallographic results are consistent with the proposed symmetry model, and good circumstantial evidence has accumulated, particularly from electron microscopic studies, indicating that the model may be generally applicable.[5] With the aid of the symmetry model as an attractive working hypothesis, various physical and chemical methods may be applied to choose the most compatible geometrical arrangement of subunits within an oligomeric protein.

The Symmetry Model. The symmetry model is based on the following postulates:

1. Chemically identical polypeptide chains in a protein oligomer are likely to be spatially equivalent.
2. If the free energy of subunit association is sufficiently large to impart a unique oligomeric structure to a protein, that structure must define a closed system of finite extent.

There are some apparent exceptions to the symmetry model, and it has been suggested that these exceptions are still compatible with evolutionary and thermodynamic principles expressed in a more relaxed symmetry model. For example, if a protein contains eight identical polypeptide chains that are not all spatially equivalent, then the next most likely situation is a model with four subunits with one type of orientation and four of another.[5]

A necessary result of the spatial equivalence of protein subunits is that they are related to one another by one or more symmetry operations. In the case of globular oligomeric proteins (typically containing 2–12 subunits) the allowable symmetry operations are limited to rotations that may be about one or more axes, all passing through one centrally located point (recall that the asymmetry of the polypeptide chain excludes the possibility of inversion or planes of symmetry). The possible geometrical arrangements for this type of symmetry are conveniently described in point group notation using the Schönfliess symbols for symmetry operations. Other more complex quaternary structures exist for certain proteins as in those repeating structures possessing line or helical symmetry (e.g., tobacco mosaic virus), but they will not be considered here.

Point Groups. Globular oligomeric proteins may belong to any of three point groups, cyclic, dihedral, or cubic. The cubic point groups may be further subdivided into tetrahedral, octahedral, or icosohedral symmetry containing 12, 24, or 60 identical subunits, respectively.

Cyclic symmetry. Cyclic symmetry is denoted by the symbol C_n. The geometrical arrangement of the subunits is such that a rotation of $360°/n$ transposes the structure into itself (i.e., it is undistinguishable from the original one). A molecule containing no symmetry belongs to the trivial point group C_1. A dimeric protein with chemically and spatially identical polypeptide chains must necessarily belong to the point group C_2. Hence, rotation of 180° about the symmetry axis produces an identical structure (Fig. 16-1). A consequence of the twofold rotation axis is that the symmetry must hold for

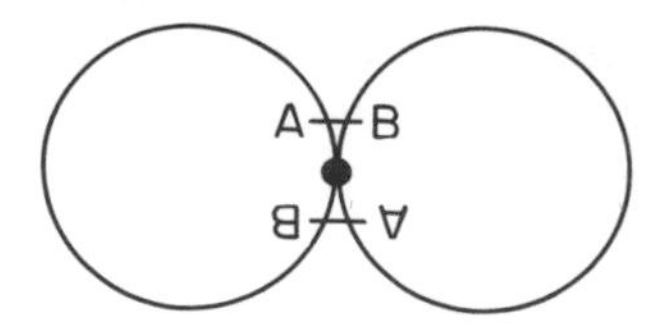

Fig. 16-1 A dimer with C_2 symmetry. The central dot locates the position of the twofold axis (perpendicular to the plane of the paper).

the amino acid residues in the contact domain between subunits, so that for each residue A on subunit 1 contacting residue B on subunit 2, there must be a residue A on subunit 2 contacting residue B on subunit 1 (Fig. 16-1). This type of contact has been termed an isologous bonding domain[1] and the formation of such a dimer is said to be an isologous association. A trimer composed of three equivalent subunits has C_3 symmetry with each subunit related to the others by a threefold axis of symmetry. The subunits are bonded to each other through a pair of complementary bonding sets (Fig. 16-2). Such an association is called heterologous.

In general, a molecule with C_n symmetry will be a closed ring containing n subunits (e.g., with a square, pentagonal, and hexagonal appearance for molecules containing 4, 5, and 6 nearly spherical subunits, respectively). If n is odd, C_n is the only symmetry possible. For even values of n (exclusive of 2), it is necessary to distinguish between cyclic and dihedral symmetry (or the cubic point group for $n = 12$, 24, or 60).

Dihedral symmetry. This point group, designated D_n, contains an *n-fold* axis with n twofold axes perpendicular to it. Two models each containing four subunits and possessing D_2 symmetry are shown in Fig. 16-3. At one extreme (Fig. 16-3*a*) the isologous association of two isologous dimers is represented in a manner that places subunits at the corners of a square (for spherical subunits). By rotating the dimers with respect to each other in that model until the subunits occupy positions at the vertices of a tetrahedron, a third set of isologous contacts are formed and the model has a tetrahedral appearance (Fig. 16-3*b*). The latter model has been termed "pseudotetrahedral" by Monod et al.,[1] who regard this structure as a logical end-point in

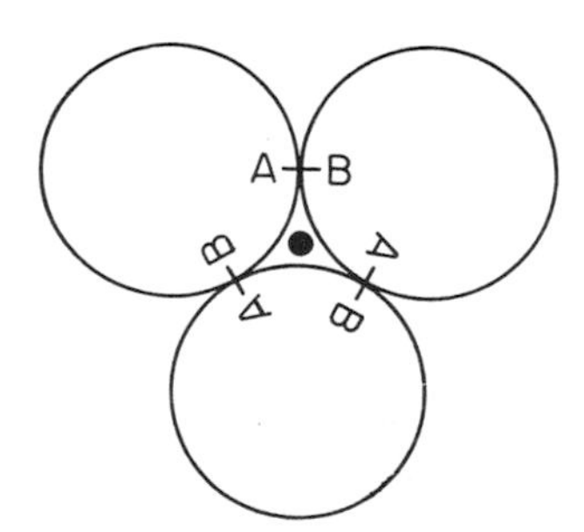

Fig. 16-2 A trimer with C_3 symmetry. The threefold axis is indicated by the central dot. In the spherical subunit model shown, the heterologous complementary bonding sets, A and B, are at an angle of 120° to each other.

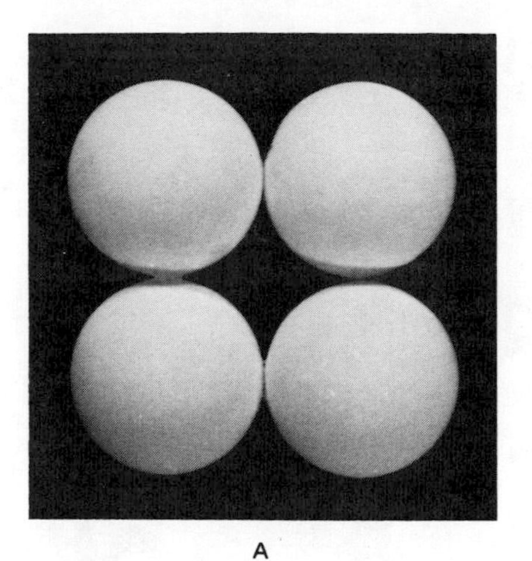

Fig. 16-3 Alternate arrangements of spherical subunits in a tetramer of D_2 symmetry. (*a*) Square presentation. (*b*) Tetrahedral appearance. Models of D_2 symmetry with an appearance between that of (*a*) and (*b*) can also be constructed.

the evolution of a D_2 tetramer. It should be noted that any method of observation, such as electron microscopy, that cannot resolve the relative orientation of roughly spherical subunits in a protein, cannot distinguish between C_4 and D_2 symmetry if the latter has a nearly planar geometry.

The dihedral point group D_n for n greater than 2 can be arranged in two general ways: by the further isologous association of isologous dimers[2] (Fig. 16-4) or by the heterologous association of isologous dimers (or the

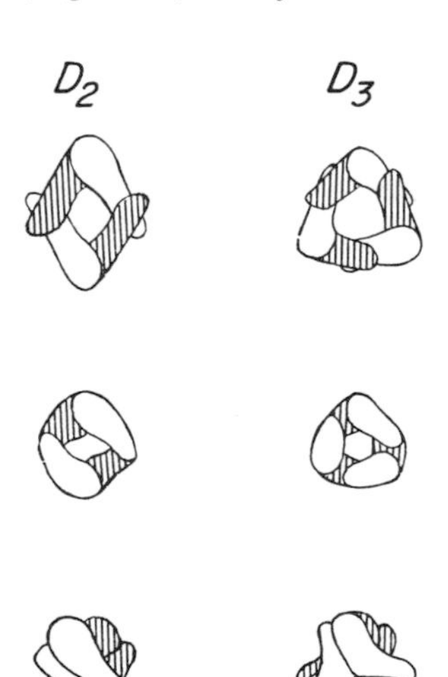

Fig. 16-4 Examples of D_2 and D_3 symmetry obtained by isologous association of isologous dimers.[2]

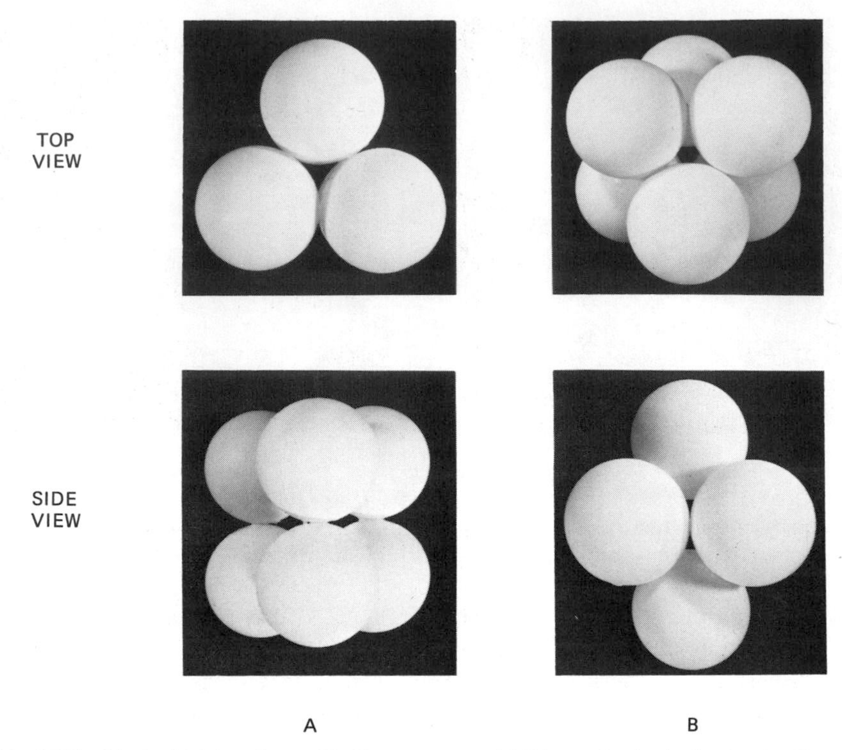

Fig. 16-5 Stacked trimer rings with D_3 symmetry. (*a*) Rings of spherical subunits directly superimposed (eclipsed) as viewed (top) along the threefold axis. (*b*) Rings staggered. Intermediate rotations of the rings with respect to one another also have D_3 symmetry.

geometrically equivalent isologous association of two identical heterologous rings). For association of rings the subunits in the rings may assume any orientation ranging from direct superposition to totally staggered, as shown in Fig. 16-5 for D_3 symmetry. The staggered form potentially has an additional bonding set per subunit and is a logical end-point in evolution from either the isologous association (giving an additional heterologous bonding set on closure) or from heterologous dimer association (giving an additional isologous association when staggered). The electron micrographs of *Escherichia coli* glutamine synthetase (cf. Fig. 13-7) may be interpreted on the basis of a model with D_6 symmetry.

Cubic symmetry. A protein containing 12 identical subunits may possess tetrahedral (T) symmetry. A variety of geometrical arrangements with T symmetry are possible. One such arrangement is the heterologous association of four heterologous trimers, each placed at the vertex of a tetrahedron. Two geometries obtained for the heterologous association of six isologous dimers

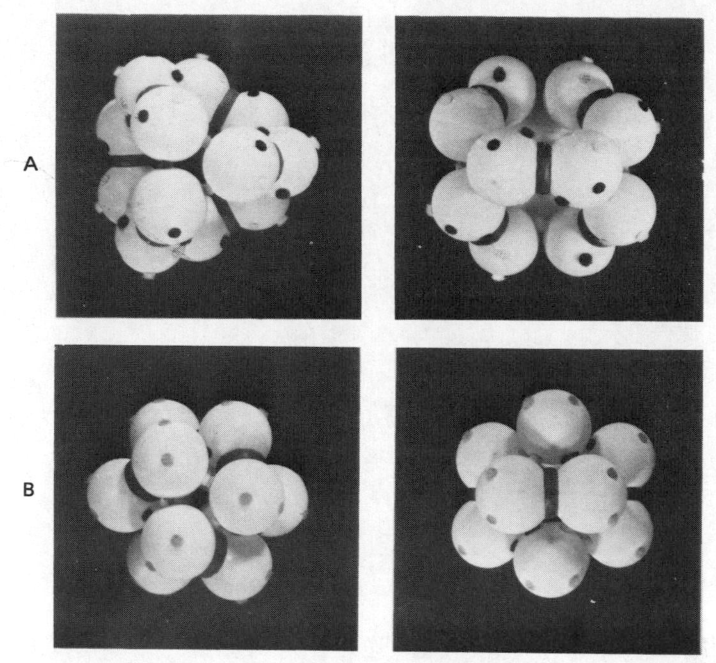

Fig. 16-6 Models showing two ways in which six isologous dimers may be arranged to give tetrahedral symmetry. (*a*) An "open" form. (*b*) The compact form. The open form can be transformed to the compact form by rotation of the dimers with respect to one another. The intermediate geometrical presentations also have tetrahedral symmetry.

are shown in Fig. 16-6. The compact form has an additional bonding set for each of the 12 subunits that are located at the vertex of an icosohedron. This compact tetrahedral model may be considered as a logical end-point in evolution to give "pseudoicosohedral" symmetry. The enzyme aspartate-β-decarboxylase is thought to have such a geometrical arrangement.[5,8]

Octahedral (O) symmetry requires 24 identical subunits. These may be arranged in a number of ways, and several evolutionary options exist for the formation of oligomeric proteins with this symmetry. The enzyme dihydrolipoyl transacetylase contains 24 identical subunits and probably has octahedral symmetry. The square presentation observed in electron micrographs of this enzyme (Fig. 16-7) is consistent with several models, for example, unresolved trimers might be placed at the corners of a cube (i.e., face of an octahedron). The further association of other proteins around this basic unit in the pyruvate dehydrogenase complex is an exciting demonstration of the role of symmetry in more complex organization.[6] A last case of icosohedral symmetry which may be mentioned is the theoretically possible oligomer of 60 identical subunits. No such structures have yet been described; however,

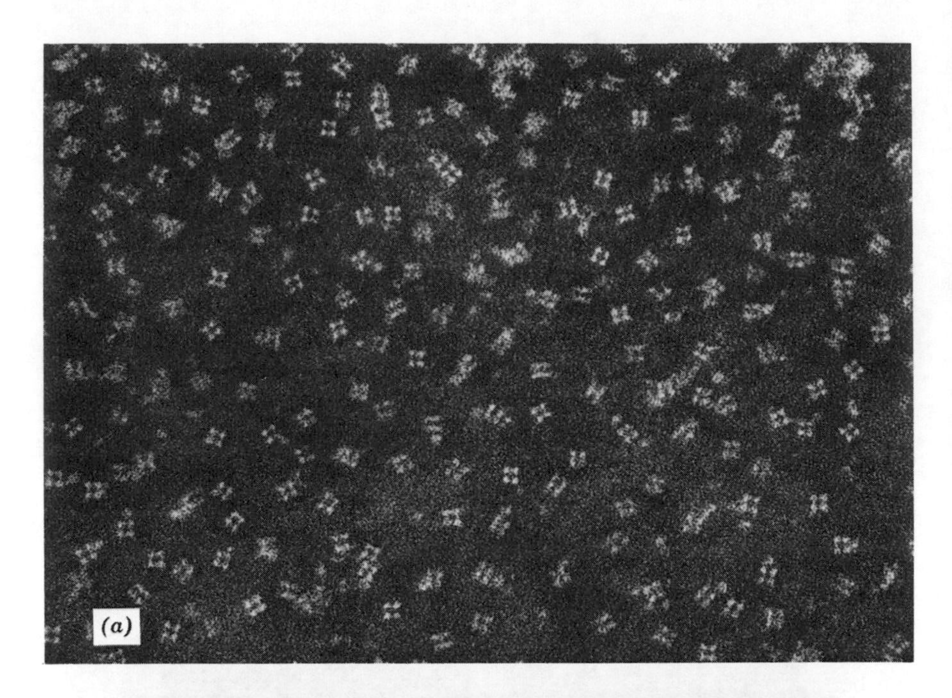

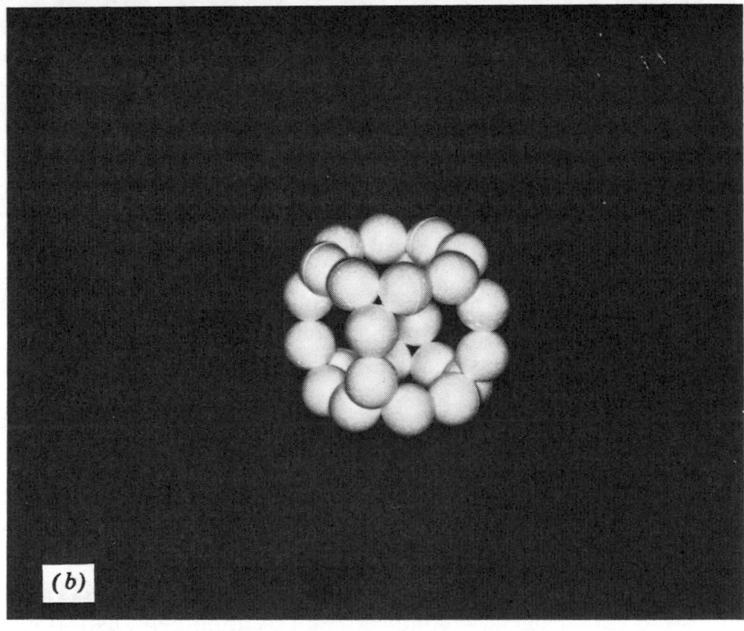

Fig. 16-7 (*a*) Electron micrograph of dihydrolipoyl transacetylase negatively stained with sodium phosphotungstate. (*b*) Model of the enzyme with 24 identical polypeptide chains. (Courtesy of Robert M. Oliver and Lester J. Reed.)

quasi-equivalent multiples of 60 protein subunits are a common occurrence in the protein shells of icosohedral viruses.[7]

Properties of Intersubunit Bonding Sets. The noncovalent forces responsible for intersubunit association are similar in nature to those responsible for the folding of polypeptide chains in the establishment of tertiary structure. Chemical or physical perturbations that cause denaturation within a polypeptide chain also promote dissociation, often under somewhat milder conditions. In some cases dissociation is observed after chemical modification of specific amino acid residues (e.g., see Reference 5) under conditions where further denaturation is not observed. Such studies are useful for choosing a compatible symmetry model.[4,5] They also illustrate that the effective removal of even a single amino acid residue from the bonding set, or steric interference with the intersubunit domain, can completely alter the observed association.

Such observations are also in accord with observed values for the free energy of subunit association that are in the range of -2 to -10 kcal/mole subunit. A detailed discussion of the energetics of intersubunit interaction can be found in the review by Klotz, Langerman, and Darnall.[4] Figure 16-8 illustrates the effect of small differences in free energy of association (ΔG°) on the monomer–octamer equilibrium of hemerythrin. It is seen that changes of the order of 1 to 2 kcal/mole are sufficient to produce a significant shift in the relative amounts of these species, most noticeably at the lower concentrations. These changes in ΔG° may be compared with estimates of about 6 kcal/mole for the ionization of a carboxyl group (e.g., of aspartic acid) and about 2.0 kcal/mole (cf., Chapter 6) for formation of an apolar bond (with no change in conformational entropy). Thus the additional stabilization afforded by only one or two apolar interactions may be sufficient to determine whether a protein is principally oligomeric or monomeric at a given concentration.

Despite small free-energy requirements, intersubunit bonding in oligomeric proteins is highly specific, and oligomers that undergo reversible dissociation will find their proper partners even if the association is carried out in a mixture of many other proteins. This specificity is most likely the result of stringent requirements for precise fit of a large number of residues, including many that make little if any contribution to the free energy. For example, all surface atoms hydrogen bonded to water in the dissociated state must be hydrogen bonded in the intersubunit bonding domain, otherwise the unfavorable free energy of hydrogen bond rupture will prohibit association.

The precise requirements for the geometry of the intersubunit bonding domain suggest that the evolution of an isologous domain would be favored

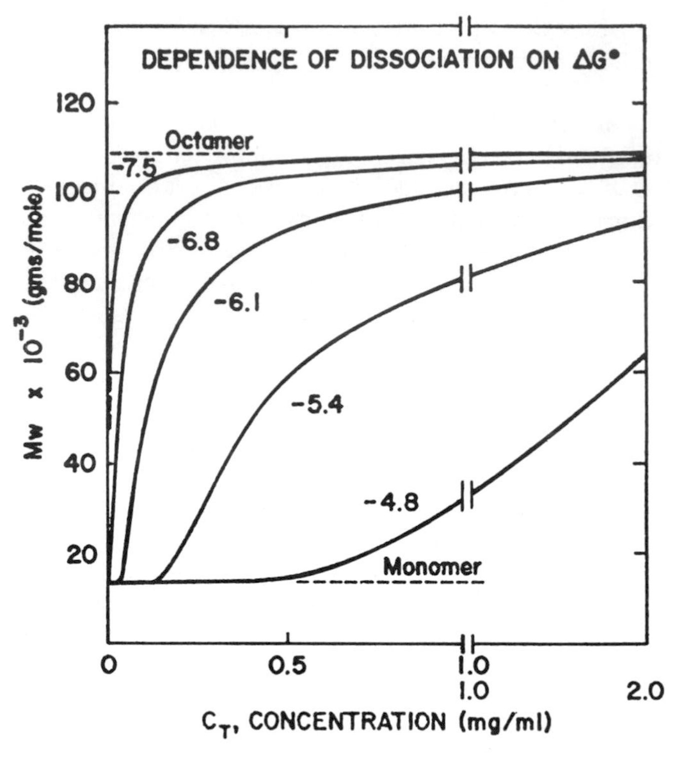

Fig. 16-8 Plots of weight-average molecular weight as a function of concentration computed for various values of the free energy change in a monomer ⇌ octamer equilibrium.[4]

over a heterologous one, for in the former case, when $\frac{1}{2}$ of the residues form a favorable fit, the rest necessarily do so because they are related by symmetry. Some experimental evidence for this concept is provided by the distribution of subunit stoichiometry compiled by Klotz and co-workers as summarized in Table 16-1. The preponderance of dimers and tetramers is immediately striking. In addition, there are few oligomers containing an odd number of subunits. Of those listed, several are the result of transverse cleavage of double ring structures and the rest are somewhat uncertain. It can be concluded that few, if any, naturally occurring oligomers possess C_3 or C_5 symmetry. Since there is no *a priori* reason to suppose that the formation of C_4 oligomers is favored over C_3 or C_5, it is likely that most tetrameric proteins have D_2 symmetry. Consequently, it appears that isologous association is highly favored. There is also evidence that the association of oligomers containing more than four subunits often occurs by the secondary association of isologous dimers and that the negative free energy $(-\Delta G)$ of formation per subunit of the isologous bond per subunit is greater than that of heterologous bonds in the oligomer.[5]

Table 16–1 Frequencies of various subunit stoichiometries based on determinations for 110 different proteins.[4]

Number of subunits	Number of proteins with designated number of subunits
2	44
3	6
4	37
5	2
6	8
7	0
8	5
9	0
10	4
12	4

The Relation of Symmetry to Biological Function. The advantage of an oligomeric protein possessing symmetry in its capacity for allosteric control was discussed and referenced in Chapter 15. Additional advantages, including stability and geometric versatility with more limited genetic information, have been postulated. With some modification the principles discussed here can be extended to more complex biological systems. The discussion in Reference 3 can serve as a fascinating starting point for additional reading in this area.

REFERENCES

1. J. Monod, J. Wyman, and J.-P. Changeux, "On the nature of allosteric transitions: A plausible model," *J. Mol. Biol.*, **12,** 88 (1965).
2. K. R. Hanson, "Symmetry of protein oligomers formed by isologous association," *J. Mol. Biol.*, **22,** 405 (1966).
3. A. Engström and B. Strandberg, Eds., *Symmetry and Function of Biological Systems at the Macromolecular Level* (Nobel Symposium Series, Vol. 11), Wiley, New York, 1969.
4. I. M. Klotz, N. R. Langerman, and D. W. Darnall, "Quaternary structure of proteins," *Ann. Rev. Biochem.*, **39,** 25 (1970).
5. R. H. Haschemeyer, "Electron microscopy of enzymes," *Adv. Enzymol.*, **33,** 71 (1970).
6. C. R. Willms, R. M. Oliver, H. R. Henney, B. B. Mukherjee, and L. J. Reed, "α-Keto acid dehydrogenase complexes. VI. Dissociation and reconstitution of the dihydrolipoyl transacetylase of *Escherichia coli*," *J. Biol. Chem.*, **242,** 889 (1967).
7. D. L. D. Caspar and A. Klug, "Physical principles in the construction of regular viruses," *Cold Spring Harbor Symp. Quant. Biol.*, **27,** 1 (1962).
8. W. F. Bowers, V. B. Czubaroff, and R. H. Haschemeyer, "Subunit structure of L-aspartate β-decarboxylase from *A. faecalis*," *Biochemistry*, **9,** 2620 (1970).

XVII Fibrous Proteins

The group of fibrous proteins is of particular interest because of the effectiveness with which a great variety of physical and chemical methods have been utilized to study their structure and function. We shall highlight some of these results.

Because of their ability to aggregate into fibers or filaments, the fibrous proteins provide the structural framework of higher organisms and are responsible, as well, for dynamic functions such as muscular contraction and blood clotting. Fibrous proteins are found in nearly all forms of life including single-celled organisms (e.g., actin and myosin in protozoa; flagellin in bacteria). Proteins of this type were apparently established very early in evolution as the basic macromolecules specialized for structure and motility in living systems. Those which have been obtained in monomeric form in solution are found to be highly elongated or anisometric in shape, and consequently show high intrinsic viscosities. In addition, a high proportion of helical secondary structure in most of these proteins produces a characteristic optical rotatory dispersion pattern. Thus measurements of viscosity or optical rotation provide a convenient means for assessing the structural integrity of these proteins in solution.

In nature most fibrous proteins occur in large fibrous structures composed of repeating units and showing varying degrees of partial crystallinity. Because of this characteristic, studies of their structure by X-ray diffraction techniques were begun long before other methods were available. An early classification system, proposed by Astbury in 1940, divided the fibrous proteins into two groups based on particular features of their fiber diffraction patterns. The first of these is the collagen group including collagen and related synthetic polypeptides. The second, the "K-M-E-F" (keratin-myosin-epidermin-fibrinogen) group, is based on a common type of diffraction pattern, the α-pattern. Some of these proteins also exhibit a β-type pattern under some conditions. Silk fibroin, which normally occurs in nature in the β-structure, is also included in the K-M-E-F group.

The characteristic X-ray α-pattern of the K-M-E-F group is similar but not identical to the diffraction pattern of an α-helix (cf. Chapter 14). For example, these patterns show a 5.1-Å reflection on the meridian, whereas the α-helix of synthetic polypeptides produces a strong off-meridional reflection at a spacing of 5.4 Å. A number of fibrous proteins give reflections consistent with the existence of superhelices composed of a number of α-helices (coiled-coils). Such an arrangement allows the α-helices to fit together in order to achieve efficient packing though repeating interactions of side chains. A model for a two-chained coiled-coil is shown in Fig. 17-1. It is not yet known, however, how many chains may participate in such superhelices or whether this form is prevalent throughout the structure of the fiber. A thorough discussion of these models and the theory of helical diffraction patterns is given by Dickerson[1] and by Holmes and Blow.[2]

1 α-KERATIN[1]

Keratin has been extensively studied because of its ready availability in naturally-occurring structures such as wool and hair. Another valuable source is the porcupine quill. Although a variety of techniques has been used, it has been primarily through X-ray diffraction and electron microscopy that we have obtained most insight into the biological organization of this protein.

The α-form of keratin is believed to consist of coiled-coils characterized by a dense 5.15 Å meridional reflection in the fiber diagram. These coils are organized into microfibrils of about 70 Å diameter. In hair the microfibrils are imbedded in a matrix of amorphous ground substance containing sulfur-rich proteins and possibly other components. The packing is hexagonal with a spacing of about 100 Å between fibrils. An electron micrograph of a cross section of an α-keratin fiber is shown in Fig. 17-2. The substructure within the microfibrils is consistent with a "9 + 2" model in which nine "protofibrils" of coiled-coils, composed of two or three α-helical strands, each are packed around a central core of two similar protofibrils. This is the same type of arrangement found in certain other biological structures such as centrioles and cilia. X-ray patterns obtained when α-keratin is stained with heavy metals indicate that the coiled-coils of the microfibrils occur in segments of about 25 Å length, connected by amorphous material.

Although the coiled-coil model has been widely accepted as a good approximation to the structure of α-fibrous proteins, some uncertainty still exists. Comparison of experimental X-ray diffraction patterns with the predictions of the simple theory for the coiled-coil model indicate that several important reflections are absent or are shifted in position. Attempts to account for the data in terms of alternative models have been reviewed by Parry.[3] For example, some features of the observed diffraction patterns can be explained in terms of straight helices. Another model, the segmented rope, has also been shown to give a good fit to the experimental data. Here, a repeating distribution of hydrophobic residues is assumed to exist in order to account for the intensity and position of critical equatorial and meridional reflections in fiber diagrams of keratin and other α-fibrous proteins. Another interesting observation in the study of the X-ray diffraction of keratin is the conversion from the α-pattern to the β-pattern that occurs when the fibers are stretched. The possibility exists that both types of structural element are always present in the natural fibers. In that case the change in diffraction pattern upon stretching may represent distortion of the α-helices which causes

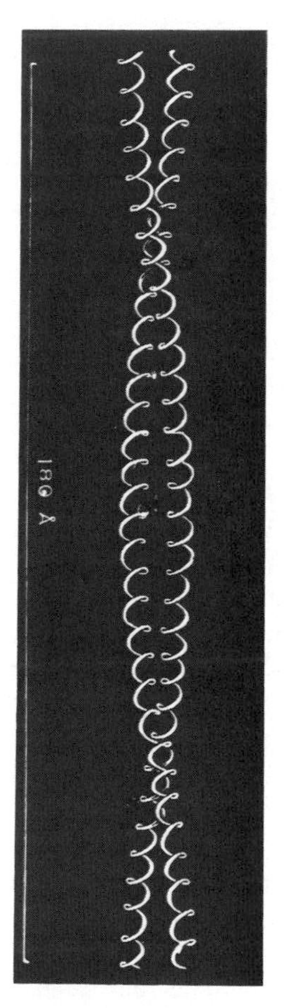

Fig. 17-1 A model of a coiled-coil consisting of two chains running antiparallel. A schematic drawing is shown to the left. On the right an atomic model made up of alternating leucine and alanine residues is given. The repeating unit is seven residues and is marked in some units by an arginine model. Hydrogen bonds are represented by spiral springs. [From C. Cohen and K. C. Holmes, *J. Mol. Biol.*, **6**, 423 (1963).]

the α-helical reflections to disappear and the β-pattern reflections to become more discernible on the X-ray film.

The study of α-keratin has demonstrated the way in which the basic units of a protein organize to form well-ordered superstructures. A characteristic feature of the α-keratin structure is the high degree of crosslinking achieved through disulfide bonding. The resulting insolubility of the material, however, has made the isolation of the basic protein subunit of keratin difficult, and little is yet known of its chemical and physical properties. Materials obtained by reduction or oxidation of the disulfide bridges have generally been heterogeneous and difficult to work with. One form that is soluble under some conditions is a protein of 640,000 molecular weight called prekeratin.[4] This may prove amenable to study by other physical methods aimed at obtaining insight into keratin structure. The chemistry of α-keratin and related substances has been covered in the review of Seifter and Gallop.[5]

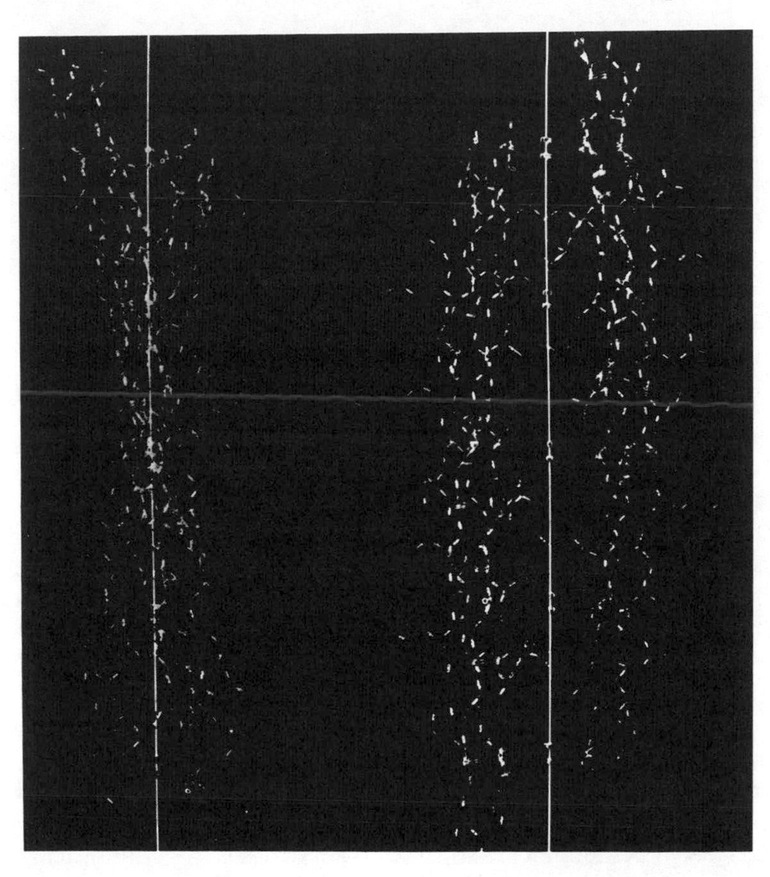

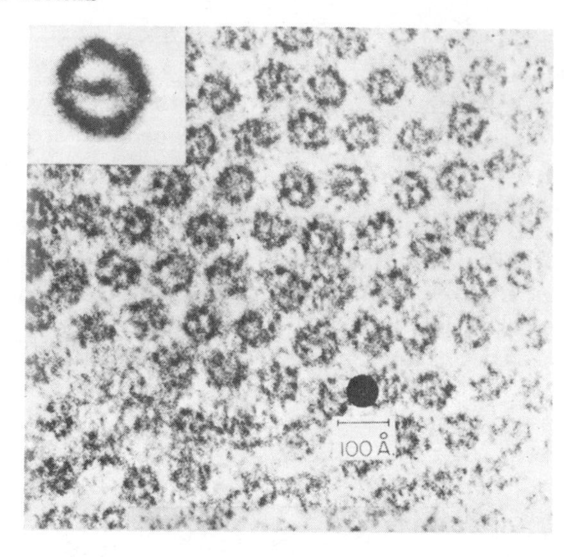

Fig. 17-2 Electron micrograph of a cross section of an α-keratin fiber stained with osmium tetroxide and lead hydroxide. Printing has been reversed so that unstained microfibrils appear dark. The black sphere indicates the size of the hemoglobin molecule for comparison. Inset shows several microfibrils superimposed. [From B. K. Filshie and G. E. Rogers, *J. Mol. Biol.* **3,** 784 (1961).]

2 MUSCLE PROTEINS

The proteins of muscle have been the object of more research aimed at elucidating structure and function than any other fibrous protein. The fascination of the material is no doubt attributable to the remarkable capacity of this protein system to produce mechanical work from chemical energy (via the hydrolysis of adenosine triphosphate, ATP).

The contractile protein actomyosin, amounting to about 75% of the fibrous protein in muscle, was first isolated in 1939. Actomyosin solutions exhibit a characteristic high viscosity and flow birefringence, indicating a high degree of asymmetry in molecular shape. The protein complex can be readily separated into its component proteins myosin and actin, with concomitant reduction in viscosity.

A third protein, tropomyosin, accounts for 5 to 10% of the protein in vertebrate striated muscle, and a number of minor components exist as well. The principal physical results on myosin, actin, and tropomyosin will be discussed; further information and thorough documentation can be found in the review of Seifter and Gallop.[5] A model for their role in muscular contraction has been presented by Huxley.[6]

2-1 Myosin

Myosin is the protein component of muscle responsible for the characteristic α-pattern observed in whole muscle with X-ray diffraction. In addition, myosin possesses ATPase activity, that is, the ability to catalyze the hydrolysis of ATP to ADP and inorganic phosphate, a reaction that yields about 8 kcal of free energy. It is this energy-producing reaction that is viewed by many investigators as the source of energy for contraction. The mechanism by which this energy transformation is achieved is still unknown.

Myosin is normally isolated from whole muscle by salt extraction and examined for homogeneity by sedimentation velocity (cf. Chapter 7). Unfortunately, difficulties in preparation of a homogeneous native material have led to conflicting results on some physical parameters of the molecule. However, it is now fairly well agreed that its molecular weight is about 510,000 and the molecule is highly asymmetric (length = 1520 Å). Almost all the hydrodynamic methods discussed in Chapters 7 and 8 for size and shape determination have been drawn on heavily in the study of this molecule, and good agreement has been found between many of these earlier studies and the more recent results from electron microscopy.

A great deal has been learned from the study of certain fragments of the molecule that can be prepared with considerable reproducibility by brief exposure to a proteolytic enzyme (trypsin, subtilisin, or chymotrypsin). These fragments are termed light meromyosin (LMM) and heavy meromyosin (HMM). Determination of yields of the fragments and their molecular weights (150,000 for LMM and 340,000 for HMM) indicated that one of each occurs in the parent myosin molecule.

Heavy meromyosin was found to possess the actin binding property and the ATPase activity of native myosin. Further tryptic digestion of heavy meromyosin has been utilized to prepare smaller fragments termed Subfragment 1 (HMM S-1) and Subfragment 2 (HMM S-2). The former, with molecular weight of about 120,000, still possesses the ATPase activity and the ability to combine with actin. Light meromyosin, on the other hand, lacks enzymatic activity but resembles myosin in that it is only sparingly soluble at low ionic strength. Viscosity measurements indicate that LMM is highly asymmetric; optical rotation studies indicate an α-helical content approaching 100%. Thus this portion of the molecule is the one responsible for the α-pattern and the fibrous character of myosin. In contrast, the other part of the molecule, which carries the active site for ATP hydrolysis and is isolated in the heavy meromyosin fragment, could be better categorized as a globular protein. A summary of physical properties of myosin and its subfragments is given in Table 17-1.

Studies of myosin by electron microscopy have revealed that the molecule is very likely a dimer. A long rodlike portion is observed with two globular heads (Fig. 17-3). Study of the parent molecule and its subfragments has led to the model shown in Fig. 17-4. A variety of physical parameters (from light scattering, sedimentation equilibrium, X-ray diffraction) support the hypothesis that the rod portion of the molecule is a two-stranded coiled coil (cf. Fig. 17-1). Thus it would appear that each helical chain terminates in

Table 17-1 Physical chemical parameters for myosin and its subfragments obtained from rabbit muscle. The fragments HMM and LMM and subfragments HMM S-1 and HMM S-2 are defined in the text; "rod" refers to the entire rodlike portion of the molecule (see Fig. 17-3). [From S. Lowey, H. S. Slayter, A. G. Weeds, and H. Baker, *J. Mol. Biol.*, **42,** 1 (1969).]

	Myosin	HMM	LMM	HMM S-1	HMM S-2	Rod
Molecular weight $\times 10^{-3}$	510 ± 10	340 ± 10	140 ± 5	115 ± 5	62 ± 2	220 ± 10
Intrinsic viscosity (dl/g)	2.1 ± 0.1	0.49 ± 0.02	1.2 ± 0.1	0.064 ± 0.01	0.4 ± 0.1	2.40 ± 0.1
Intrinsic sedimentation coefficient $\times 10^{13}$	6.4 ± 0.1	7.2 ± 0.1	2.9 ± 0.1	5.8 ± 0.1	2.7 ± 0.1	3.4 ± 0.1
Rotatory dispersion constant, b_0	400 ± 10	320 ± 10	630 ± 10	230 ± 10	610 ± 10	660 ± 10
% α-Helix	57	46	90	33	87	94

the globular portion that can be broken off as HMM S-1. Some additional low molecular weight subunits occur in the globular portions, as shown by the fine zigzag lines in Fig. 17-4. These subunits, comprising about 8% of the total protein, can be dissociated from the heavy-chain core of myosin by a variety of methods including a reversible dissociation by the use of concentrated salt solutions.[7] ATPase activity is restored upon recombination of the light and heavy portions.

The organization of myosin molecules into the filaments that occur in muscle is typical of the self-assembly processes found among all fibrous

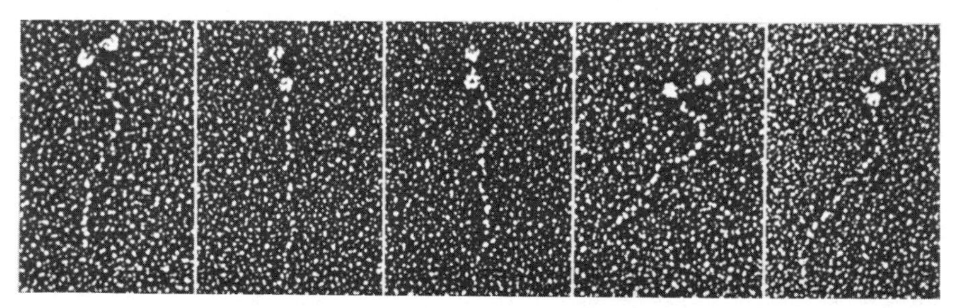

Fig. 17-3 Electron micrograph of typical molecules of rabbit muscle myosin sprayed on freshly cleaved mica and shadow cast with platinum. Two lobes are observed in the head of the molecule. [From H. S. Slayter and S. Lowey, *Proc. Nat. Acad. Sci., U.S.*, **58,** 1611 (1967).]

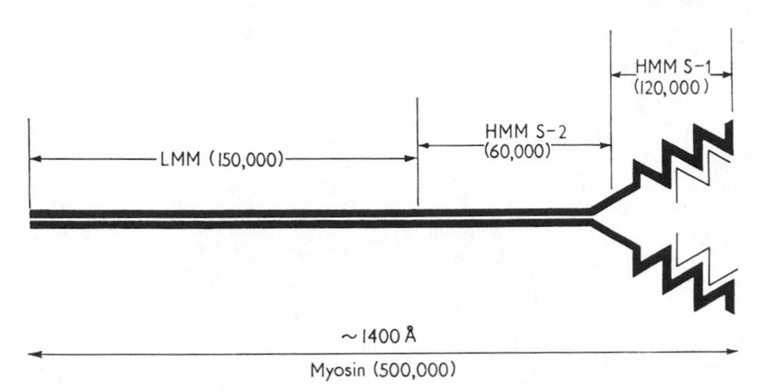

Fig. 17-4 Schematic representation of the myosin molecule showing regions corresponding to light meromyosin (LMM) and the subfragments of heavy meromyosin (HMM) and their molecular weights. The two large polypeptide chains total 400,000 Daltons; there are four light chains (shown by thin lines in the HMM S-1 segments) with molecular weights around 20,000. [S. Lowey and D. Risby, *Nature*, **234,** 81 (1971); diagram from S. Lowey, H. S. Slayter, A. G. Weeds, and H. Baker, *J. Mol. Biol.*, **42,** 1 (1969).]

proteins. Myosin filaments obtained by mechanical disruption of muscle tissue have a length of about 1.5 μ and a diameter of 160 Å. They often show projections clustered around the ends and a smooth zone in the middle (Fig. 17-5). Such structures can also be obtained from purified myosin. In the buffered 0.5 M salt solutions in which myosin is usually studied, the protein exists almost entirely in the 500,000 molecular weight form. Upon reduction of the ionic strength, the molecules rapidly assemble to form filamentous structures, as shown in Fig. 17-6*a*. The mode of aggregation of the monomers is illustrated in Fig. 17-6*b*.

2-2 Actin

Actin can be isolated from muscle in a fibrous form referred to as F-actin. F-actin exists as long thin filaments that appear to consist of two strands wound around one another in a right-handed helix.[8] At low ionic strengths, F-actin which contains 1 mole of ADP per polypeptide chain disaggregates into subunits of molecular weight about 44,000 (globular or G-actin). G-actin containing bound ATP can be readily reaggregated *in vitro* to form filaments of the fibrous form by the addition of neutral salts. In the process, ATP is dephosphorylated to ADP. Divalent cations (Ca^{2+} or Mg^{2+}) also play a role in the polymerization process.

The role of the bound adenine nucleotides in relation to the properties of actin is still not entirely clear. In aqueous solvents their removal is accompanied by decreases in specific optical rotation, suggestive of partial denaturation, and the resulting protein is incapable of polymerization to F-actin.[9]

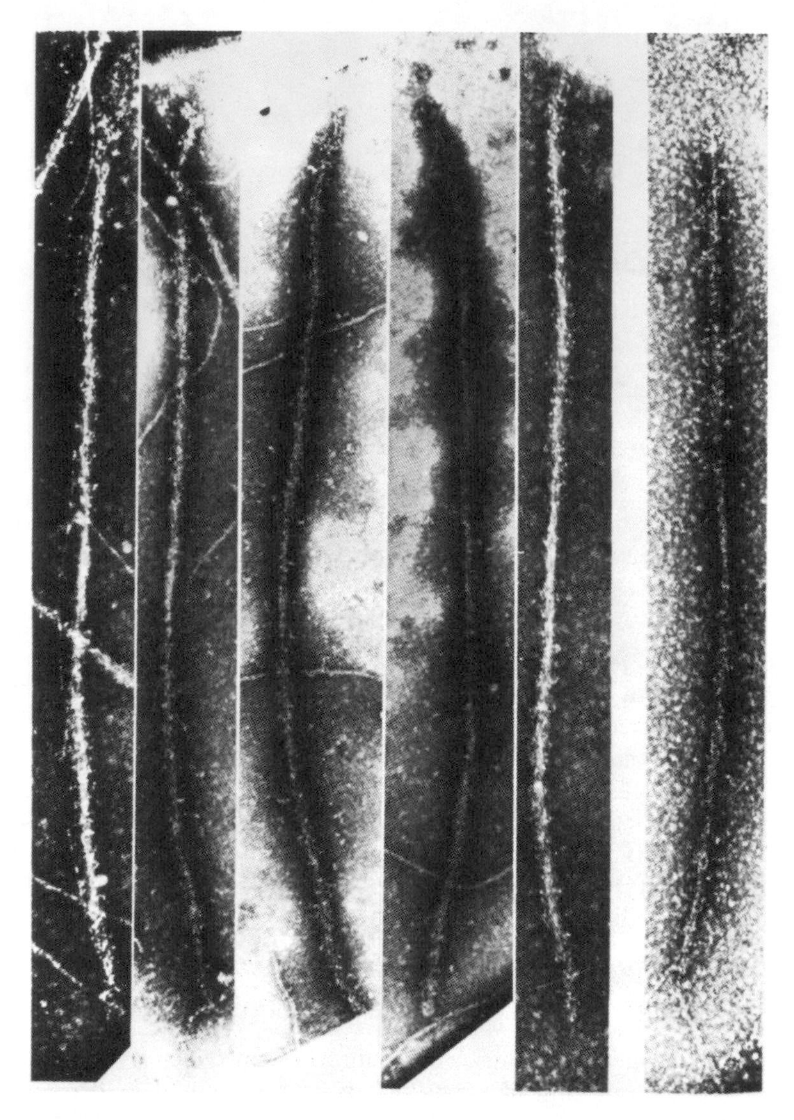

Fig. 17-5 Myosin filaments obtained from homogenized muscle tissue. The thick filaments show a characteristic bridge-free zone in the center and projections clustered at each end. [From H. E. Huxley, *J. Mol. Biol.*, **7**, 281 (1963).]

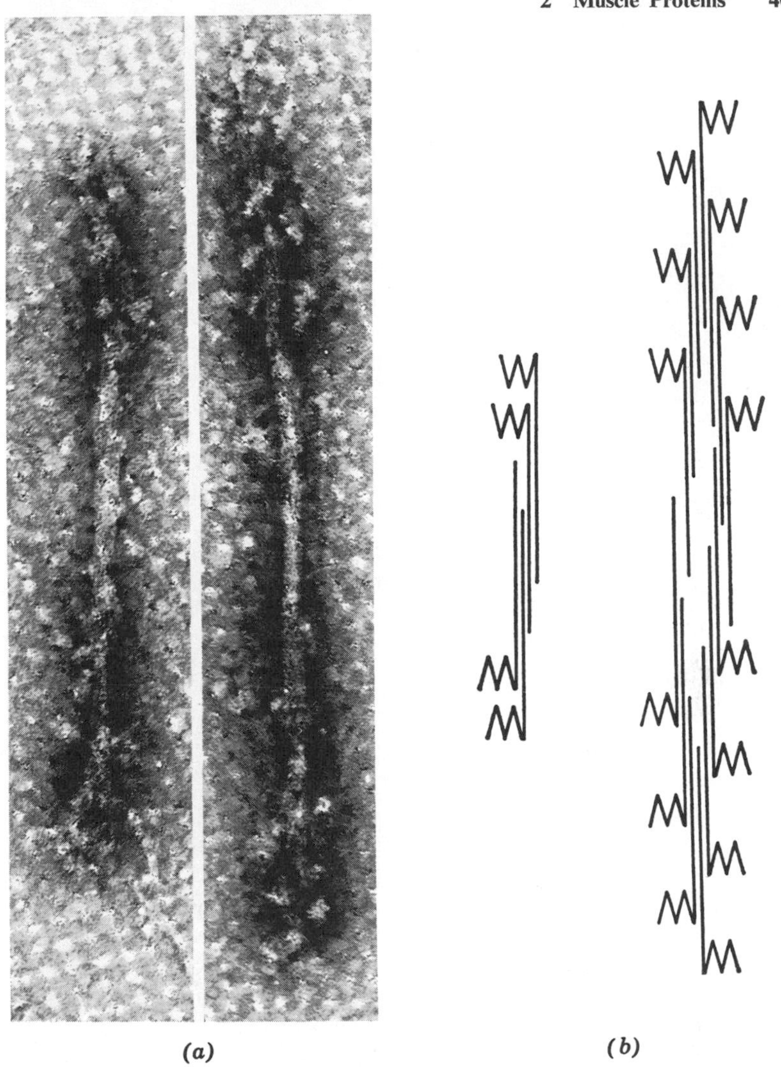

Fig. 17-6 (*a*) Electron micrograph of reconstituted myosin filaments. (*b*) Schematic model of myosin aggregation. [Courtesy of H. E. Huxley.]

Under special conditions, however, as in concentrated sucrose solutions, the protein retains its ability to polymerize. In general, the presence of ATP favors aggregation, and polymerization occurs readily even at 1°C upon addition of KCl and Mg^{2+}. However, ADP-actin is regarded to be the predominant form *in vivo*.

The polymerization process may be followed by viscosity measurements,

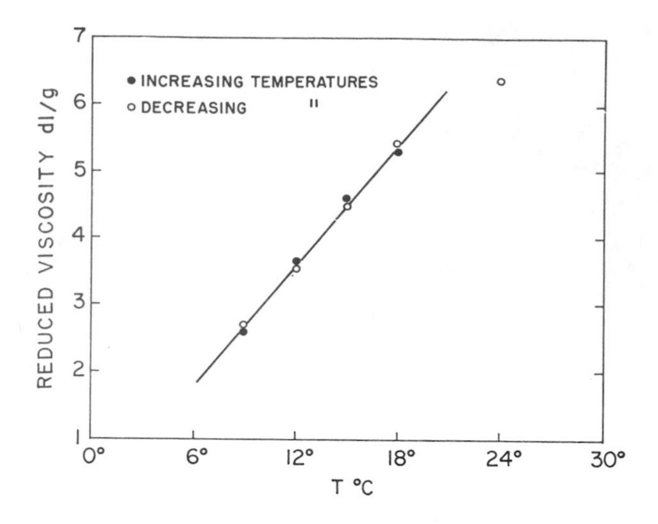

Fig. 17-7 Temperature dependence of the reversible polymerization of G-ADP-actin, as followed by viscosity measurements. Protein concentration 0.9 mg/ml in 2 m*M* $MgCl_2$. 100% polymerization (F-actin) is achieved at 29°. (Courtesy of Robert J. Grant.)

provided the actin preparaton has been carefully purified to remove tropomyosin (see below) which contributes markedly to intrinsic viscosity. As shown in Fig. 17-7, aggregation of G-ADP-actin by Mg^{2+} increases with increasing temperature, suggesting the involvement of hydrophobic interactions. Under these conditions the reaction is fully reversible, but in the presence of tropomyosin or orthophosphate the protein is stabilized in the polymerized form.

The interaction of F-actin filaments with myosin and subfragments of myosin has received intensive study. Electron micrographs, when analyzed by a three-dimensional reconstruction technique, reveal a stoichiometric combination of molecules of subfragment I (the "head" subunit) with G-actin units in the F-actin structure.[10] The tilt of the myosin fragments with respect to the filament axis produces a characteristic "arrowhead" pattern.

2-3 Tropomyosin

Tropomyosin is the third major protein of muscle that appears to occur in all species. It is sometimes designated tropomyosin B to distinguish it from another (actually very different) muscle protein, tropomyosin A or paramyosin, that occurs in invertebrate muscle. In many ways this protein resembles the LMM portion of myosin. It produces an α-pattern in X-ray diffraction and shows a high optical rotation consistent with as much as 90% helical content. Studies with light-scattering and hydrodynamic methods indicate a molecular weight of about 70,000 and possibly two subunits per molecule.[11] Its overall structure is considered to be a two-chain coiled-coil of about 400 Å length. Tropomyosin is unusual among fibrous proteins in that it forms para- (or semi-) crystalline fibers with a very high degree of crystallinity. The molecules are found to align in a head-to-tail manner in such crystals.[12]

3 FIBRINOGEN AND FIBRIN[13]

Fibrinogen is a soluble blood protein that is synthesized in the liver of vertebrates and accounts for about 3% of the total protein of plasma. Under the action of thrombin, an enzyme that is activated during the process of blood clotting, fibrinogen is converted to the aggregated fibers of the fibrin clot. Extensive studies have been made on its structure and the changes that occur during clotting. Although species differences in amino acid sequence exist, the fundamental properties of the molecule appear to be the same for all vertebrates. The discussion below deals primarily with bovine and human fibrinogens, which have been studied in the greatest detail.

3-1 Physical and Chemical Properties of Fibrinogen

Fibrinogen has a molecular weight of about 340,000. Molecular dimensions obtained by hydrodynamic and light-scattering methods are of the order of 400 to 500 Å in length and 50 to 100 Å in width; however, the exact values are still in question, since measurements by different techniques have not always agreed.[14] Heterogeneity in preparations due to proteolytic digestion, denaturation, and/or aggregation is probably responsible for the discrepancies in physical parameters determined by various methods. It is also possible that the hydrodynamic model (usually a prolate ellipsoid of revolution) used to obtain these values does not adequately represent the frictional characteristics of the real molecule. Electron micrographs of convincing clarity have thus far been obtained for fibrinogen only by shadowing methods (Fig. 17-8). The general appearance of the image has been interpreted in terms of a three-beaded structure for the molecule.

Native fibrinogen can be dissociated with various denaturing agents after reduction or other treatment for disulfide bond rupture, and weight-average molecular weight is found to drop to $\frac{1}{6}$ of its original value. Identification of the separated polypeptide chains has shown that fibrinogen contains three pairs of polypeptide chains, confirming earlier results from *N*-terminal analysis. By convention the three types of chains are denoted α, β, and γ. Although very similar in molecular weight, the three types differ

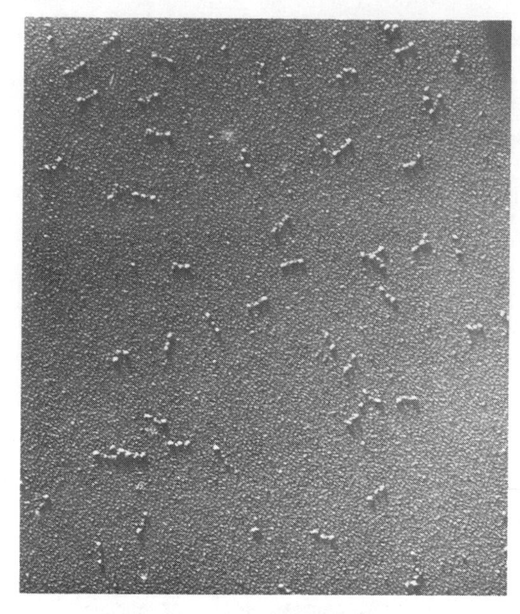

Fig. 17-8 Electron micrograph of shadowed bovine fibrinogen. 115,000×. [From C. E. Hall and H. S. Slayter, *J. Biophys. Biochem. Cytol.*, **5**, 11 (1959).]

sufficiently to be separable in the form of sodium dodecyl sulfate complexes by gel electrophoresis. Large portions of the *N*-terminal sequences of the chains have been determined. Each chain is found to contain a small carbohydrate component; together these sugars account for about 3% of the weight of fibrinogen.[15]

The arrangement of the three pairs of polypeptide chains in the native molecule is still unclear. Physical studies have been consistent with an elongated molecule having a 2-fold rotation axis perpendicular to its long axis, and with the *N*-terminals of the α-chains at opposite ends of the molecule.[16] A schematic diagram based on this model and the electron micrograph of Fig. 17-8 is shown in Fig. 17-9. The *N*-terminals of each set of α, β, and γ chains are grouped together on the basis of the finding of interchain disulfide bonds in that region of the sequences. Evidence that conflicts with this model has been obtained, however. Cyanogen bromide digestion of native fibrinogen gives rise to a molecular fragment of about 25,000 daltons that contains six *N*-terminal peptides, two from each type of chain, connected by disulfide bridges.[17] If not due to spurious disulfide interchange, these results indicate that all six *N*-terminals of the native molecule must be grouped closely together, possibly in the center of the molecule, or all at one end with symmetry like that of the myosin structure (Fig. 17-4).

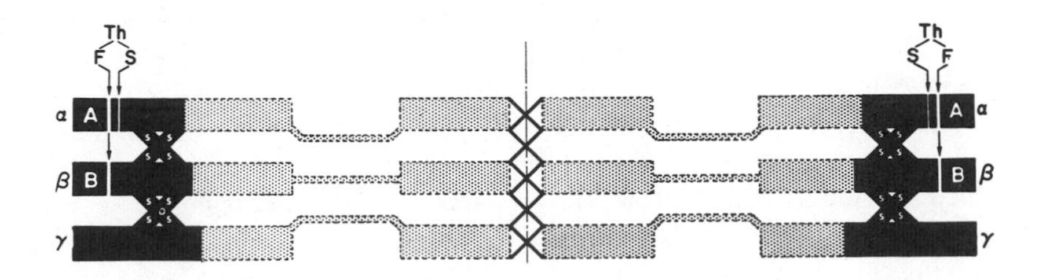

Fig. 17-9 A possible arrangement of the six polypeptide chains of fibrinogen. Cleavage sites by thrombin (Th) in clotting are indicated by arrows. [From B. Blombäck, M. Blombäck, A. Henschen, B. Hessel, S. Iwanaga, and K. R. Woods, *Nature*, **218,** 130 (1968).]

3-2 The Fibrinogen–Fibrin Conversion

Early theories of blood clotting and the historical development of our present concepts have been reviewed.[14,18,19] The final stage of clotting involves the thrombin-catalyzed conversion of fibrinogen to fibrin. This process begins with the hydrolysis of four arginyl-glycyl bonds in fibrinogen (F) to produce fibrin monomer (f) and four peptides, two of fibrinopeptide A and two of fibrinopeptide B. The reaction can be written:

$$\mathrm{F} \xrightarrow{\text{thrombin}} \mathrm{f} + 2\mathrm{A} + 2\mathrm{B}$$

Fibrinopeptide A (bovine) is a chain of 19 amino acids and includes the *N*-terminal glutamic acid residue present in the α-chain of the original fibrinogen molecule. The B fibrinopeptides are derived from the β-chains of fibrinogen and are 21 amino acid residues in length. The *N*-terminal residue of bovine fibrinopeptide B, which does not react with *N*-terminal reagents, has been identified as pyroglutamic acid (i.e., a cyclic form in which the α-amino group takes part in a peptidelike bond with the γ-carboxyl group of the side chain). The peptides are characterized by a high proportion of di-carboxylic amino acid residues in all species examined. An unusual amino acid, tyrosine-*O*-sulfate, is found in the B-peptide in some species.

The release of the A fibrinopeptide by thrombin proceeds more rapidly than the release of the B peptide, and is obligatory for subsequent clotting. Following peptide release, the activated molecules undergo association, ultimately leading to the characteristic ropelike mesh structure of the clot. A very similar clot is produced by a snake venom enzyme (amusingly termed "reptilase") that splits off only the A peptides. The process appears to be identical except for a possible reduction in lateral association of the activated monomers in the polymeric structures of the clot. Fibrin clots show a characteristic band pattern in the electron microscope and a strong 230 Å repeat by X-ray diffraction; however, the molecular organization of the clot is still debated. Some interesting possibilities are discussed by Bang.[20]

The interactions responsible for the aggregation of fibrin monomers in clot formation have not yet been clarified. When clotting is carried out with purified fibrinogen and thrombin preparations, the aggregation is fully reversible. Such solvents as 3 *M* urea or dilute acetic acid are capable of dissolving the clot and returning fibrin to the monomeric form. Some type of hydrogen bonding appears likely, although its exact nature is still in question. Another contributing factor is probably apolar bonding. The solubilizing capability of solvents that weaken apolar bonds suggest at least partial involvement of interactions of that type in the polymerization process.

Under physiological conditions the polymerization of fibrin is not reversible; this is the result of an additional enzymatic mechanism which produces covalent crosslinkages between fibrin molecules. The enzyme involved has been termed Factor XIII (fibrin-stabilizing factor) or plasma transglutaminase. After activation by thrombin, it catalyzes a transpeptidation reaction between the ε-amino groups of lysine residues and the γ-carboxyl groups of glutamic acid residues of adjoining fibrin molecules.[21] Further details on this and other aspects of fibrinogen research may be found in References 13 and 19.

4 COLLAGEN[22–25]

As the principal fibrous constituent of structural tissues such as skin, tendon, cartilage, and bone, collagen is the most abundant protein in the animal kingdom. It has been estimated that between 25 and 35% of the total protein of the mammalian body is in the collagen family. Collagen frequently occurs in association with elastin, a protein of similar properties, and is found imbedded in an extracellular matrix of mucopolysaccharides, water, other proteins, and small molecules. Because of its prevalence in nature and its unusual chemical and physical properties, it has been the object of intensive research.

4-1 Structure and Chemistry

Collagen, as ordinarily isolated from sources such as skin, bone, cornea, or tendon, shows a distinctive chemical composition. Approximately 33% of all amino acid residues is glycine, and approximately 25% consists of proline and hydroxyproline. The primary structure of the protein consists for the most part of repeating triplets of the form Gly-X-Y, where X and Y may be any of the amino acid residues that occur in collagen (although X is frequently proline). This unique composition forms the basis for the unusual structure and physical properties of this protein. The total content of proline and hydroxyproline is relatively constant in collagens throughout nature from flatworms to humans; however, the proportion that occurs as hydroxyproline varies widely. The significance of the hydroxyproline content and the factors that determine which proline residues in the polypeptide chains are to be hydroxylated are still a puzzle. Another rare amino acid occurring in collagen is hydroxylysine, formed by hydroxylation of lysyl residues. Hydroxylating enzymes for proline and lysine have been identified and isolated; they apparently act on the newly synthesized collagen (*protocollagen*) after (and possibly during) polypeptide chain synthesis. Small amounts of carbohydrate (galactose and glucosylgalactose) are found in *O*-glycosidic linkage to hydroxylysine isolated from collagen.

Examination of connective tissue sections by electron microscopy shows collagen fibrils organized into successive sheets lying at right angles to each other. The resultant structure, as shown in Fig. 17-10*a*, resembles plywood. Intact collagen fibrils can be obtained by disruption of such tissues. They show a characteristic periodicity of about 640 Å, consistent with a model in which adjacent molecules are staggered by approximately one-fourth of their length (Fig. 17-10*b*).

The basic collagen molecule (*tropocollagen*) is isolated by extraction of connective tissue (e.g., skin) with a cold neutral salt solution or a cold acid solution. Such preparations from mammalian sources spontaneously reaggregate into fibrils at neutral pH when the temperature is raised to body temperature (37°). The process is initially reversible, although with time the fibrils become stabilized and can no longer be disaggregated by dropping the temperature. Figure 17-11 shows the time course of the thermal gelation of collagen as determined by the simple measurement of opacity in a Klett colorimeter. After 7 days only about 25% of the fibrils of normal collagen are capable of disassembling at 5°. In contrast, a much greater degree of reversibility is found for fibrils of an abnormal collagen obtained from animals suffering from lathyrism, a chemically-induced connective tissue disease. Fibril stabilization is now recognized to be due to crosslinks involving aldehyde groups that are absent in lathyritic collagen. Crosslinking of native collagen can be blocked by treatment with aldehydic reagents; it can be stabilized in reconstituted fibrils by chemical reduction.[26]

As with other fibrous proteins, the self-assembly of collagen into fibrils is its most striking and least understood property. In addition to the native type fibril of 640 Å period that can be obtained *in vitro*, collagen is found to form two other distinct types of superstructure. One of these, the fibrous-long-spacing (FLS), shows a major period of about 3000 Å, coinciding with

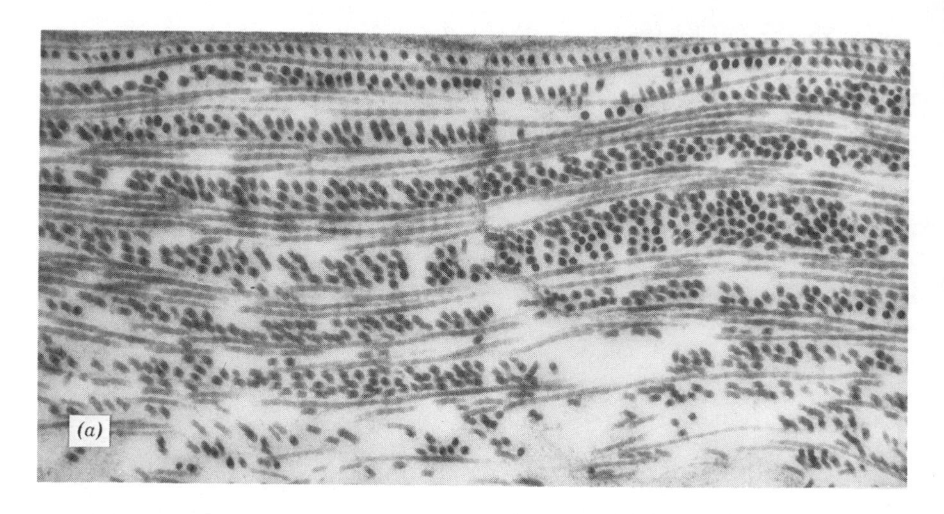

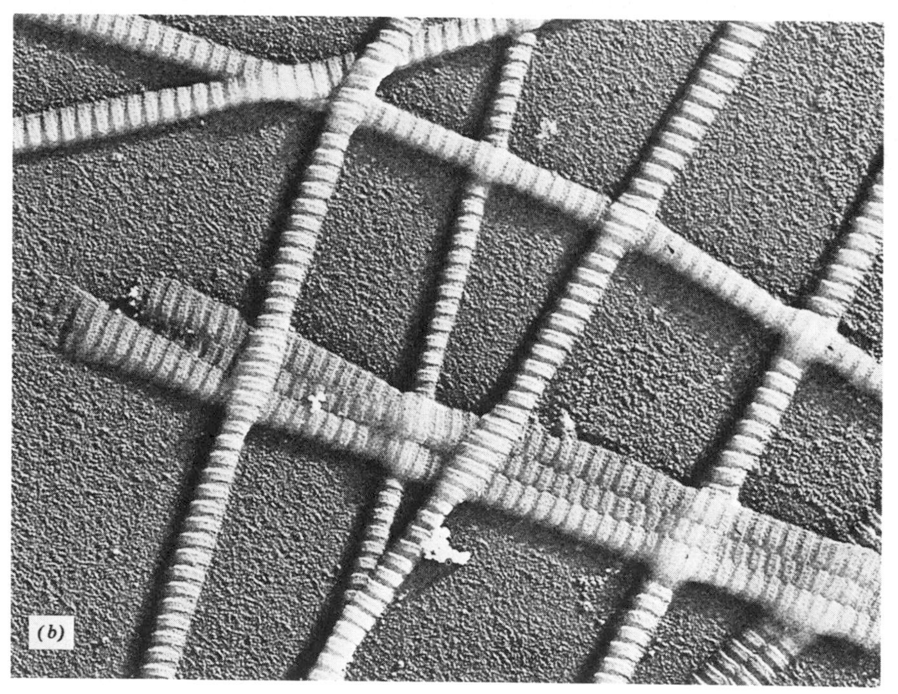

Fig. 17-10 (*a*) Plywood-like organization of sheets of collagen fibrils in tadpole cornea. The fibrils in one layer are at right angles to those in the next. 31,500×. (Electron micrograph by Marie A. Jakus) (*b*) Intact collagen fibrils obtained from skin and shadowed with chromium, showing characteristic 640 Å periodicity. 36,000×. (Courtesy of Jerome Gross.)

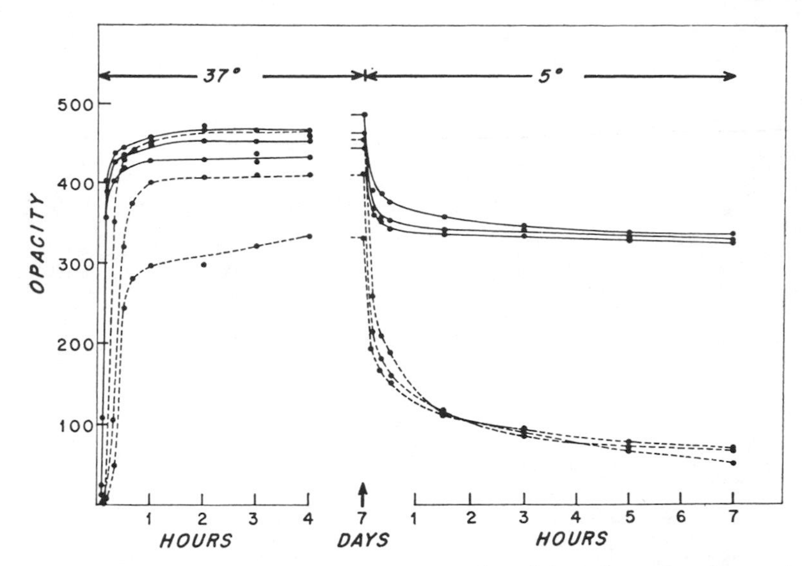

Fig. 17-11 The temperature-dependent aggregation of several samples of normal collagen (solid lines) and lathyritic collagen, a type lacking intramolecular crosslinks (broken lines). Fibril concentration is assayed by solution opacity. Reversibility of the association upon cooling to 5°C (at right) differs markedly in the two preparations. [From J. Gross, *Biochem. Biophys. Acta*, **71,** 250 (1963).]

the length of the collagen molecule. The FLS structure as revealed in the electron microscope after phosphotungstic acid staining is shown in Fig. 17-12*a*. Another type of collagen aggregation called segment-long-spacing (SLS) is induced by the addition of adenosine triphosphate (or certain other acidic materials) to solutions of collagen monomers. The collagen then precipitates in small crystallites of about 3000 Å length (Fig. 17-12*b*). In the SLS all molecules have the same polarity and there is no overlap. Figure 17-12*c* shows the fine structure revealed by phosphotungstic acid staining of native type reconstituted fibrils.

Collagen fibrils have been extensively studied by X-ray diffraction methods for determination of the basic architecture of the collagen molecule. The wide angle X-ray diffraction pattern is consistent with the existence of a helical triple-chained structure; a meridional spacing of 2.91 Å corresponds to the distance between residues along the helical axis. The collagen pattern is readily distinguishable from that of the single chain α-helix with its characteristic 1.5 and 5.4 Å meridional spacings. The size of the collagen molecule (285,000 molecular weight) and its dimensions as observed in the electron microscope (about 3000 Å length and 15 Å width) support the helical rod model. The generally accepted structure of the molecule is comparable to the

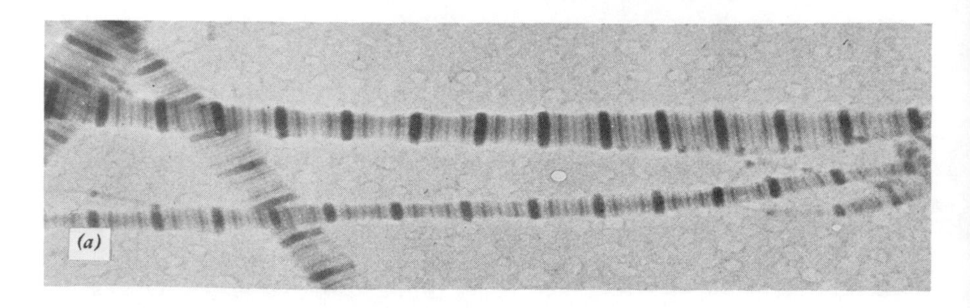

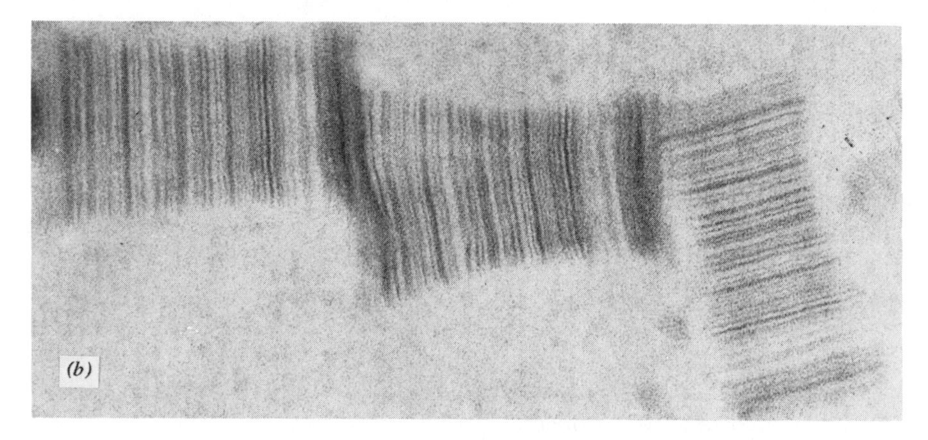

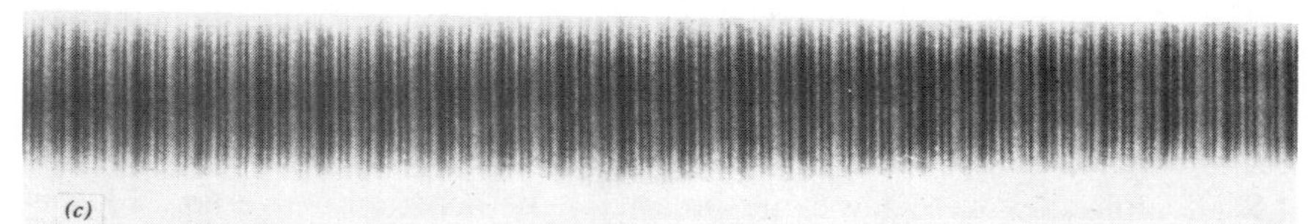

Fig. 17-12 Electron micrographs of three types of fibrous structures formed from collagen monomers in solution. (*a*) Fibrous-long-spacing. 27,500×. (*b*) Segment-long-spacing. 150,000×. (*c*) Reconstituted native-type fibril. 74,000×. (Courtesy of Jerome Gross.)

polyglycine II or poly-L-proline II structures; three parallel helical polypeptide chains are stabilized in the superhelix by virtue of one hydrogen bond per three residues.[27] This model can accommodate all amino acid sequences in collagen, including the sterically restricted -Gly-Pro-Hypro- sequence. Studies on a variety of polytripeptides [e.g., $(\text{Gly-Pro-Pro})_n$] have led to improved coordinates for the model.[24] Attention has also been given to an alternate triple-chained model with two hydrogen bonds for every three residues.[28] Although permitting possibly increased stabilization through

H-bonding, this model suffers from rather close van der Waals' contacts between some atoms (see discussion in Reference 1) and has not been confirmed by the model polypeptide studies. The controversy over collagen models, however, has been helpful in the refinement of model building methods and in focusing attention on the criteria used (e.g., the atomic van der Waals' radii, bond angles).

Soluble collagen as usually extracted is easily denatured to single chains (α-chains) of molecular weight about 95,000, crosslinked dimers (β-form) of molecular weight about 190,000, and a small proportion of crosslinked triple-stranded molecules (γ). Most collagens are found to contain two types of polypeptide chains, $\alpha 1$ and $\alpha 2$, in proportions given by the formula $(\alpha 1)_2\alpha 2$. The two types of chains are very similar and each appears to be capable of forming a collagenlike structure *in vitro* alone, that is, $(\alpha 1)_3$ or $(\alpha 2)_3$. However, $(\alpha 1)_2\alpha 2$ forms preferentially in mixtures.[29] Discovery of more than one distinct molecular species of collagen in a single organism now necessitates the use of a more complicated notation scheme. Chick cartilage has been found to contain two different chains of the $\alpha 1$ type: Type I which occurs in a collagen of normal composition, that is, $[\alpha 1(I)]_2\alpha 2$; and a second, Type II, which occurs in a collagen of composition $[\alpha 1(II)]_3$.[30] Forms of the type $[\alpha 1]_3$ may also exist in lower organisms, as well as molecules in which all three chains differ, that is, $\alpha 1\alpha 2\alpha 3$.[24] An additional complication is that any preparation may exhibit microheterogeneity due to variability in the hydroxylation of prolyl and lysyl residues. A growing amount of sequence information, primarily on cyanogen bromide peptides (reviewed in Reference 24), will aid in the identification and characterization of the various molecular forms of collagen.

The proportions of α, β, γ, and sometimes higher order species in a collagen preparation depends on conditions of extraction. Apparently, as collagen matures *in vivo*, crosslinking of chains, both intramolecular and intermolecular, is increased, and extraction of monomers into aqueous solutions becomes more difficult. Crosslinking appears to involve conversion of certain lysyl and hydroxylysyl residues to aldehydes, followed by Schiff-base formation (by reaction with an available lysyl amino group) or by aldol condensation between aldehydes on adjacent chains. The importance of these bonds to the rigidity and strength of the fibrils is demonstrated by the rubbery connective tissue and skeletal malformations of lathyritic animals whose collagen lacks crosslinking capacity. Other types of unusual covalent bonds, which may exist in some collagens, have been reviewed by Harding.[31]

Most of the data previously amassed on collagen has referred to extracellular collagen, which makes up the bulk of the material isolated from sources such as skin and bone. However, in the cell collagen appears to be synthesized in the form of polypeptide chains of 120,000 to 125,000 molecular weight. When the resultant triple-chained *procollagen* is isolated, it is found to form SLS aggregates *in vitro* like those of Fig. 17-12*b* except for the presence of an extension of about 130 Å at the *N*-terminal end.[32] It is very likely that procollagen is the form in which collagen is transported from the collagen-synthesizing cells into the extracellular matrix where formation of tropocollagen, crosslinking, and fibril organization occurs.

4-2 Degradation by Specific Collagenases

Highly specific and controlled degradation of collagen is required for the remodeling of connective tissues that occurs in many natural processes of development or reorganization. The application of physical and chemical methods to the collagen molecule has proved invaluable in identifying specific collagenases and elucidating their mechanism of action. For example, in the action of collagenase from tadpole fin on collagen, it was found that products of molecular weights 200,000 (TC^A) and 70,000 (TC^B) were formed from the tropocollagen (TC) of calf skin. Figure 17-13 illustrates the sedimentation equilibrium results for the three species. Viscosity data were consistent with the rigid rod structure for all three species; identical optical

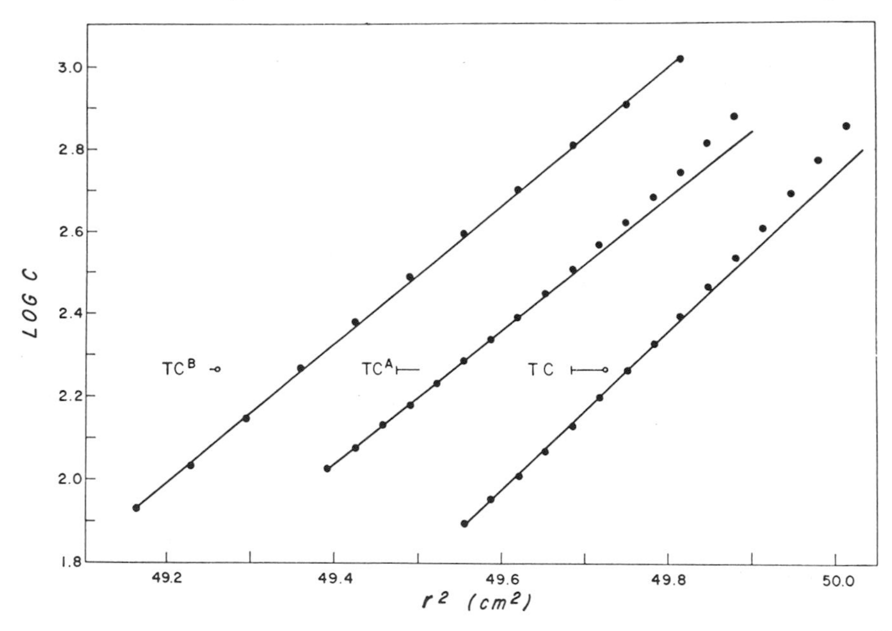

Fig. 17-13 Sedimentation equilibrium analysis for the two collagen fragments TC^B (molecular weight 70,000) and TC^A (molecular weight about 200,000) obtained by the action of tadpole collagenase on calf skin tropocollagen TC (molecular weight 285,000). [From T. Sakai and J. Gross, *Biochemistry*, **6,** 518 (1967).]

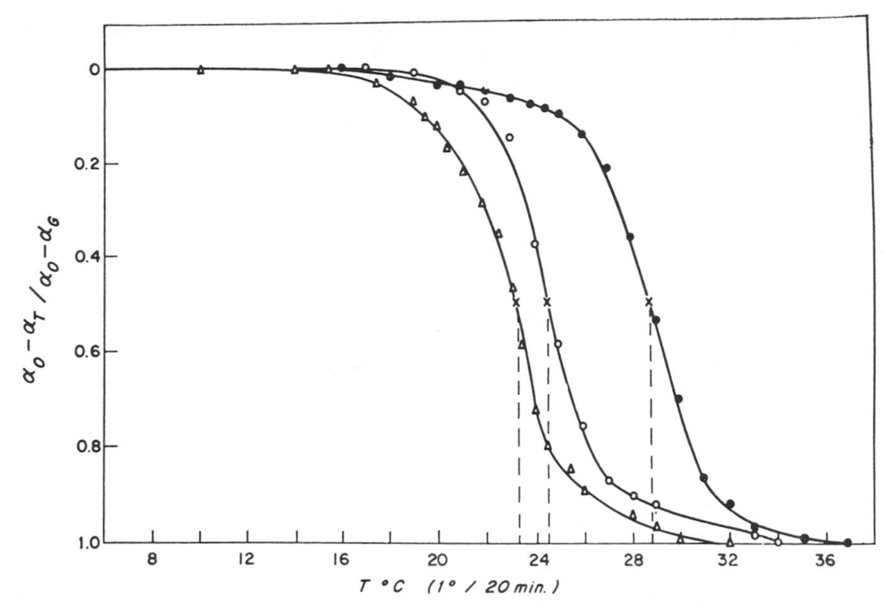

Fig. 17-14 Thermal denaturation, as followed by optical rotation, of tadpole collagen TC (closed circles) and of the fragments produced by the action of tadpole collagenase: TC^B (triangles) and TC^A (open circles). The fractional change in rotation at 365 nm is plotted. α_o, α_T and α_G refer to rotation values at 15°, T, and 40°, respectively. Crosses indicate the positions of 50% denaturation; the corresponding melting temperatures may be read on the abscissa. [From T. Sakai and J. Gross, *Biochemistry*, **6,** 518 (1967).]

rotatory dispersion profiles indicated that little change in secondary structure resulted from the enzymatic action. The postulate that the enzyme acted only to cleave the molecule at a single site about one-quarter of the distance from the carboxyl terminals was supported by observation of SLS aggregates formed by the fragments.[33] Denaturation studies indicated a possible physiological significance of the cleavage. As shown in Fig. 17-14, the fragments denature (and thus become susceptible to ordinary proteolytic digestion) at a lower temperature than the native molecule. The same methods are being applied to the detection and characterization of specific collagenases from a number of other tissues that undergo remodeling processes.

REFERENCES

α-Keratin

1. R. E. Dickerson, "X-ray analysis and protein structure," in *The Proteins*, 2nd ed., Vol. 2, H. Neurath, Ed., Academic Press, New York, 1964.

2. K. C. Holmes and D. M. Blow, "The use of X-ray diffraction in the study of protein and nucleic acid structure," *Meth. Biochem. Anal.*, **13,** 113 (1965).
3. D. A. D. Parry, "A proposed conformation for α-fibrous proteins," *J. Theor. Biol.*, **26,** 429 (1970).
4. A. G. Matoltsy, "Prekeratin," *Nature*, **201,** 1130 (1964).

Muscle Proteins

5. S. Seifter and P. M. Gallop, "The structure proteins," in *The Proteins*, 2nd ed., Vol. 4, H. Neurath, Ed., Academic Press, New York, 1966.
6. H. E. Huxley, "The mechanism of muscular contraction," *Science*, **164,** 1356 (1969).
7. L. C. Gershman and P. Dreizen, "Relationship of structure to function in myosin. I. Subunit dissociation in concentrated solutions," *Biochemistry*, **9,** 1677 (1970).
8. J. Hanson and J. Lowy, "The structure of F-actin and of actin filaments isolated from muscle," *J. Mol. Biol.*, **6,** 46 (1963). R. H. Depue, Jr., and R. V. Rice, "F-actin is a right-handed helix," *J. Mol. Biol.*, **12,** 302 (1965).
9. M. S. Lewis, K. Maruyama, W. R. Carroll, D. R. Kominz, and K. Laki, "Physical properties and polymerization reactions of native and inactivated G-actin," *Biochemistry*, **2,** 34 (1963).
10. P. B. Moore, H. E. Huxley, and D. J. DeRosier, "Three-dimensional reconstruction of F-actin, thin filaments and decorated thin filaments," *J. Mol. Biol.*, **50,** 279 (1970).
11. E. F. Woods, "Molecular weight and subunit structure of tropomyosin B," *J. Biol. Chem.*, **242,** 2859 (1967).
12. D. L. D. Caspar, C. Cohen, and W. Longley, "Tropomyosin: crystal structure, polymorphism and molecular interactions," *J. Mol. Biol.*, **41,** 87 (1969).

Fibrinogen and Fibrin

13. "Fibrinogen: structural, metabolic and pathophysiologic aspects," *Thromb. Diath. Haemorrhag., Suppl.* **39** (1970).
14. H. A. Scheraga and M. Laskowski, Jr., "The fibrinogen-fibrin conversion," *Adv. Protein Chem.*, **12,** 1 (1957).
15. L. Mester, "Structure et role des fractions glucidiques des glycoproteines impliquees dans la coagulation du sang," *Bull. Soc. Chim. Biol.*, **51,** 635 (1969).
16. A. E. V. Haschemeyer, "Charge distribution of fibrinogen as determined by transient electric birefringence studies," *Biochemistry*, **1,** 996 (1962).
17. K. R. Woods, M. S. Horowitz, and B. Blombäck, "Effect of thrombin on the molecular weights of N-terminal fragments of human fibrinogen," *Thromb. Res.*, **1,** 113 (1972).
18. E. W. Davie and O. D. Ratnoff, "The proteins of blood coagulation," in *The Proteins*, Vol. 3, H. Neurath, Ed., Academic Press, New York, 1965.
19. B. Blombäck, "Fibrinogen to fibrin transformation," in *Blood Clotting Enzymology*, W. H. Seegers, Ed., Academic Press, New York, 1967.
20. N. U. Bang, "Ultrastructure of the fibrin clot," in *Blood Clotting Enzymology*, W. H. Seegers, Ed., Academic Press, New York, 1967.

21. L. Lorand, N. G. Rule, H. H. Ong, R. Furlanetto, A. Jacobsen, J. Downey, N. Oner, and J. Bruner-Lorand, "Amine specificity in transpeptidation. Inhibition of fibrin cross-linking," *Biochemistry*, **7,** 1214 (1968).

Collagen

22. A. J. Bailey, "The nature of collagen," in *Comprehensive Biochemistry*, Vol. 26B, M. Florkin and E. H. Stotz, Eds., Elsevier, Amsterdam, 1968.
23. G. N. Ramachandran, Ed., *Treatise on Collagen*, Vol. 1, *Chemistry of Collagen*, Academic Press, New York, 1967.
24. W. Traub and K. A. Piez, "The chemistry and structure of collagen," *Adv. Protein Chem.*, **25,** 243 (1971).
25. M. E. Grant and D. J. Prockop, "The biosynthesis of collagen," *New Eng. J. Med.*, **286,** 194, 242, 291 (1972).
26. M. L. Tanzer, D. Monroe, and J. Gross, "Inhibition of collagen intermolecular cross-linking by thiosemicarbazide," *Biochemistry*, **5,** 1919 (1966). M. L. Tanzer, "Intermolecular cross-links in reconstituted collagen fibrils," *J. Biol. Chem.*, **243,** 4045 (1968).
27. A. Rich and F. H. C. Crick, "The molecular structure of collagen," *J. Mol. Biol.*, **3,** 483 (1961).
28. G. N. Ramachandran, V. Sasisekharan, and Y. T. Thathachari, "Structure of collagen at the molecular level," in *Collagen*, N. Ramanathan, Ed., Wiley-Interscience, New York, 1962.
29. C. Tkocz and K. Kühn, "The formation of triple-helical collagen molecules from α1 or α2 polypeptide chains," *Eur. J. Biochem.*, **7,** 454 (1969).
30. E. J. Miller and V. J. Matukas, "Chick cartilage collagen: A new type of α1 chain not present in bone or skin of the species," *Proc. Nat. Acad. Sci., U.S.*, **64,** 1264 (1969).
31. J. J. Harding, "The unusual links and cross-links of collagen," *Adv. Protein Chem.*, **20,** 109 (1965).
32. P. Dehm, S. A. Jimenez, B. R. Olsen, and D. J. Prockop, "A transport form of collagen from embryonic tendon: Electron microscopic demonstration of an NH_2-terminal extension and evidence suggesting the presence of cystine in the molecule," *Proc. Nat. Acad. Sci., U.S.*, **69,** 50 (1972).
33. J. Gross and Y. Nagai, "Specific degradation of the collagen molecule by tadpole collagenolytic enzyme," *Proc. Nat. Acad. Sci., U.S.*, **54,** 1197 (1965).

Index of Scientists

Subject Index